Religion and Biopolitics

Mirjam Weiberg-Salzmann • Ulrich Willems
Editors

Religion and Biopolitics

Springer

Editors
Mirjam Weiberg-Salzmann
Cluster of Excellence "Religion and Politics"
University of Münster
Münster, Germany

Ulrich Willems
Institute for Political Science
University of Münster
Münster, Germany

ISBN 978-3-030-14579-8 ISBN 978-3-030-14580-4 (eBook)
https://doi.org/10.1007/978-3-030-14580-4

This Springer imprint is published by the registered company Springer Nature Switzerland AG
The registered company address is: Gewerbestrasse 11, 6330 Cham, Switzerland

Acknowledgements

At this point, it is our more than pleasant duty to express our gratitude to all those who participated in the preparation of this book through to its publication. First of all, we would like to thank the contributors to and participants of our international conference on "Religion and Biopolitics" in Münster in October 2015, which laid the foundation for this book. We would also like to gratefully acknowledge those who made the book financially possible, namely, the Deutsche Forschungsgesellschaft, the Cluster of Excellence "Religion and Politics in Pre-Modern and Modern Cultures," and the Centre for Advanced Study in Bioethics, University of Münster, Germany. Our special thanks also go to Ruth Langer for her support in organizing the conference; our chair secretary Kate Backhaus, who supported us in proofreading and communication with all parties involved; and our student assistant Paul Schnase, for the editorial work. And last but not least, we would like to thank all those who actively supported us on our way from Springer Verlag to the printing stage: Johannes Glaeser and Jessica Fäcks.

Münster, Germany — Mirjam Weiberg-Salzmann
Münster, Germany — Ulrich Willems
December 21, 2018

Contents

About the Authors

Monika Boppert is full professor for Moral Theology and director of the Seminar of Moral Theology of the Faculty of Catholic Theology, University of Muenster. From 2012 until 2015, she was professor for Theological Ethics and director of the Institute for Social Ethics at the Faculty of Theology, University of Lucerne, Switzerland, and research associate/assistant professor at the Institute for History and Ethics of Medicine of the Medical Faculty, University of Heidelberg, from 2001 until 2012. From 1995 until 2000, she was Ph.D. fellow and later on scientific coordinator of the interdisciplinary Postgraduate College "Ethics in the Sciences and Humanities" at the International Centre for Ethics in the Sciences and Humanities (IZEW) of the University of Tübingen, Germany. Bobbert is a member of several commissions and committees for applied ethics.

Wybo Dondorp is an Associate Professor of Biomedical Ethics at Maastricht University, the Netherlands, Research Schools CAPHRI and GROW. He is a member of the Health Council of the Netherlands, past chair of the Task Force Ethics & Law of the European Society of Human Reproduction and Embryology (ESHRE), and past coordinator of the Special Interest Group on Ethics of the International Society of Prenatal Diagnosis and Fetal Therapy (ISPD). He participated in the research team for the last evaluation of the Dutch Embryos Act (2012) and has published on different aspects of this legislation, mostly in the *Dutch Journal of Health Law*. His international publications are mainly on the ethics of reproductive medicine, genomics, and screening.

John H. Evans is Professor of Sociology at the University of California, San Diego, USA. He earned his Ph.D. from Princeton and spent some time as postdoctoral fellow at Yale. He has been a professor at the University of California since 1998. His research focuses on religion, culture, politics, and science. His first book *Playing God? Human Genetic Engineering and the Rationalization of Public Bioethical Debate* (Chicago UP 2002) won the Distinguished Book Award from the Religion Section of the American Sociological Association. His other books are *Contested Reproduction: Genetic Technologies, Religion and Public Debate* (Chicago UP 2010),

The History and Future of Bioethics: A Sociological View (Oxford UP 2012), *What is a Human? What the Answers Mean for Human Rights* (Oxford UP 2016), and *Morals Not Knowledge: Recasting the Contemporary U.S. Conflict between Religion and Science* (California UP 2018).

Anja Hennig is Scientific Assistant at the Chair of Comparative Politics at the Faculty of Cultural Studies at European University Viadrina in Frankfurt (Oder). She received her Ph.D. at the same university in 2011. Core areas of her research are the shifting borders between religion and politics before the background of cultural pluralization and secularization, the comparative analysis of morality politics and religion in Catholic societies, the relationship between biopolitics, religion, and democracy, and most recently religious–right-wing alliances in gender-related conflicts in Europe. She has written the book *Moralpolitik und Religion. Bedingungen politisch-religiöser Kooperation in Polen, Italien und Spanien* (Ergon-Verlag 2012) and coedited with Jeffrey Haynes, *Religious actors in the public sphere: Means, Objectives, and effects* (Routledge 2011). Recent and forthcoming publications by Hennig include "Morality politics in a Catholic democracy: a hard road towards liberalization of gay rights in Poland" in J. Haynes (ed.), *Religion and politics in Europe, the Middle East and North Africa: Secularization, democracy and citizenship* (Routledge, 2010), 208–225, and "Religion vs. economy? The regulation of artificial reproduction techniques in Italy, Poland, and Spain." in T. Beckers and Pascal Siegers (eds.), *The Boundaries of Life* (Springer, forthcoming).

Göran Hermerén is Professor Emeritus of Medical Ethics, Lund University, Sweden. From 2002 to 2011, he was the president of the European Group on Ethics in Science and New Technologies, Brussels, Belgium, and is currently a member of the National Council on Medical Ethics, Stockholm, Sweden, and chair of the permanent working group for science and ethics of ALLEA (All European Academies). He has served on many governmental commissions in Sweden, as a referee for international journals, and is involved in several ongoing EU-funded research projects. He has published books and papers on research ethics, ethics of stem cell research, the goals of medicine, and priority setting in health care and contributed to many encyclopedias. Some of his recent publications are "The Principle of Proportionality Revisited: Medicine, Health Care and Philosophy" 15(4): 373–382, (2012). "Definitions of fraud and misconduct revisited." *Jahrbuch für Wissenschaft und Ethik*, Vol. 18. (de Gruyter 2013: 85–121); "Human stem-cell research in gastroenterology: experimental treatment, tourism and biobanking." *Best Practice & Research Clinical Gastroenterology*, Vol. 28 (2014: 257–268); "Risk screening, testing and diagnosis: ethical aspects." *International Encyclopedia of Social and Behavioral Sciences*. 2nd ed, ed. by James Wright (Elsevier 2015: 698–705); and "Werte." In: *Handbuch der Bioethik*. Ed. Dieter Sturma, Bert Heinrichs (J.B. Metzler 2015: 165–174).

Mats Johansson is a philosopher and an associate professor in medical ethics at Lund University, Sweden. His research targets various issues in research ethics and clinical ethics. His work often focuses on the ethical and regulatory challenges of proxy decision making. Johansson is currently also working as Lund University's advisor in matters concerning research ethics and research ethics regulation and serves as a member in the university's ethics council. He has been recruited for conducting ethics analysis on local, regional, and national level. Some (relevant and recent) publications include Hug, K. and M. Johansson, "Challenges to informed consent in first-in-human trials involving novel treatments: A case study of Parkinson's disease." *J Parkinsons Dis* (forthcoming); Hug, K. and M. Johansson, "The ethics of withdrawal: the case of follow-up from first-in-human clinical trials." *Regen Med* 12(1) 25–36 (2017); Broström, L. and M. Johansson, "Involving children in non-therapeutic research: on the development argument." *Med Health Care Philos* 17(1): 53–60 (2014); Johansson, M. and L. Broström, "Does peer benefit justify research on incompetent individuals? The same-population condition in codes of research ethics." *Medicine, health care, and philosophy* 15: 287–294 (2012); and Hug, K., G. Hermeren, and M. Johansson, "Withdrawal from biobank research: considerations and the way forward." *Stem Cell Rev* 8 (4): 1056–65 (2012).

D. Gareth Jones has been a Professor of Anatomy and Structural Biology at the University of Otago, Canada, and Deputy Vice-Chancellor. Prior to this, he held positions at the University of Western Australia and University College London. He is a Visiting Fellow at St Edmunds College, Cambridge, UK, and an Adjunct Professor at Liverpool Hope University. He is Deputy Chair of the New Zealand Government's Advisory Committee on Assisted Reproductive Technology. Research interests include human anatomy and ethics (especially the uses of human cadavers and tissues), neuroethics (e.g., neuroimaging and dementia), stem cell technology (induced pluripotent stem cells), artificial reproductive technologies (developmental issues, enhancement, and therapy), and bioethics and theology. Jones has authored many articles and books, among them, *The Peril and Promise of Medical Technology* (Peter Lang, 2013), *Bioethics: When the Challenges of Life Become too Difficult* (ATF Press, 2007), and *Designers of the Future: Who Should Make the Decisions?* (Monarch Books, forthcoming).

Kevin McGovern is a Catholic priest who was ordained in 1988. He holds a Diploma in Applied Science from the Queensland University of Technology, Australia, and a Licentiate of Sacred Theology from Weston Jesuit School of Theology in Boston, USA. From 1997 to 2006, he lectured in Christian ethics and moral theology at the Brisbane College of Theology, Australia. From 2007 until 2016, he was the director of the Caroline Chisholm Centre for Health Ethics. He is a member of the Department of Moral Theology and Canon Law at Catholic Theological College in Melbourne and an Adjunct Lecturer at Australian Catholic University. In 2011, he was appointed by the Australian Government as a member of the Heerey Review of Australia's cloning and embryo research laws. Since 2012,

Fr McGovern has been a member of the Australian Health Ethics Committee (AHEC).

Jennifer Merchant obtained her Ph.D. in Political Sciences from Sciences Po Paris and now teaches at the Université de Paris II. Her research privileges a multidisciplinary approach at the crossroads of political science, law, gender studies, and bioethics in Anglo-American and European countries. She is currently working on two projects: (1) analyzing from a comparative perspective (France/USA) policies relative to gender and health care and research and (2) comparative public policy analyses of the framing of human genome editing (HGE) and their impact on the future of human reproduction. She is a member of the *Inserm* Ethics Committee and of the Institut Universitaire de France and was a member of the U.S. National Academy of Sciences Committee on HE. Her publications include *Procréation et politique aux Etats-Unis, 1965–2005* (Paris, Berlin, 2006), "Assisted Reproductive Technology in the United States: Towards a National Regulatory Framework," and "Privacy, Autonomy and Public Policy: French and North American Perspectives" in the special issue on "Solidarity and Autonomy: Two Conflicting Values in English and French Health Care and Bioethics Debates," Marie Gaille & Ruth Horn (eds.), *Theoretical Medicine and Bioethics*, 37(6): 503–516 (December 2016) and her coauthorship of *De la Constitution aux droits: les fondements juridiques des droits des femmes aux Etats-Unis* (with Vincent Michelot) (ENS Editions, 2018).

Ingrid Metzler is a postdoc at the Department of Science and Technology Studies of the University of Vienna, Austria. Her research focuses on the intersections of biomedicine and democratic orders, in particular in the area of genetics and genomics in human reproduction.

Frank O'Keeffe received his Masters of Bioethics from the John Paul II Institute for Marriage and Family in Melbourne, Australia, in 2016. Previously, he obtained a Bachelor of Laws (LLB) from La Trobe University, Melbourne, Australia, in 2013 and was admitted to the legal profession before the Victorian Supreme Court in 2015. His main research interests are procreation and assisted reproductive technology in Australia and Christian ethics.

Massimiliano Passerini graduated from the University of Perugia, Italy, in Philosophy and History. He wrote his thesis on political religion and gnosis in the work and thought of Eric Voegelin. At the University of Münster, Germany, he completed a second degree in Romance Philology with a thesis on the political passions of the young Ugo Foscolo. He worked as a research assistant at the Cluster of Excellence "Religion and Politics" at Münster University in a project on religion and biopolitics. As part of this project, he researched the influence and argumentation of religious and political actors in public biopolitical debates and normative processes in Italy, focusing especially on embryonic stem cell research, cloning, and pre-implantation diagnostics.

He is now a lecturer in contemporary history and Italian language and culture at the Universities of Münster and Osnabrück.

Anna Pichelstorfer is working as a university assistant (prae doc) at the Department of Science and Technology Studies of the University of Vienna in Austria. She is interested in questions of science, law, and democracy, in particular in public debates and policymaking around biomedical technologies.

Jacob Dahl Rendtorff is Associate Professor of Ethics at the Department of Social Sciences and Business at Roskilde University in Denmark. He received his Ph.D. at the University of Copenhagen in 1999 and his doctoral degree in Administration and Social Sciences of the Roskilde University in 2010. Rendtorff's current main research interests are business ethics, corporate social responsibility, corporate legitimacy, philosophy of management, and stakeholder theory. Bioethics, biotechnology, innovation, and nanotechnology are also issues of research. He is the author of, among others, the books *Bioetik og ret: Kroppen mellem person og ting* [Bioethics and law. The body between person and thing] (Gyldendal 1999), *Basic Ethical Principles in European Bioethics and Biolaw* (Center for Ethics and Law 2000), *Responsibility, Ethics and Legitimacy of Corporations* (Copenhagen Business School Press 2009), *French Philosophy and Social Theory: A Perspective for Ethics and Philosophy of Management* (Springer 2014), and *Cosmopolitan Business Ethics: Towards a Global Philosophy of Management* (Routledge 2018).

Nathalie Schiffino-Leclercq is Professor of Political Science at the University of Louvain (Belgium). She is a guest researcher at the Université Laval (CAPP, Canada) and at the Université Saint-Louis (CReSPO, Brussels). She earned a Ph. D. in Government and Public Administration from Louvain. Areas of expertise include policy analysis, crisis management and risk regulation, as well as theories of democracy. Major works include *Organizing after crisis: The challenge of learning*, Public Action Series (Peter Lang 2015), *Crises politiques et démocratie en Belgique* L'Harmattan (2003), and "Uncertainty and cleavages at stake: Do the Belgian Constitutions of 1831 and 1993 stabilize political power?" in L. Imbeau and S. Jacob (eds.), *Behind a veil of ignorance? Power and uncertainty in constitutional design,* Studies in Public Choice, vol. 32 (Springer 2003: 121–133).

Mirjam Weiberg-Salzmann is Senior Research Fellow at the Cluster of Excellence "Religion and Politics" in Münster. She received her Ph.D. in Political Sciences in 2009. Her research areas are international politics, comparative government, and peace and conflict studies. She focuses on the topics religion and politics, violence and conflicts in multiethnic and multireligious states, democracy, democratization and demography of developing countries, and biopolitics. Her regional focus lies on South and South-East Asia, the USA, Great Britain, Ireland, Italy, and Norway. Weiberg-Salzmann is the author of *Die Dekonstruktion der Demokratie durch die Kultur* (Nomos 2011), *Ursachen und Erklärungen gewaltförmiger Konflikte in Demokratie der Dritten Welt* (Universität Rostock 2008), and *Zwischen Bürgerkrieg* *und friedlicher Koexistenz. Interethnische Konfliktbearbeitung* (Transcript Verlag 2007) and has coedited with T. Salzmann and V. Skirbekk *Wirtschaftspolitische Herausforderungen des demographischen Wandels* (VS 2010). Some of her recent articles include "The Radicalisation of Buddhism in the Twentieth and Twenty-first Centuries" in *Politics, Religion & Ideology*, 15/2 (2014), 283–307, and "Ending Ethnic Civil Wars by Negotiations or by Military Defeat?" in *Journal of Asian and African Studies*, 50/2 (2015), 135–159.

Guido de Wert is Professor of Biomedical Ethics at Maastricht University, the Netherlands, Research Schools CAPHRI and GROW. He is a member of the Ethics Committee of the European Society of Human Reproduction and Embryology (ESHRE) and of the Public and Professional Policy Committee (PPPC) of the European Society of Human Genetics (ESHG). He served on the Central Committee (CCMO) that reviews all embryo research protocols in the Netherlands. He participated in both evaluations of the Dutch Embryos Act (2006, 2012). His international publications include "Treatment of the Embryo: Facts and Moral Questions." H. M. Vroom et al. (eds.), *Looking beneath the Surface: Medical Ethics from Islamic and Western Perspectives* (Rodopi, 2013; 145–174).

Ulrich Willems is Professor of Politics at the Institute of Political Science at the Westfälische Wilhelms Universität Münster and Principal Investigator at the Cluster of Excellence "Religion and Politics." He earned his Ph.D. in Politics at TU Darmstadt in 1996 and completed his habilitation at the University of Hamburg in 2008. His core areas of research are political theory, theory of democracy, political sociology, the political and judicial constitution of pluralistic societies, politics and religion, associations, the political system, and socially regulative policy, especially biopolitics. Willems has written the book *Entwicklung, Interesse und Moral: die Entwicklungspolitik der Evangelischen Kirche in Deutschland* (Leske + Budrich 1998). He has edited and coedited a number of publications, such as *Moderne und Religion. Kontroversen um Modernität und Säkularisierung* (Transcript 2013), *Politik und Kontingenz* (Springer VS 2012), and *Interessenvermittlung in Politikfeldern* (VS 2009). Some of his recently published articles include "Religion und politisches System in der Bundesrepublik Deutschland" in *Zeitschrift für Religion und Gesellschaft*, 1/2 (2011), 313–318; "Reformbedarf und Reformfähigkeit der Religionspolitik in Deutschland" in *Zeitschrift für Politik*, 55/1 (2008), 65–86; and "Moderne und Religion nach dem Zusammenbruch der Säkularisierungstheorie: Does anything go?" in D. Pollack (ed.), P. L. Berger: *Nach dem Niedergang der Säkularisierungstheorie. Mit Kommentaren von Detlef Pollack, Thomas Großbölting, Thomas Gutmann, Marianne Heinbach-Steins, Astrid Reuter und Ulrich Willems* sowie einer *Replik von Peter L. Berger*, Münster: Centrum für Religion und Moderne, 2013, 29–33.

Yvonne Kathrin Zelter, Mag. Theologiae, is scientific assistant of Monika Bobbert at the Seminar for Moral Theology of the Faculty of Catholic Theology, University of Münster, Germany. Her dissertation project about ethics of genome editing

(CRISPR/Cas9) is situated at the Catholic Faculty (Prof. Dr. Dr. Matthias Beck) of the University of Vienna, Austria. After her theological studies in Trier and Vienna from 2008 until 2015, she completed a traineeship at a grammar school in Vienna and a journalistic training at the Catholic Media Academy Vienna. Her research focus lies on biomedical ethics with special regard to questions concerning the beginning of human life and on interreligious and intercultural ethics.

Part I
Introduction

Chapter 1
Challenging the Political: Religious Actors and Religious Arguments in Liberal Democracies

Mirjam Weiberg-Salzmann and Ulrich Willems

Abstract Biopolitical questions are nationally and internationally important and there is profound moral-ethical disagreement about whether technologies—such as embryo research or cloning—should be allowed or forbidden. Among the social groups that strongly shape the character of these debates are many religious organizations and actors. But the legitimacy and the functionality of such religious interventions into the "secular" state are both politically and theoretically controversial. The controversy centers on the liberal theory that public debates—in particular decision-making processes—should be restricted to arguments and reasons which are in principle accessible, intelligible, and agreeable to all citizens. And "religious" arguments and reasons do not fulfill these standards. Despite this theoretical debate, there is a lack of empirical knowledge about the extent, nature, and consequences of the use of religious arguments. Therefore, how religious traditions and actors formulate their bioethical positions and the kinds of reasons they give to validate their positions are of particular interest. Moreover, how do religious actors influence politics, e.g., in which ways and under what conditions do religious positions and arguments achieve influence in political decisions?

1 Introduction: Religion and the Formation of Public Policy

In general, the debates concerning the relationship between politics and religion within the political science have been conclusively decided: the differentiation into a secular (political) and clerical/sacred (religious) sphere and the institutional separation of both powers have seemed to be irreversible. The historical experience of modernization, secularization, and democratization going hand in hand in Western countries demonstrated that the separation of religion and politics was a precondition of successful state foundation. At the same time, it was assumed that conflict and violence would increase if religion and politics intermingled. It is a prevalent consensus that the separation of church and state and accordingly a secular state

M. Weiberg-Salzmann (✉) · U. Willems
University of Münster, Münster, Germany
e-mail: mirjam.weiberg-salzmann@uni-muenster.de; ulrich.willems@uni-muenster.de

M. Weiberg-Salzmann, U. Willems (eds.), *Religion and Biopolitics*,
https://doi.org/10.1007/978-3-030-14580-4_1

(in terms of neutrality toward religious communities and convictions) was an essential (albeit insufficient) precondition for stable democracies.

Recent partial revision of this strict separation can basically be attributed to growing inconsistencies between empiricism and theory. At least in the USA, the theoretical discussion about the admissibility of religious arguments into public dialogue has never completely come to a standstill. It used to be that liberal arguments tended to classify democracies according to how their laws gave expression to a separation of religion and politics, so that some were viewed as compatible with liberal standards and others were not. Today's theoretical debate is, however, conducted in a more sophisticated way. The majority of theorists who used to argue along strictly secular-liberal lines have now moved over to wanting to limit religious arguments in public dialogue instead of excluding them. Thus, there are only a few on the side of the pluralist critics who would support an unconditional admission of religious arguments into public dialogue, and who in particular would allow exclusively religious reasons for a decision. Moreover, in non-Western countries as well as within Europe, all these presuppositions could easily be challenged since separation between church and state was often not nearly as clear as theoretically claimed. In many non-Western countries, religious institutions assumed important welfare, legitimizing and integrating functions. Religious convictions continued to be a strong motive for many actors, and quite frequently one could find religious elements in constitutions and laws. Despite empirical evidence pointing to the importance of religion for political and social processes, academia primarily regarded religion as a symbolic or instrumental resource and as an independent or structuralizing element only at the margins. In the case of institutions as well as concepts, separation was often not as clear as theoretically demanded. However, general statements may be made only to a limited extent because of the very specific relationship between religion and politics in different countries and traditions. Even between seemingly similar countries in the same geographic area (Germany–Switzerland–Austria; USA–Canada; Denmark–Sweden–Norway) or religious setting (Australia–New Zealand; France–Italy–Spain; Netherlands–Belgium), the presence, acceptance, and effectiveness of religious arguments are extremely different. Even though empirical research recorded these heterogeneities in great detail, the liberal separation paradigm continued to dominate the normative and theoretical discourse. Cases that did not fit into this scheme were merely considered as specific rather than representative instances. The existing debate about the role played by religion in modern democracies moreover reveals a focus on questions about the influence religious actors may have on politics ("what influence, if any, exists?") and its structural and institutional prerequisites ("what elements of the system enable/encourage such influence?"). However, questions concerning how (religious) argumentation is patterned, how the content of arguments is structured, and about the discursive environment has essentially been marginalized.

Our focus is on the relationship between politics and religion in Western democracies, in particular on the role of religious actors and the discursive forms and reasons underlying religious arguments in political discussions. The starting point is the continuing debate on separating state and church in liberal-democratic states and, associated with this, on the legitimacy and functionality of religious interventions in the "secular" state. In the past, the theoretical discussion about the influence of religion has been characterized by two positions: on the one hand, there are the representatives of a liberal concept of democracy (e.g., Rawls, Habermas, or Audi), who are willing to concede no space for religious arguments, or at most a merely subordinate significance, in the political, public sphere, and who favor a strict separation between secular-public and private-religious spheres. In this view, only a religiously neutral state can best meet the citizens' demands and mediate conflicts in a pluralist society. Religious arguments, however, are too particularistic and contain within them potential conflicts. On the other hand, there are the representatives of a pluralistic concept of democracy (e.g., Wolterstorff, Quinn, and Bader), who consider a complete separation of the spheres both impossible and unnecessary. According to them, the neutrality of liberalism bears an epistemological limitation of, inter alia, religious arguments and contains the unjustified insinuation that religious arguments were not generally intelligible and particularly conflict prone.

We have chosen the controversial issue of biotechnology to analyze the role of religious actors and arguments in forming public policy. Biological and life sciences are a suitable object of study for several reasons: progress in the life sciences has made human nature accessible to technical intervention to a hitherto unknown degree, thus putting more and more biopolitical topics on the political agenda. Furthermore, there is a deep-rooted moral and ethical disagreement about whether such technologies ought to be allowed or prohibited. Finally, yet importantly, it is religious organizations that are among the social groups decisively shaping these debates. The rapid scientific progress in these fields caused considerable concern among many religious actors because it affects core aspects of their religious beliefs. The biopolitical debates are of vital interest, since they address key aspects of how human identity is defined: what is it that makes us human beings in the first place? For the answers to this fundamental question, religious beliefs play a fundamental role. Thus, the public is both highly sensitive and responsive to the arguments religious actors employ. Depending on country and tradition, the religious actors hold different positions, make use of different reasoning, play different roles in decision-making processes, and have different impacts on political settlements. In comparing the political debate about and the regulation of embryo research, in particular pre-implantation genetic diagnoses, therapeutic and reproductive human cloning, and human stem cell research in different countries, we try to get closer to the questions raised above. Overall, we will show that the theoretical discussion (about the function and status of religion/religious arguments) fails to play a sufficient role and that the structure of a religion's arguments depends less on a general institutional arrangement than on the specific manifestations of religion and the sociocultural context. In the end, we try to answer the question of whether the democratic state should (or should not) renounce all religious substantiation of political decisions in this field.

2 The Institutional Separation of Religion and Politics

In the debates on modernity, there was a clear answer to the question about what kind of role religion would play in politics: none. At least in Western countries, religion was to be dismissed into the private sphere.[1] After centuries of conflicts between the religious and political systems, it seemed irreversible that each had finally developed its own legitimations, institutions, and membership rules. The Peace of Westphalia not only put the seal on confessionalization but also on the subordination of religion to the political sovereign.[2] Thus, the first step toward subjectivizing belief and an individualized inner religion was taken. The rise of nation states in Europe and later on their democratization led to the irrevocable weakening of the churches' secular power.[3] Research on political transformation and democratization, which emerged after the fall of the Iron Curtain, and debates concerning the "fourth wave of democratization" claimed that the separation of church and state was an important (even though not sufficient) precondition for stable democracies.[4] Although religion was also accorded a positive and workable potential in political opinion- and decision-making processes, it was considered most beneficial if religion remained within the spheres of private life or civil society. And the legitimacy of political decisions, too, was seen as greatest if they were based on generally intelligible, secular argumentation thus accessible to all citizens. Religious arguments were meant to be generally excluded from the political decision-making process to ensure an egalitarian inclusion of all members of a community in civil society. In ethnically and religiously fragmented societies, there was a particular risk of the exploitation of

[1]Francois Foret and Xabier Itcaina (2012) make critical use of the concept of multiple modernities, in which secularization is only more or less intrinsic. Hans Joas (2014) challenges the thesis that modernization leads to secularization and proposes a future in which modernity and faith can be mutually enriching.

[2]Current approaches consider secularization as a process of transformation (with separation, interaction, and co-implication) in which religion and politics are jointly contested and reshaped (Scherer 2013).

[3]Four aspects have prevailed in the secularization debate: they assume the decline, disappearance, privatization, and transformation of religion (compare Casanova 1994). For a detailed exploration of the history, the sociology, America's exceptionalism, and the revival of world religions, Bryan (2010) and Taylor (2014).

[4]On the third wave of democratization, compare Huntington (1991). The third wave of democratization designates the spread of democratic rule beginning in Southern Europe at the end of the 1970s. The first wave from 1828 to 1926 democratized, among others, the USA and Mexico and the second wave from 1943 to 1962 Nazi and Fascist regimes (e.g., Germany, Italy, Japan) as well as colonies that were granted independence (e.g., India). Marie A. Eisenstein (2012) specifies that in a liberal democracy, one can pursue an absolute separation of church and state, but one can neither ask for nor pursue an absolute separation of religion and politics. Critical of the position that secularization is a necessary condition for democracy (Losonczi 2015).

religion leading to long-lasting conflicts and violence (Anderson 2004, 2009; Dark 2000; Hildebrandt and Brocker 2005; Nehushtan 2015).[5]

Although some form of institutional separation is basically undisputed,[6] there has been criticism of the assumptions that religion that no longer had a legitimizing role in politics was outdated, and that no policies or bills should be pursued that cannot be secularly legitimized, since in pluralistic societies this could be considered unjustified discrimination against religious citizens. Yet recently the more far-reaching question is being asked of whether "the existence of the liberal and secular state is based on assumptions which the state itself cannot guarantee"—both insofar as liberal societies depend on their citizens being committed to common values and therefore using their freedom reasonably, but also in the sense of how the political sphere is extrapolated, legitimized, and delimited at all (Böckenförde 2004; Stein 2001; Sigurdson 2010; Moore 2009; Stüwe 2018; Bernt Torvild 2018).[7] Both questions have become all the more urgent because the modern state is more and more frequently in the position of deciding conflicts that need to be regulated (e.g., abortion, reproductive medicine, cloning, euthanasia) but which do not have fixed moral and ethical standards. Open questions are: which actors and agents should participate in these processes of negotiation and decision-making, and what shall their arguments involve? Do religious traditions dispose of specific moral resources in answering exactly these questions or is it precisely the cultural and ideological pluralism of society which prohibits drawing on religious traditions?

2.1 Religious Arguments in Liberal Democracies

Since the 1990s an intense normative debate has developed within political theory about the status of religious arguments in the public spheres of liberal democracies. The controversy centers on liberal theorists' thesis that democratic public debate should be conducted with arguments and reasons which are either secular or are acceptable to all citizens, at least in principle. Liberal theorists maintain that "religious" arguments and reasons are not legitimate justifications because legal coercion

[5]Compare the previous studies by Casanova (1994). For the positive role of religion in politics and its capacity for peace-making, Banchoff (2008). For an overview of religion in world politics, Rowe (2012) and Esposito (2008).

[6]Generally, the separation of church and state in Western democracies is not contested. Controversial issues concern its institutional structure (e.g., a separation, cooperation, established church or endorsed church model) as well as informal practices (the actual influence of the church, support of religious communities). Benyamin Neuberger (2011) argues that the general trend in liberal Western democracies is one of convergence of the different models toward a closer cooperation between church and state. On alternative approaches regarding individually negotiated forms of separation, compare Alfred Stepans' "twin tolerations" (2001).

[7]David S. Gutterman and Andrew R. Murphy (2016), for example, emphasize the political importance of religion in regard to the individual by arguing that questions of religion are questions of personal, social, and political identity.

is only legitimate if the decisions are being made on the basis of a shared (moral) justification. Or, more precisely, the normative theoretical political debate about the status of religious arguments in the public sphere of liberal democracies is (at least mainly) not about the institutional or legal regulation of the relationship between church and state, but about defining what can be morally expected of religious people's public statements and actions when they are acting in their roles as citizens. The debate is focused on either defending or refuting the hegemonic liberal thesis that the public dialogue in democracies needs to be conducted either solely or at least always with secular arguments and reasons, or at least with those ones basically acceptable to all citizens, for legal coercion was legitimate only if decisions were taken on the basis of shared (moral) justifications—however, this restriction, according to the liberal theorists' views, removes "religious" arguments and reasons from the pool of necessary and sufficiently legitimate justifications (Willems 2003).[8] These liberal positions, however, are not monolithic, but differ considerably from each other. The liberal mainstream is occupied by authors whose position might be called moderately exclusivist. They do not usually plead for a complete exclusion of religious arguments from public debate, but claim that religious arguments cannot constitute the only underpinning of regulations that are reinforced with legal coercion. These positions differ regarding the specific provisions they make for how religious arguments should be used, regarding how specific arguments are categorized as religious or nonreligious, and regarding in which spheres religious arguments should be restricted.

Robert Audi distinguishes between religious and secular arguments on the grounds of accessibility and on that of their suitability for consent (Audi 2008). Since religious arguments are neither universally intelligible nor universally acceptable, they do not qualify as justifications for coercive measures. Instead, he recommends that the principles of secular reason and secular motivation be applied to all political decisions. Although religious arguments are not excluded from the political process, he demands that religious citizens should have to put forward secular reasons in addition to religious ones. Audi further assumes that religious citizens in liberal democracies are generally "bilingual" (in terms of their religious and secular convictions) and are able to balance their religious and secular reasons and motivation. By contrast, John Rawls draws the crucial dividing line between comprehensive doctrines (the "Good") of religious and secular origin on the one hand and a narrow set of universally acceptable, political principles of justice on the other (Rawls 1993; 1999). If they are to serve as justifications, this small set of universally acceptable, political principles of justice must exist independently from the doctrines of the Good. Moreover, Rawls limits the area in which religious and nonreligious comprehensive doctrines (the Good) are not able to develop any capacity of justification to fundamental questions of justice and of the development of society's basic

[8]Willems, Ulrich. (2003). Religion als Privatsache? Eine kritische Auseinandersetzung mit dem liberalen Prinzip einer strikten Trennung von Politik und Religion. *Politik und Religion*, eds. Michael Minkenberg und Ulrich Willems, 88–112. Wiesbaden: Westdeutscher Verlag.

structure. However, religious citizens may resort to religious arguments in those debates as long as they also put forward (within an adequate time frame) arguments derived from basic political principles of justice. Regarding the legitimacy of religious arguments, Jürgen Habermas differentiates between the political institutions that draw up or implement generally binding decisions and the informal public (Habermas 2008a, b). Habermas assumes that religious experiences and beliefs have access to a reservoir of moral sensitivities. Therefore, secular citizens can be expected not to deny a priori that religious arguments might contain reasonable content. Religious citizens are entitled to put forward their religious arguments in public dialogue. However within political institutions, only publicly accepted or secular reasons are legitimate. Positions that are religiously justified therefore must be translated into secular language. The translation of religious content into secular language constitutes a cooperative mission of religious and nonreligious citizens. According to Habermas, however, there is no clear-cut diving line between secular and religious reasons. All citizens together establish it within the political process.

There have been a number of objections raised against these liberal or deliberative positions on the use of religious arguments in politics, (Wolterstorff 1997, 2003; Wolterstorff and Cuneo 2016; Audi and Wolterstorff 1997; Quinn 1997, 2006; McConnell 1999; Walzer 1998, 2015; Bader 1999, 2007; Taylor 1998; Grotefeld 2000, 2006; Willems 2003). Audi's distinction between religious and secular reasons is opposed on the grounds that neither is every religious argument inaccessible nor is every secular argument acceptable: religious arguments may be quite plausible and open to rational criticism, whereas secular utilitarian arguments may be highly divisive and controversial. Rawls's claims are challenged by doubts about whether there in fact exist enough indisputable principles in pluralistic societies that can reasonably be expected to have broad consent. These doubts appear especially justified if we, like Taylor, assume that the supposedly free-standing principles of justice only exist in versions influenced by comprehensive doctrines of Good. Similar objections are raised against the procedural, discursive generation of principles of consent that Habermas assumes. A criticism that applies equally to all three theories is that religious people will be disproportionally affected (compared to secular citizens) by the envisaged restrictions on public debate. But what kind of alternatives do the critics of liberal or deliberative positions suggest for the use of religious arguments in public? Some authors—such as Nicholas Wolterstorff, Philip L. Quinn, and Michael W. McConnell—do not perceive any reasons for religion to receive special treatment. Specific restrictions on the use of religious arguments are neither necessary nor justifiable, they argue. Other authors, such as Richard Bellamy, Michael Walzer, and Veit Bader, agree on the need for regulation but argue that it not be restricted to religious arguments alone. On the one hand, their proposals not only target the use of religious and nonreligious arguments but also their impact and the range. On the other hand, they suggest that conditions and incentives for mutually acceptable negotiated solutions should be created.

Regarding this normative theoretical debate about the role of religious arguments, it seems remarkable that it is conducted at all. For apart from anecdotal evidence, there exists no systematic empirical knowledge about the extent, nature, and consequences of the use of religious arguments in public debate. We do not even yet know whether religious stakeholders accept the proposed restrictions, criticize them as unfair, or even completely ignore them in practice. We also do not know if productive ways to deal with pluralism have emerged in the practice of liberal democracies beyond the horizon of liberal theory, and which still await the decoding of their normative content.

3 The Development of the Biopolitical Debate

The use of modern biotechnologies like human embryonic stem cell research, human cloning, and pre-implantation genetic diagnosis has provoked heated political debates in almost all countries.[9] While in the 1960s, questions concerning human life were still discussed within a religious context, over time the discussion has been largely secularized. The scientific community and industry frequently demand that biopolitical issues are to be separated from religious issues and that the discussion ought to focus solely on medical facts. The focus often lies on utilitarian and functional viewpoints as opposed to moral issues. The positions found in the public debate are, however, more diverse. Religious ideologies have played an important role in the debates on abortion and stem cell research, among others, although some people might not have even been aware that they had been arguing on the basis of religious worldviews (Schlieter 2005; Lauritzen 2005; The President's Council on Bioethics 2015; Dodds and Ankeny 2018).

Religious thinkers and traditions have been involved in bioethical discourse since the 1950s. However, in the course of time the manner as well as the extent of their interventions has changed. From a historical viewpoint, bioethics are rooted in religion, not least since religion was an integral part of medical ethics. Medical ethics have a long-standing tradition dating from the Middle Ages of being based on the sacraments and theological virtues. In Western countries, Christianity has, furthermore, influenced the founding of hospitals and the treatment of the sick. Taking the origin of bioethics into account, it becomes obvious that the majority

[9]Pre-implantation diagnostics is used as part of in vitro fertilization (IVF). By examining the embryo's cells before implanting it in the uterus, it helps detect the embryo's genetic predispositions (e.g., for gene defects or certain disorders). Embryonic stem cell research makes use of the developmental potential of so-called pluripotent cells—undifferentiated cells that can develop into any type of body cells—from IVF-generated embryos that are destroyed during the procedure (Bryant and Schwartz 2008; Devolder 2015; Pereira 2016). In the political debate, human cloning refers to the method of somatic cell nuclear transfer, which was used to generate the sheep "Dolly" in 1996. As part of this technique, the nucleus of a zygote—the first embryo phase—is replaced by the genetic material of a somatic cell, and the cell is then stimulated to behave like an embryo or, more precisely, like a zygote. The term reproductive cloning is used when the embryo is implanted into a uterus and carried to term—as in the case of "Dolly." In therapeutic cloning, the embryo is destroyed in order to obtain embryonic stem cells that carry the genetic information of another organism and that can subsequently be used for therapeutic purposes (compare Seidel Jr. 2003, p. 452; Korobkin and Munzer 2007, pp. 7–25; Jensen 2014).

of this first generation of bioethicists were clergy or religious men. However, a counter-trend emerged as early as the 1960s that was set to eliminate any transcendent-religious argumentation. While on the one hand many less-religious specialists started taking over the field, many theologians and religious ethicists gave in to society's demand for secularization. Science, law, philosophy, and the educational system demonstrate similar tendencies toward driving religion into the private sphere. Since most bioethicists received their education in this academic context, they turned toward secularization just as their educational institutions did. At the same time, the 1960s was a period of inner-theological controversies. Since the turn of the century, Protestant denominations had faced questions concerning the demystification of the Holy Scriptures, Protestant liberalism, the "Social Gospel Movement," and theologies of "God's Death"; modernism was its Catholic counterpart (Crockett 2013). The contraception controversy was the most prominent incident with regard to bioethics and moral theology, especially in the context of the release of the encyclical *Humanae Vitae* in 1968. Although the controversy was a Catholic phenomenon, the discussion was widely followed in society as a whole. The encyclical generated a reaction diametrically opposed to what had been expected from it. Laymen and parts of the clergy were disappointed after its publication. This feeling of discontentment with the Church's teachings laid the foundation for the so-called "New Morality" and "Proportionalism," which spread in many seminars and theological faculties. This was of particular importance to bioethics, since Proportionalism tended to emphasize the consequences and life situation of a particular action, in order to morally evaluate it (Hallett 1995). Taken to an extreme, this argumentation was capable of justifying any biopolitical measure, with the reasoning that it brought about more positive than negative consequences (the end justifies the means). Looking at prominent institutions and leading bioethicists (e.g., in the USA) shows that many disillusioned Christians constituted those who broke away from clerical structures in order to found alternative secular institutions and consequently renounce or deprioritize theological input.[10] Clergymen who wanted to appeal to a wider audience were another one of the reasons for the secularization of discourse. Restricting religious discourses developed out of the necessity of using a moral language that was appropriate for the pluralist public and the interdisciplinary nature of the topic. Religious actors no longer only addressed their own members, with whom they could refer to a common set of argumentation structures and meanings. Instead, the immediate audience comprised a heterogeneous group of scientists and professionals, who did not share the same moral values, let alone the same religious concepts. Within this setting, it was necessary to translate the religious views of a particular tradition into secular language. The formal separation of church and state made it difficult for religious actors to use

[10]Examples are for instance André Hellegers, who later founded the "Kennedy Institute of Ethics" in Georgetown. Hellegers, a gynecologist, was a member of the Pontifical Commission for Birth/Population Control. Daniel Callahan, publisher of the "Commonwealth Magazine," was appalled by the encyclical as well, and became the cofounder of the Hastings Center. The Kennedy Institute and the Hastings Center both had great influence on bioethics.

"religious" arguments, so far as institutionalized forms of bioethical discourse were concerned (e.g., in preparing or assisting political decisions). The use of commonly understood language and nonreligious argumentation was required in order to have their own viewpoints recognized in the documents and reports of the various bioethics commissions (Myskja 2009). However, since the end of the 1980s at the latest, a new countermovement that criticized the secularization of the biopolitical and bioethical discourse emerged (Stein 2007, p. 313; Tollefsen 2008; Hurlbut 2017).

Within the individual countries, biopolitical and bioethical discourses developed very differently with regard to the agents involved, arguments used, and institutional forms. While in some countries hardly any or only very tentative public debates about stem cell research and PID took place and were limited to the parliament or else the topic was consciously ignored by the governments, in other states debates have taken place in the media and in parliament on a federal level. Ethics commissions and expert panels were installed in almost all the countries, although here the extent and the form of consultation differed greatly, too. We have selected 16 countries in order to compare the biopolitical debate. All countries are similar with regard to systems and religion (Christian plural democracies), yet at the same time vary when it comes to the confessional structure of the dominant religious traditions, courses of historical conflicts, the institutionalized relationship between church and state, the extent of religious pluralism, and the way the terms state and nation are defined. The selected states can be divided into six main groups, classified according to the historical pattern of the relationship between religion and enlightenment (Martin 1978): First, Protestant states with high religious pluralism, low anti-clericalism, a low status for the clergy, religiously influenced civil religion, strict separation between church and state, and without a religious party, e.g., USA. Second, Protestant-English states with moderate religious pluralism, low anti-clericalism, a middling status for the clergy, religiously influenced civil religion, ties maintained between state and church, and without a religious party, e.g., Great Britain, Canada, Australia, and New Zeeland. Third, Protestant-Scandinavian states with low religious pluralism, low anti-clericalism, a high status for the clergy, secularly influenced civil religion, and smaller religious parties, e.g., Denmark and Norway. Fourth, mixed-confession states with a high level of religious pluralism (duopoly), low anti-clericalism, a high status for the clergy, religiously influenced civil religion, a partially maintained connection between state and church (partial separation), and influential religious parties, e.g., Netherlands, Germany, Switzerland, and Belgium. Fifth, Catholic-Latin states with low religious pluralism, distinct anti-clericalism, a low status for the clergy, a religious as well as secular civil society, the separation of church and state, and strong religious parties, e.g., France, Italy, Spain, and Austria. Sixth, Catholic states with little religious pluralism, low anti-clericalism, a high status for the clergy, a religious civil society, close connection of state and church and influential religious parties, as well as an integration of nationalistic religion against outside threats, e.g., Ireland and Poland.

3.1 Discursive Strategies of Religious Actors Within the Bio-political Debate

While the influence of religion on politics has been studied intensely in recent years, less is known about the discursive strategies employed by religious actors in political debates. However, many fundamental assumptions and justifications may be deduced from the general discussion about the relationship of the religious and political (democratic) systems. Among the following factors, we expect three to have the most important impact on the discursive strategies of religious actors within the biopolitical debates in different countries and traditions: the degree of secularization, the (non-)existence of a hegemonic religious tradition, and the specific content of a religious tradition.

The first factor is the degree of secularization. We posit that the more secularized a society is, the less likely it is that religious actors will resort to religious arguments. Arguments in democratic political debates primarily have the function of convincing other political actors of one's positions. Thus, in general, the more a society is secularized, the less sense it makes to adduce religious arguments. However, one can imagine the following two exceptions from this rule: First, even in a secularized society there may be a compelling rationale for employing religious arguments if religious actors, who are a minor part of a wider political coalition, aim to mobilize their constituency. A religious-secular split may also be conducive to the use of religious arguments despite a high degree of secularization. Furthermore, religious arguments also make sense if religious actors do not primarily want to have their positions accepted but rather strive to contest symbolically the privatization of religion (Casanova 1994, p. 225; Gabriel 2009; Achterberg et al. 2009; Halman and Petterson 2002). Furthermore, one may assume in the second conjecture that religious vitality has an influence on the use of religious arguments. The more intense the believers' belief, the more sense religious arguments make in view of the democratic premise (instrumental-rational), because under these circumstances they not only create resonance (as in the case of a low secularization level), but may also develop a considerable mobilization potential, since the value of religion is higher in private life. The probability of the use of religious arguments increases especially when a low secularization level and high religious vitality occur together. The more society holds onto traditional values and the more positive the attitude is toward a combination of religion and politics, the more probable is the influence of religious arguments on public concerns (Fink 2007, p. 177; Minkenberg 2002, p. 135). From the first and second factors follows a further assumption, namely that Catholic countries may show a stronger tendency to use religious argument in public than majoritarian Protestant countries, for Catholic countries are characterized by a low level of secularization and a higher level of religious vitality. Furthermore, the Catholic Church also expects its members to adhere to its guidelines and instructions in politics. Protestantism does not recognize any ethical mandate of the church which might bind an individual Christian in his political activity.[11] The next factor is the existence or nonexistence of a hegemonic religious tradition. In the case of a religious monopoly or hegemony, actors of the monopolistic or hegemonic religious tradition can appeal to a wider audience that is favorably inclined toward religious arguments. Moreover, in the case of a religious monopoly or hegemony, it is likely that elements of the monopolistic or hegemonic religious tradition will constitute a major part of a country's civil religion or public philosophy. Religious arguments from the monopolistic or hegemonic religious tradition can then be easily linked to a country's civil religion or public philosophy and be granted legitimacy.[12]

In addition to the religious framework, a religion's size and vitality, as well as the different contextual traditions of argumentation within Christianity, also play a role in the use of religious arguments. Crucially, religious traditions differ greatly concerning their views on the relationship between faith and reason. The Catholic Church has taught that morality is part of natural law and can therefore be discerned by reason alone. For Catholics natural law thus constitutes a resource for arguing in a secular manner. Correspondingly, the Catholic Church has no need to rely on the use of religious arguments. The use of nonreligious argumentation resources may even have a more wide-reaching effect than solely in its circle of believers (democratic premise). Therefore, it is more probable that the Catholic Church will seldom have recourse to religious arguments. Within Protestantism, one has to differentiate between several traditions. Lutherans adhere to the doctrine of the two kingdoms of God. Since in Lutheran teaching the earthly kingdom of a secular ruler is granted a substantial degree of autonomy, religious arguments prove ill-suited to debates on secular affairs. In addition, some Lutheran theologians even deny the existence of a separate Christian ethic that would be distinct from a universal ethic of humanity. Calvinism and sects of the Anglican Church are more "state-distant" than Lutheranism and give greater importance to autonomy. Reformed churches maintain that the doctrine of the Lordship of Christ applies to all spheres of life and deny any autonomy to earthly affairs and its laws. Other evangelical churches denounce the autonomy of earthly affairs and its laws in an even more radical fashion, advancing a literal interpretation of the Bible. With their strict adherence to scripture they have the lowest variability in traditions of argument. Hence, one may expect especially them to use religious arguments. This tendency increases when the religious traditions reject religion's restriction to the private life and tend to have conservative or orthodox views. One can expect that the Catholic and Lutheran Churches use overtly religious arguments less frequently (if at all) than reformed or evangelical churches. Yet, one further needs to consider that religious traditions are also internally divided by the

[11]"The most obvious is the difference between the notable more religious and Catholic countries of Southern Europe and the less religious countries of the Protestant North" (Davie 1999, p. 71; Kongregation für die Glaubenslehre 2002; Graf 2009, p. 15; Wiemeyer 2009, p. 32; Leitner 2002, pp. 371, 377).

[12]Nonetheless, different studies have shown different effects of religious pluralism (Stark and Finke 2000; Chaves and Cann 1992; Hamberg and Pettersson 1994; Unger 2006, p. 105).

degree to which they accept or reject the liberal doctrine of a, more or less strict, separation between religion and politics.[13]

Furthermore, smaller (possibly more radical) groups tend to attain public validation by using religious arguments, whenever their interests are concerned. Moreover, in addition to traditional argument, one must pay attention to the formal (hierarchical) organization of a religious tradition. Centralized organizations, such as the Catholic Church (with an international structure and the Pope's authority), can probably control the more radical members of their clergy better; it therefore becomes more likely that religious arguments will be used somewhere within a religious tradition, the less hierarchically structured it is (Moltmann 1997; Martin 1999; Silomon 2009; Bahle 2002).

Additionally, certain relationship patterns between church and state favor the use of religious arguments in public discussions: religious political elites and religious parties promoted religious arguments in the past if they resonated favorably with their corresponding voters. Altogether it is valid to say that, as long as the majority of the political elite remain members of one of the two Christian churches, one may continue to expect religious argument to resonate with the political elite (Willems 2001, 2004; Heron 2000). In the past, a country's institutional structure was often seen as a crucial variable in the process of mobilizing social/religious movements. However, it is debatable in which direction any such causation points. We assume that a close relationship between church and state is an expression of the acceptance and legitimacy of religious traditions and therefore of religious arguments, which thus in turn promotes the use of religious arguments (Tatalovich and Smith 2003, p. 96; Monsma 2000, p. 82).

Whether religious agents are involved in the public sphere depends not only on a country's religious and political structure but also on the subjects to be discussed: from an insider's perspective, the threat to core beliefs demands that religious agents respond. At the same time, there is a relatively great "external" acceptance of the church's engagement in certain political fields (such as human rights, war and peace, euthanasia, marriage, and family). This is especially true in countries where the church is seen as an alternative to repressive political systems. Under these circumstances, the use of religious arguments is especially likely when the reservoir of nonreligious arguments to enforce religious positions is limited (Langendörfer 2003; Silomon 2009; Pollack 2002). Moreover, discourses on a given topic frequently connect to already existing traditions. Therefore, it is easier for religious agents to put forward religious arguments when the dominating discourse can be connected to religious arguments. Hence religious positions may include aspects of a general position. Religious arguments may also be included with the aim of establishing a counter-discourse (Gottweis and Prainsack 2002). Finally, all the factors can be brought together into one argument: the historical course of conflicts between enlightenment and religion fundamentally influences the use of religious arguments and the specific way the church intervenes politically and acts as a lobby (Martin

[13]For the different positions within the Christian groups, Bruce (2006). For the rising influence of Evangelical organizations in politics, Freston (2006).

1978, 2005; Casanova 2007; Warner 2002; Lübbe 2004). In regard to evaluating how conflicts unfold, we expect that countries which subordinate religion demonstrate less religious activity, whereas more religious arguments may be expected in contexts of conflict or collaboration.

4 Characteristics of Religious Arguments

Liberal theory will not permit religious arguments (at least in the political decision-making process), a restriction justified by their largely not being generally intelligible and negotiable—unlike secular or scientific arguments. The foundation of the liberal, democratic state is to refer to arguments which are based on "a common human reason" (Habermas 2005, p. 125; Barbato 2011). Political decisions are only then legitimate when they are formulated in a language comprehensible and acceptable to all citizens. Apparently religious arguments do not belong in this category. On the one hand, this suggests that a dichotomous relationship between religious and secular arguments exists and that both forms of argumentation can be clearly distinguished from one another and on the other hand, that religious arguments are not comprehensible to people who do not share their epistemological premises (Habermas 2005, 2007, 2008a, b; Winandy 2011). So what are religious arguments and how do they differ, for example, from secular arguments? Which theoretical and empirical characteristics distinguish religious arguments? In general, religious arguments are understood as all arguments which refer to God, to theological positions, or to religious authorities and institutions. This common definition covers the majority of religious arguments, but says little about their theoretical status and their demarcation against other arguments (e.g., secular). Although this definition is a reasonably workable one for research, the question remains as to whether it encompasses all categories of religious arguments. It is thinkable, for example, that due to cultural shifts religious arguments were adopted as neutral arguments and that speakers are not aware of their religious character when using them (e.g., human dignity, respect for life). Some religious arguments have become common knowledge in Western societies, without one necessarily being aware of their biblical roots.[14] It is furthermore conceivable that for certain reasons religious arguments are being avoided by individual speakers, e.g., they reduce their argumentation down to "a common human reason" in the form of ethical and legitimate considerations in order to make their arguments also comprehensible to nonbelievers. The core of argumentation appears to be genuinely theological, but the form is secular. And how does one deal, for instance, with Catholic natural law? The reference to natural law, in the normative natural sense, cannot be classified as a

[14]Rainer Prätorius (2005) presents the example of social programs of American churches. Hans-Georg Ziebertz (2016) refer to the relation between religion and human rights. Compare also Jedan 2013.

directly religious argument, but could be classed as such by the intention, because the intention is to justify divine law with the help of a seemingly objective order, even though the moral concepts are defined in advance.

4.1 Epistemological Status and Empirical Operationalization

The question of whether one should approve religious arguments in public politics is closely related to that of the (epistemic) status of religious arguments. Religious arguments are assigned a different cognitive and logical status than secular arguments. Not only the truth content but also the acceptance of religious arguments as "knowledge" is in dispute. According to Habermas, religious convictions within a certain (religious) discursive context are debatable and criticizable (and therefore limitedly "true"), but in discourses operating according to general rationality an unclear, indefinite residue remains. Furthermore, in modern societies the Bible and the church have lost their universal authority. Consequently, claims of belief are reduced to historical truths, no longer containing any immediate and absolute logical truth which could have validity for all members of the state (Appiah 2008; Mehta 2008).[15] If Habermas's reasoning is correct that religious arguments do not have a place in public discourse since their accuracy (and truth) cannot be proven, or rather because they lack a shared inter-subjective sense, these might be a specific characteristic of religious arguments by which one may define them more closely. Let us first test the claim that religious beliefs (in contrast to secular) are only limitedly true: generally it applies to all categories of arguments that if a logical conclusion (deduction or argumentation) follows the premise (assumption), then the conclusion is true if the premises are true (deductive arguments). The second possibility is the inductive argument, which claims that the individual case is the norm. Here the rule also applies that it is rational to take the conclusion as being true if the premises are true. But what makes statements true? Truth can, as with Habermas, be the result of a discourse ("true is what withstands the discourse"). Other philosophical approaches determine truth criteria according to Pragmatism ("true is what is useful") or Realism ("true is what is rationally accepted"), etc. All theories about truth introduce one condition, e.g., consensus, tradition, innovation, and praxis, under which an argument becomes true, false, or unclassifiable. Since a whole range of plausible truth theories exist, which each have their legitimacy in certain areas/levels, it cannot be conclusively proven when or whether an argument is true. An argument becomes true when the theory's criteria are met; this applies to secular as well as religious arguments (Quinn 1991, pp. 324, 340).

[15]For a discussion of rationality, meaning, and truth of religious beliefs, Sweet (2007) and Evans and Manis (2009). Joppke (2015) tries to define and demark religion through the history of social and political theory (pp. 5–85).

In close relation to the question of truthfulness is the question of whether religious conviction can be classed as knowledge. By definition knowledge is a "true, justified opinion, provided that the opinion itself is not derived from false assumptions and the premises, from which the conclusions are derived, are correct" (Landwehr 2007). It is generally argued that a historical knowledge does not exist, but that it has to be considered in its respective cultural or historical context. This applies to every kind of argument.[16] Given this concept of knowledge, the claim that religious arguments do not count (as knowledge) in political decision-making due to their allegedly doubtful sources, their limited rational justification and the lack of inter-subjective reliability must be relativized, as these objections are also valid for other belief structures such as political ideologies. Accordingly, knowledge criteria can insufficiently separate religious from other arguments.[17]

Another possibility would be to define religious arguments by their meaning. In this case, it is not a question of whether a religious argument is true (verification), but whether it could be true (verifiable). An argument's meaning, namely its logical priority, comes before the question of truth. For if one cannot specify the meaning of a statement, the question about the truth is superfluous. To ask about its meaning is to ask about the conditions that make a statement true (Just 1975). Looking at the status of religious arguments from different philosophical approaches (such as logical atomism, logical positivism (or rather empiricism), critical rationalism, the ordinary language school, post-analytical and continental philosophy), almost all modern attempts originate from using positivist methods of verification to determine the status of religious arguments. Empirical perceptions/verifiable facts, or the verification and falsification of either a group-specific or open discourse, are used as criteria to differentiate between reasonable and nonsensical, provable and unprovable, scientific and nonscientific, meaningful and meaningless, and useful and harmful religious statements. These criteria are assumed to be self-evident and thus lack any justification of their own. This is surprising, because it directly contradicts the theory's claim that it is not based on the validity of any absolute authority, either in the moral or sociopolitical areas. Ultimately, it was religion's claim to be absolute that was the reason to separate religious arguments from public discourse. Despite their own requirements, certain methodologies have become linked to values which for their part are not empirically controllable/verifiable. The belief in unlimited human sensibility and rationality seems to be especially beyond any criticism (Schaeffler 1983; Stiver 2004).

[16]Claudia E. Haupt (2015) shows that even the concept of neutrality is a notoriously ambiguous one that required a detailed historical and socio-legal understanding of the context.

[17]Nukhet A. Sandal (2012) addresses religion as a specific form of knowledge produced by an "epistemic community" which he defined with Haas as a network of professionals with recognized expertise and competence in a particular domain and an authoritative claim to policy-relevant knowledge within that domain or issue area. Gregory W. Dawes (2016) identifies four particular sources of knowledge which religious believers appeal to: by way of signs, by acquaintance, by way of discursive rational arguments, and by way of knowledge that depends on testimony. He states that insofar as religious utterances are assertive, they make claims about the origins of nature of the world and our place within it. These claims are broadly comparable to those made by science. Even if they may be not of a form that allows them to be tested experimentally, they do assert what are thought to be matters of fact.

We also can try to address the definition of religious arguments from an empirical point of view: how can religious arguments be usefully defined for research? Audi relies on four criteria by which he distinguishes religious arguments from secular arguments (Audi 2008). First, an argument is essentially religious if its content refers to a divine or religious authority/rule, a religious source/scripture, or a religious leader. The question of whether the use of religious content as an addition to a wider argumentation can be called religious is more difficult. Audi concludes that these religious arguments would be of secondary order, but of great relevance because they are being prescribed an important justificatory role. The second criterion is explicitly related to the manner in which an argument is justified. An argument is religious if its premises and/or its conclusion cannot be understood or accepted independently of religious considerations (scripture, authority, etc.). Not all arguments with religious content therefore have to be considered religious if they only represent an addition to a primarily moral argument and do not justify the moral itself (i.e., when the premise and the conclusion can also be justified and understood without the religious argument). Correspondingly, not all religious arguments must have religious content. That is the case in regard to natural law or to ethical conclusions that may not currently display religious references, but whose content is derived from religion and whose underlying substantiation cannot be justified without the religious derivation. The motivation of an argument can be seen as the third criteria: an argument is religious if its proponent displays a pronounced motivation to enforce religious goals with a religious justification regardless of whether the content of the argument is overtly religious or not.[18] The fourth criterion relates to the historical genealogy of the argument: in this case an argument is religious if one is able to trace it back to at least one religious proposition. The argument may be persuasive on its own or through its reference to a religious source, but the power of persuasion ultimately depends on the addressee. Although Audi's statements demonstrate a few enlightening assumptions about the nature of religious arguments, ultimately his criteria are less clear than they may appear at first. In nearly every case, the boundaries between religious and nonreligious arguments are fuzzy or heavily dependent on the judgment of the observer. As Audi himself admits, without a thorough knowledge of the religion in question, it is impossible to classify an argument as "religious." One example of these blurred boundaries is the argumentation of the Catholic Church and its use of nature and the everlasting natural law as a point of reference for ethical evaluation: upon closer examination, natural law proves to be a bogus secular argument. The reference to natural law in a normative natural sense cannot be directly classified as a religious argument, but its intention is religious since it tries to justify divine law with the help of a seemingly objective order, although its moral concepts had already been predefined. The divine order is therefore not the result of objective natural deduction, but rather an objective natural order is suggested that confirms the divine law in hindsight. Moreover from a

[18]It is arguable that the motivation of a religious person is already included in their patterns of behavior, even if they do not explicitly mention it (Seibert 2009).

methodological point of view, it is questionable whether it is even logically possible to derive normative statements and moral evaluations from a descriptive assessment of nature. It is also questionable whether natural law with its Catholic characteristics (understood as a natural, eternal moral law, which prescribes every human's aims, rights, and obligations) is a neutral or even a secular point of reference for an ethical evaluation of morally controversial questions. Neither the postulate of an eternally valid natural law—since the position of the Church regarding important issues has changed multiple times throughout the past centuries—nor the independence of the interpreting authority is convincing, since Catholic doctrine interprets natural law as binding and authentic from the perspective of Christian-Catholic ideas on order and morality (compare Gomez-Lobo 2002; Haakonssen 1992; Oderberg and Chappell 2004).

However, compared to other research studies dealing with religious arguments, Audi offers a detailed definition. Most authors avoid making a definition, conducting the analyses with the typical religious markers,[19] with reference to Audi (Riordan 2004, p. 186), by demarcating secular characteristics or simply by using the form of definition that US Supreme Court Justice Potter Steward once applied to pornography: I know it when I see it.[20] Bayer is skeptical concerning the possibility of any general theory of religious argumentation/arguments because criteria fail to demarcate between religious and secular/scientific arguments. His most important criterion of religious arguments is their function (semantics and pragmatics) (Beyer 2004,

[19]Examples for religious arguments are "otherworldly sources" or "metaphysical"; for secular arguments "innerwordly" or "postmetapysical,"; see Baumeister (2011), p. 230.

Andre Bächtinger et al. (2013) using as an indicator, "...whether a statement contains an explicit 'religious marker' such as reference to God, Jesus, or the Bible. In addition, we also create an indicator that measures the degree of explicit religiosity" (p. 106). David Yamane (2000) examines the "...extent to which religious groups employ specifically religious legitimations in their public testimony before the state legislative committees." He mentions six categories of religious legitimation: reference to the Bible; teaching and tradition of the church; ecclesiastical documents or leader; personal authority (p. 181). David Domke and Kevin Coe (2008) analyze the "language of the faithful" of different American presidents (p. 30). They differentiate between two types "of religious communication": "invocations of God" and "invocations of faith." "To invoke God is to make direct reference, often by name, to a supreme being. Examples include any mention of God, Christ, the Creator, Providence, the Almighty, or the like. Such language is the most explicit type of religious communication that any person, including a U.S. president, can use. Invocations of faith are more subtle, but not necessarily less important. To invoke faith is to use terms that over time have become laden with spiritual—especially Christian—meaning, such as Scripture, blessing, heaven, faith, mission, pray, and so on. Such language does not inevitably carry religious meaning for all listeners, but there is a vocabulary of faith embedded in American culture that conveys religious sentiments to anyone listening for such cues—and millions are doing so" (p. 31). Gregory P. Magarian (2011) emphasizes two classes of religious arguments "[...two particular varieties of religious argument-arguments that rest on claims of divine revelation, and arguments based on fealty to religious authorities-cast serious doubt ...]" (p. 144).

[20]Michelle Dillon (1996) demarcates a secular type of legitimation, which subdivides into scientific authority, legal authority, social-economic authority, and personal experience (p. 28). Rhys Williams and N.J. Demerath (1991) propose another, third category of cultural legitimation (civil-religious), which is built upon the subcategories law, morals, and national/civic identity (p. 56).

p. 10). Some political case studies have also tried to isolate the characteristics of religious arguments, i.e., specific language and sentence forms, sociological and psychological functions, as well as the transmission of theological worldviews. These classifications were especially applied in the area of political rhetoric and the characterization of political regime types or historical personalities. Instead of a comprehensive theory, often only a few characteristic features are mentioned: words and quotes/allusions/paraphrases that can clearly be assigned to a religion; elements of language, which stem from religious practices ("brother and sister"); stylistic elements (parables, stereotypical nature); and a typical intonation or a religious context (Dube 2004). The emphasis of most works, besides the language's characteristics, lies in the meaning and purpose of the use of religious arguments. Sociological works which look at the specifics of religious communication are also scarcely helpful, as criteria such as the "tropical character" of religious language, the "production of a surplus of impressions," or "communicative social events" do not suffice for a clear empirical classification (Tyrell et al. 1998). Linguistics has also attempted to find possible delimitation criteria. An argument, from a linguistic perspective, is definitely religious if the provided context is (expressly) religious (in language, situation, or behavior). If this context is not provided, then one cannot definitely define the argument as being religious if it has no further religious-semantic characteristics. A religious argument can also only be called religious if members of society recognize it as being religious. At the same time, the meaning of an argument does not need to be the same for the producer and the recipient (Kaempfert 1972, p. 32). This conclusion contains not only a potential criterion (the intention and the reception), but also a problem, as one cannot see the inside of a speaker/listener. Throughout the linguistic approach, semantic, pragmatic, and contextual conditions also become apparent alongside functional logic. Furthermore, some studies provide a gradation of religious arguments according to their content and use or the level of secularization.[21] The linguistic or functional marking of religious arguments has justifiably been rejected in religious studies as being reductive. Theologians lament that the situation and the proposition of religious arguments have especially been neglected. According to them, religious arguments are a realization of a religion's pattern of behavior and interaction. Religious arguments are a religion's situation-related linguistic actualizations, religiousness being its distinct attribute (Weyel and Gräb 2006; Dalferth 1981; Peterson 1998; Pleines 2008). This finding leads to two conclusions: Firstly, that there cannot be a general definition for religious arguments, but that they must always be thought of in reference to a specific tradition. The comprehensive definition of an argument as religious (be it according to content, motivation, or reception) must be preceded by an analysis of the respective religion/denomination. Secondly, a religious argument depends on the situation in which it is expressed. Consequently, it cannot definitively be concluded that an argument is

[21]Kaempfert (1971) distinguishes between several forms of secularization of religious arguments that complicate the division. He mentions as forms "religious re-interpretations," "crypto religious uses," and "secular re-interpretations."

religious, but only under which circumstances it could be. Whether an argument can be assigned to a religion (and is seen as religious by that religion) is determined by the religious tradition (Mockabees 2001, p. 685; Smith 2012, p. 479; Hilpert 2010, p. 199).

Our overview has confirmed the assumption that a simple definition of religious arguments as "arguments that are based on God, on theological positions or religious authorities and institutions" is insufficient. A closer definition of religious arguments can accordingly only be made through an interdisciplinary approach. Furthermore, all conclusions indicate that there cannot be a conclusive definition of religious arguments. Religious arguments depend on their tradition's historical development, the context, the function they serve, and the situation in which they are applied. Within this framework, a graduated situating of arguments as more or less religious seems necessary. Methodologically, this causes problems: there exists not only no definition of religious arguments which conclusively states when an argument is religious, but also no simple criteria which provide conditions under which arguments could be religious. It is also impossible to obtain any criterion for differentiation from the logical or epistemological status of religious arguments, as it results in there being no end to the eternally receding questions about the causes and reasons for our convictions, in which an unquestionable (moral) decision does not become apparent. This applies both to religious and all other arguments. With previous theories it is impossible to define the status of religious arguments, as many points of criticism are equally applicable to the construction of any worldview (ideology) with a universal claim of validity.

Finally, we suggest a two-level definition of religious arguments: the parameters of the first level (content, source, intention, belief, convention, validity) determine whether an argument could be termed a religious argument in the first place. In the evaluation of the material, one must state which respective criterion of definition the agents refer to. The labels "content" and "source" are relatively easy to identify as they point to obvious markers in the text. The two criteria of "convention" and "validity" refer to the reception of the argument and the social identity of the actor and can be traced by exploring their background within the religious tradition and society. Neither have an underlying quality, in semantic-linguistic terms, nor can they be defined by the effect their use has on a community. The two criteria of "intention" and "belief" are difficult to operationalize. The religious significance can only be fully understood by reconstructing the argument's articulation process and its genesis. Whether the religious argument is deemed permissible in the wider public discourse primarily depends on the second level, which is the level of gradation: i.e., is it purely religious, not clearly religious, does it have an additional nonreligious dimension, does it use religion in nonreligious contexts, and is there a secular transfer? It is all these factors that primarily decide the admissibility of religious arguments in public discourse. Except for the case of secular transfer, all these gradations are considered problematic in liberal democratic theory. With this additional classification of arguments according to religious content, the different modes of argumentation and the underlying strategies of religious agents, as well as their acceptance in society, shall be recorded in a more detailed manner (Table 1.1).

Table 1.1 Characteristics of religious arguments

First level: Definition of religious arguments *Religious arguments...*
1. ...refer to religious content, e.g., (quoted or paraphrased) statements from a religious text or a divine commandment of a certain religious tradition (content).
2. ...call upon (quoted or paraphrased) statements of religious authorities or institutions (source) to justify their opinions.
3. ...are motivated in the sense that their reference to content/their connection to the content or the concepts about a religion is not coincidental or unintentional (intention of the speaker/conscious use).[a]
4. ...are motivated in the sense that their reasoning about the content of or ideas about a religion come from an (internal or external/instrumental) conviction (conviction/motivation).
5. ...must be recognized, in their reception by society, as belonging to a religious tradition (conventionality).[b]
6. ...must be (recognized and) accepted as connected to certain statements and reasoning as religious tradition by the members of that society (validity).[c]
Second level: Classification of religious arguments *The religious content of religious arguments can be defined as...*
1. ...purely religious arguments (quote and paraphrase of "content" and "source").
2. ...ambiguously religious arguments with facultative components (ad hoc created arguments by single speakers as religious reinterpretation or pseudo-religious use).
3. ...nonreligious arguments which are used in a religious context (applicable religious transformation, e.g., natural law).[d]
4. ...secularized adaptation/transfer (absorption into general terminology, e.g., human rights).

[a]Provided context is (expressly) religious (e.g., in language, situation, or behavior)
[b]An argument can be recognized as religious by external parties
[c]Argument reflects the position of a religion's member
[d]Are made based on specific religious convictions

References

Achterberg, Peter, et al. 2009. A Christian cancellation of the secularist truce? Waning Christian religiosity and waxing religious deprivatization in the west. *Journal for the Scientific Study of Religion* 48 (4): 687–701.

Anderson, John, ed. 2004. *Religion, democracy and democratization.* Abingdon OX: Routledge, Special issue 11(4).

———. 2009. Does god matter, and if so whose god? In *Routledge handbook of religion and politics*, ed. Jeffrey Haynes, 192–210. London: Routledge.

Appiah, Kwame Anthony. 2008. Causes of quarrel: what's special about religious disputes? In *Religious pluralism, globalization, and world politics*, ed. Thomas Banchoff, 41–64. New York: Oxford University Press.

Audi, Robert. 2008. *Religious commitment secular reason.* Cambridge: Cambridge University Press.

Audi, Robert, and Nicholas Wolterstorff. 1997. *Religion in the public square: the place of religious convictions in political debate.* Lanham, MD: Rowman & Littlefield.

Bächtiger, André, et al. 2013. Religious reasons in the public sphere: an empirical study of religious actors' argumentative patterns in Swiss direct democratic campaigns. *European Political Science Review* 5 (1): 105–131.

Bader, Veit. 1999. Religious pluralism: secularism or priority for democracy? *Political Theory* 27 (5): 597–633.
———. 2007. *Secularism or democracy? Associational governance of religious pluralism*. Amsterdam: Amsterdam University Press.
Bahle, Thomas. 2002. Staat, Kirche und Familienpolitik in westeuropäischen Ländern. Ein historisch-soziologischer Vergleich. In *Politik und Religion*, ed. Michael Minkenberg and Ulrich Willems, 391–411. Wiesbaden: Westdeutscher Verlag. Sonderheft 33 der Politischen Vierteljahresschrift.
Banchoff, Thomas, ed. 2008. *Religious pluralism, globalization, and world politics*. New York: Oxford University Press.
Barbato, Mariano. 2011. Auf der Suche nach einer postsäkularen "Lingua Franca"? Die postsäkulare Gesellschaft von Jürgen Habermas am Beispiel der parlamentarischen Debatte zur Spätabtreibung. In *Religion zwischen Zivilgesellschaft und politischem System: Befunde-Positionen-Perspektiven*, ed. Antonius Liedhegener and Ines Jaqueline Werkner, 56–77. Wiesbaden: VS.
Baumeister, Andrea. 2011. The use of "public reason" by religious and secular citizens: the limitations of Habermas conception of the role of religion in the public realm. *Constellations* 18 (2): 222–243.
Bernt Torvild, Oftestad. 2018. *The catholic church and liberal democracy*, Routledge studies in modern history. New York: Routledge.
Beyer, Klaus. 2004. *Religiöse Sprache: Thesen zur Einführung*. Münster: LIT-Verlag.
Böckenförde, Ernst-Wolfgang. 2004. Die Entstehung des Staates als Vorgang der Säkularisation. In *Kirche und christlicher Glaube in der Herausforderung der Zeit. Beiträge zur politisch-theologischen Verfassungsgeschichte 1957-2002*. Münster: LIT-Verlag.
Bruce, Steve. 2006. Did Protestantism create democracy? In *Religion, democracy and democratization*, ed. John Anderson, 3–20. New York: Routledge.
Bryant, Peter J., and Philip H. Schwartz. 2008. Stem cells. In *Fundamentals of the stem cell debate: the scientific, religious, ethical, and political issues*, ed. Kristen Renwick Monroe, Ronald Miller, and Jerome Tobis, 10–36. Berkeley, CA: University of California Press.
Casanova, José. 1994. *Public religions in the modern world*. Chicago: University of Chicago Press.
———. 2007. Die religiöse Lage in Europa. In *Säkularisierung und die Weltreligionen*, ed. Hans Joas and Klaus Wiegandt, 322–357. Frankfurt am Main: Fischer Taschenbuch Verlag.
Chaves, Mark, and David E. Cann. 1992. Regulation, pluralism, and religious market structure. *Rationality and Society* 4 (3): 272–290.
Crockett, Clayton. 2013. *Religion and politics after liberalism: radical political theology*. New York: Columbia University Press.
Dalferth, Ingolf U. 1981. *Religiöse Rede von Gott*. München: Christian Kaiser.
Dark, Ken R., ed. 2000. *Religion and international relations*. London: Macmillan.
Davie, Grace. 1999. Europe: the exception that proves the rule? In *The desecularization of the world: resurgent religion and world politics*, ed. Peter L. Berger, 65–83. Washington: Ethics and Policy Center.
Dawes, Gregory W. 2016. *Religion, philosophy and knowledge*. Dunedin: Palgrave Macmillan.
Devolder, Katrien. 2015. *The ethics of embryonic stem cell research (issues in biomedical ethics)*. Oxford: Oxford University Press.
Dillon, Michelle. 1996. Cultural differences in the abortion discourse of the Catholic Church: evidences from four countries. *Sociology of Religion* 57 (1): 25–36.
Dodds, Susan, and Rachel A. Ankeny, eds. 2018. *Big picture bioethics: developing democratic policy in contested domains*, The international library of ethics, law and technology. Cham: Springer.
Domke, David, and Kevin Coe. 2008. *The god strategy: how religion became a political weapon in America*. Oxford: Oxford University Press.
Dube, Christian. 2004. Religiöse Sprache in Reden Adolf Hitlers. Analysiert anhand ausgewählter Reden aus den Jahren 1933-1945. Dissertation. Philosophische Fakultät der Christian-Albrechts-Universität zu Kiel.

Eisenstein, Marie Ann. 2012. The theory and the practice of political tolerance in the U.S. Society. In *Religion, politics, society and the state*, ed. Jonathan Fox, 105–129. New York: Oxford University Press.

Esposito, John L., Darell Fasching, and Todd Lewis, eds. 2008. *Religion and globalization: world religions in a historical perspective*. New York: Oxford University Press.

Evans, C. Stephen, and R. Zachary Manis. 2009. *Philosophy of religion: thinking about faith*. Illinois: IVP Academic.

Fink, Simon. 2007. *Forschungspolitik zwischen Innovation und Lebensschutz: Die Determinanten von Embryonenforschungspolitiken im internationalen Vergleich*. Baden-Baden: Nomos.

Foret, François, and Xabier Itçaina, eds. 2012. *Politics of religion in Western Europe: modernities in conflict*. London: Routledge.

Freston, Paul. 2006. Evangelical protestantism and democratization in contemporary Latin America and Asia. In *Religion, democracy and democratization*, ed. John Anderson, 21–41. London: Routledge.

Gabriel, Karl. 2009. Phänomene öffentlicher Religion in Europa. In *Religion und Öffentlichkeit: Probleme und Perspektiven*, ed. Mariano Delgado, Ansgar Jödicke, and Guido Vergauwen, 11–40. Stuttgart: Kohlhammer.

Gomez-Lobo, Alfonso. 2002. *Morality and the human goods: an introduction to natural law ethics*. Washington, DC: Georgetown University Press.

Gottweis, Herbert, and Barbara Prainsack. 2002. Religion, Bio-Medizin und Politik. In *Politik und Religion*, ed. Michael Minkenberg and Ulrich Willems, 412–432. Wiesbaden: Westdeutscher Verlag. Sonderheft 33 der Politischen Vierteljahresschrift.

Graf, Friedrich Wilhelm. 2009. Christen im demokratischen Verfassungsstaat. *Aus Politik und Zeitgeschichte* 14: 15–20.

Grotefeld, Stefan. 2000. Self-restraint and the principle of consent: some considerations of the liberal conception of political legitimacy. *Ethical Theory and Moral Practice* 3 (1): 77–92.

———. 2006. *Religiöse Überzeugungen im liberalen Staat: Protestantische Ethik und die Anforderungen öffentlicher Vernunft*. Stuttgart: Kohlhammer.

Gutterman, David S., and Andrew R. Murphy. 2016. *Political religion and religious politics: navigating identities in the United States*. New York: Routledge.

Haakonssen, Knud. 1992. Natural law theory. In *Encyclopedia of ethics*, ed. Lawrence C. Becker and Charlotte B. Becker. New York: Garland.

Habermas, Jürgen. 2005. *Zwischen Naturalismus und Religion: Philosophische Aufsätze*. Frankfurt a.M.: Suhrkamp.

———. 2007. Die öffentliche Stimme der Religion. Säkularer Staat und Glaubenspluralismus. *Blätter für deutsche und internationale Politik* 12: 1441–1446.

———. 2008a. Die Dialektik der Säkularisierung. *Blätter für deutsche und internationale Politik* 4: 33–46.

———. 2008b. *Between naturalism and religion: philosophical essays*. Cambridge: Polity Press.

Hallett, Garth L. 1995. *Greater good: the case for proportionalism*. Washington, DC: Georgetown University Press.

Halman, Loek, and Thorleif Pettersson. 2002. Religion und Politik in der zeitgenössischen Gesellschaft: Differenzierung oder Entdifferenzierung? Eine komparative Analyse von EVS/WVS-Umfragedaten aus 38 Ländern. In *Politik und Religion*, ed. Michael Minkenberg and Ulrich Willems, 303–322. Wiesbaden: Westdeutscher Verlag. Sonderheft 33 der Politischen Vierteljahresschrift.

Hamberg, Eva, and Thorleif Pettersson. 1994. Religious markets: denominational competition and religious participation in contemporary Sweden. *Journal for the Scientific Study of Religion* 33: 205–216.

Haupt, Claudia E. 2015. *Religion-state relations in the United States and Germany*. Cambridge: Cambridge University Press.

Heron, Emma. 2000. Die soziale Theologie New Labours oder Der Appell an Moral, Gemeinschaft und Gewissen. In *Religion und Politik zwischen Universalismus und Partikularismus*, ed. Heinz-Dieter Meyer, Michael Minkenberg, and Ilona Ostner, 49–80. Wiesbaden: VS Verlag.
Hildebrandt, Mathias, and Manfred Brocker, eds. 2005. *Unfriedliche Religionen: Das politische Gewalt- und Konfliktpotential von Religionen*. Wiesbaden: VS Verlag für Sozialwissenschaften.
Hilpert, Konrad. 2010. Religion in den bioethischen Diskursen Deutschlands. In *Religion in bioethischen Diskursen: Interdisziplinäre, internationale und interreligiöse Perspektiven*, ed. Friedemann Voigt, 187–214. Berlin: De Gruyter.
Huntington, Samuel P. 1991. *The third wave: democratization in the late twentieth century*. Norman: University of Oklahoma Press.
Hurlbut, Benjamin. 2017. *Experiments in democracy: human embryo research and the politics of bioethics*. New York: Columbia University Press.
Jedan, Christoph. 2013. Overcoming the divide between religious and secular values: introductory essay. In *Constellations of value: European perspectives on the intersection of religion, politics and society*, ed. Christoph Jedan, 1–15. Vienna: LIT Verlag.
Jensen, Eric A. 2014. *The therapeutic cloning debate: global science and journalism in the public sphere*. New York: Routledge.
Joas, Hans. 2014. *Faith as an option: possible futures for Christianity*. Stanford: Stanford University Press.
Joppke, Christian. 2015. *The secular state under siege: religion and politics in Europe and America*. Cambridge: Polity Press.
Just, Wolf-Dieter. 1975. *Religiöse Sprache und analytische Philosophie: Sinn und Unsinn religiöser Aussagen*. Stuttgart: Kohlhammer.
Kaempfert, Manfred. 1971. *Säkularisierung und neue Heiligkeit: Religiöse und religionsbezogene Sprache bei Friedrich Nietzsche*. Berlin: Erich Schmidt Verlag.
———. 1972. Religiosität als linguistische Kategorie? *Über einige allgemeine Eigenschaften religiöser Texte. Linguistica Biblica*: 31–53.
Kongregation für die Glaubenslehre. 2002. Lehrmäßige Note zu einigen Fragen über den Einsatz und das Verhalten der Katholiken im politischen Leben (Verlautbarungen des Apostolischen Stuhls 158), eds. Bonn: Sekretariat der Deutschen Bischofskonferenz.
Korobkin, Russel, and Stephen R. Munzer. 2007. *The stem cell century: law and policy for a breakthrough technology*. New Haven, CT: Yale University Press.
Landwehr, Achim. 2007. Wissensgeschichte. In *Handbuch Wissenssoziologie und Wissensforschung*, ed. Rainer Schützeichel, 801–813. Konstanz: UVK.
Langendörfer, Hans. 2003. Politik ist nicht alles: Über die politische Präsenz der Kirchen. In *Religion und Politik: Ergebnisse und Analysen einer Umfrage*, ed. Bernhard Vogel, 41–65. Freiburg: Herder.
Lauritzen, Paul. 2005. Stem cells, and human rights: implications for a Posthuman future. *The Hastings Center Report* 35 (2): 25–33.
Leitner, Sigrid. 2002. Katholizismus und Sozialpolitik: Zur Entstehung der Sozialversicherung in Kontinentaleuropa. In *Politik und Religion*, ed. Michael Minkenberg and Ulrich Willems, 369–390. Wiesbaden: Westdeutscher Verlag. Sonderheft 33 der Politischen Vierteljahresschrift.
Losonczi, Peter, and Walter van Herck, eds. 2015. *Secularism, religion, and politics*. London: Routledge.
Lübbe, Herrmann. 2004. Der neue Groß-Terror und die Gebete des amerikanischen Präsidenten. In *Religion und Politik: Zur Theorie und Praxis des theologisch-politischen Komplexes*, ed. Manfred Walther, 362–368. Baden-Baden: Nomos.
Margarian, Gregory P. 2011. Religious argument, free speech theory, and democratic dynamism. *Notre Dame Law Review* 86 (1): 119–181.
Martin, David. 1978. *A general theory of secularization*. Oxford: Blackwell.
———. 1999. The evangelical upsurge and its political implications. In *The desecularization of the world: resurgent religion and world politics*, ed. Peter L. Berger, 37–49. Washington: Ethics and Policy Center.

———. 2005. *On secularization: towards a revised general theory*. Aldershot: Ashgate.

McConnell, Michael W. 1999. Five reasons to reject the claim that religious arguments should be excluded from democratic deliberation. *Utah Law Review* 3: 639–657.

Mehta, Pratap Bhanu. 2008. On the possibility of religious pluralism. In *Religious pluralism, globalization, and world politics*, ed. Thomas Banchoff, 65–88. New York: Oxford University Press.

Minkenberg, Michael. 2002. Staat und Kirche in westlichen Demokratien. In *Politik und Religion*, ed. Michael Minkenberg and Ulrich Willems, 115–138. Wiesbaden: Westdeutscher Verlag. Sonderheft 33 der Politischen Vierteljahresschrift.

Mockabees, Stephan T. 2001. Measuring religious commitment among Catholics and Protestants: a new approach. *Journal for the Scientific Study of Religion* 40 (4): 675–690.

Moltmann, Jürgen. 1997. *Gott im Projekt der modernen Welt. Beiträge zur öffentlichen Relevanz der Theologie*. Gütersloh: Christian Kaiser.

Monsma, Stephan V. 2000. Kirche und Staat in den USA und der Bundesrepublik Deutschland: Auf dem Weg zur Konvergenz? In *Religion und Politik*, ed. Heinz-Dieter Meyer, Michael Minkenberg, and Ilona Ostner, 81–107. Opladen: Leske+Budrich.

Moore, Scott H. 2009. *The limits of liberal democracy: Politics and religion at the end of modernity*. Downers Grove, IL: IVP Academic.

Myskja, Bjørn K. 2009. Rationality and religion in the public debate on embryo stem cell research and prenatal diagnostics. *Medical Health Care and Philosophy* 12: 213–224.

Nehushtan, Yossi. 2015. *Intolerant religion in a tolerant-liberal democracy*. Portland: Hart Publishing.

Neuberger, Benyamin. 2011. Cooperation between church and state in liberal democracies. In *Religion, politics, society and the state*, ed. J. Fox, 11–26. Boulder, CO: Paradigm Publishers.

Oderberg, David S., and Timothy Chappell, eds. 2004. *Human values: new essays on ethics and natural law*. New York: Palgrave.

Patrick Riordan, S.J. 2004. Permission to speak, religious arguments in public reason. *The Heythrop Journal* 45 (2): 178–196.

Pereira, Lygia V. 2016. *Stem cells: promise and reality*. Singapore: World Scientific Publishing Company.

Peterson, Michael. 1998. *Reason and religious belief: an introduction to the philosophy of religion*. Oxford: Oxford University Press.

Pleines, Jürgen-Eckardt. 2008. *Glauben oder Wissen: Analyse eines Dilemmas*. Hildesheim: Georg Olms Verlag.

Pollack, Detlef. 2002. Das Verhältnis von Religion und Politik in den postkommunistischen Staaten Ostmittel- und Osteuropas und seine Auswirkung auf die Vitalität des religiösen Feldes. In *Politik und Religion*, ed. Michael Minkenberg and Ulrich Willems, 435–455. Wiesbaden: Westdeutscher Verlag. Sonderheft 33 der Politischen Vierteljahresschrift.

Prätorius, Rainer. 2005. Marginalisiert in der Mitte. Das politische Geschick des “mainline” Protestantismus. In *God bless America: Politik und Religion in den USA*, ed. Manfred Brocker, 68–88. Darmstadt: Wissenschaftliche Buchgesellschaft.

Quinn, Philip L. 1991. Epistemic parity and religious argument. *Philosophical Perspectives* 5: 317–341.

———. 1997. Political liberalisms and their exclusions of the religions. In *Religion and contemporary liberalism*, ed. Paul J. Weithman, 138–161. Notre Dame: University of Notre Dame Press.

———. 2006. *Essays in the philosophy of religion*. Oxford: Clarendon Press.

Rawls, John. 1993. *Political liberalism*. New York: Columbia University Press.

———. 1999. The idea of public reason revisited. In *John Rawls: collected papers*, 573–615. Cambridge: Harvard University Press.

Rowe, Paul S. 2012. *Religion and global politics*. Ontario: Oxford University Press.

Sandal, Nukhet A. 2012. Churches in "troubles": a knowledge based analysis of clergy involvement in conflict. In *Religion, politics, society, and the state*, ed. Jonathan Fox, 163–186. New York: Oxford University Press.

Schaeffler, Richard. 1983. *Religionsphilosophie*. Freiburg: Verlag Karl Alber.

Scherer, Matthew. 2013. *Beyond church and state: democracy, secularism, and conversion*. Cambridge: Cambridge University Press.

Schlieter, Jens. 2005. Bioethik, Religion und Kultur aus komparativer Sicht. Zur Bedeutung und Methodik einer neuen Fragestellung. *Polylog – Zeitschrift für interkulturelles Philosophieren* 13: 7–14.

Seibert, Christoph. 2009. Inklusion von Religion im politischen Diskurs-eine irreführende Fragestellung? Überlegungen zur Verhältnisbestimmung von öffentlicher Vernunft und Religion bei John Rawls und Jürgen Habermas. In *Religionen und Demokratie: Beiträge zu Genese, Geltung und Wirkung eines aktuellen politischen Spannungsfeldes*, ed. Ines-Jacqueline Werkner, Antonius Liedhegener, and Mathias Hildebrandt, 29–52. Wiesbaden: VS Verlag.

Seidel, George E., Jr. 2003. Cloning: scientific background. In *Encyclopedia of bioethics*, vol. 1, 447–454. New York: Macmillan.

Sigurdson, Ola. 2010. Beyond secularism? Towards a post-secular political theology. *Modern Theology* 26 (2): 177–196.

Silomon, Anke. 2009. Widerstand von Protestanten im NS und in der DDR. *Aus Politik und Zeitgeschichte* 14: 33–38.

Smith, Steven G. 2012. The roar of the lion, the taste of the salt: on really religious reasons. *Religious Studies* 48 (4): 479–496.

Stark, Rodney, and Roger Finke. 2000. *Acts of faith: explaining the human side of religion*. Berkeley, CA: University of California Press.

Stein, Tine. 2001. "Im Bewusstsein der Verantwortung vor Gott...": Christliches Menschenbild und demokratischer Verfassungsstaat. In *Säkularisierung und Resakralisierung in westlichen Gesellschaften*, ed. Mathias Hildebrandt, 185–201. Wiesbaden: Westdeutscher Verlag.

———. 2007. *Himmlische Quellen und irdisches Recht: Religiöse Voraussetzungen des freiheitlichen Verfassungsstaates*. Campus Verlag: Frankfurt a.M.

Stepan, Alfred. 2001. The world's religious systems and democracy: crafting the "twin tolerations". In *Arguing comparative politics*, ed. Alfred Stepan, 213–253. Oxford: Oxford University Press.

Stiver, Dan R. 2004. *The philosophy of religious language: sign, symbol and story*. Oxford: Blackwell.

Stüwe, Klaus. 2018. *Religion und Politik in der freiheitlichen Demokratie [Religion and politics in liberal democracy]*. Berlin: Duncker & Humblot GmbH.

Sweet, William. 2007. Religious belief, meaning and argument. *Studies in Religion* 36 (1): 41–64.

Tatalovich, Raymond, and T. Alexander Smith. 2003. *Cultures at war: moral conflicts in western democracies*. Peterborough: Broadview Press.

Taylor, Charles. 1998. Modes of secularism. In *Secularism and its critics*, ed. Rajeev Bhargava, 31–53. New Delhi: Oxford University Press.

———. 2014. How to define secularism. In *Boundaries of toleration*, ed. Charles Taylor and Alfred C. Stepan, 59–78. New York: Columbia University Press.

The President's Council on Bioethics. 2015. *Human dignity and bioethics*. Platform: CreateSpace Independent Publishing.

Tollefsen, Christopher O. 2008. *Biomedical research and beyond: expanding the ethics of inquiry*. New York: Routledge.

Turner, Bryan S., and S. Bryan. 2010. Secularization. In *SAGE key debates in sociology*, vol. I–IV. London: Sage.

Tyrell, Herrmann, Volkhard Krech, and Hubert Knoblauch, eds. 1998. *Religion als Kommunikation*. Würzburg: Ergon.

Unger, Frank. 2006. Die USA: Supermacht wiedergeborener Christen? In *Macht, Religion, Politik: Zur Renaissance religiöser Praktiken und Mentalitäten*, ed. Margarete Jäger and Jürgen Link, 99–112. Münster: Unrast Verlag.

Walzer, Michael. 1998. Drawing the line: religion and politics. *Soziale Welt* 49 (3): 295–307.
———. 2015. *The paradox of liberation: secular revolutions and religious counterrevolutions*. New Haven, CT: Yale University Press.
Warner, Carolyn. 2002. Die katholische Kirche als politischer Akteur in Italien, Frankreich und Deutschland. In *Politik und Religion*, ed. Michael Minkenberg and Ulrich Willems, 279–299. Wiesbaden: Westdeutscher Verlag. Sonderheft 33 der Politischen Vierteljahresschrift.
Weyel, Birgit, and Wilhelm Gräb, eds. 2006. *Religion in der modernen Lebenswelt: Erscheinungsformen und Reflexionsperspektiven*. Göttingen: Vandenhoeck & Ruprecht.
Wiemeyer, Joachim. 2009. Das Engagement von Christen in politischen Parteien. *Aus Politik und Zeitgeschichte* 14: 27–32.
Willems, Ulrich. 2001. Religionspolitik in der Bundesrepublik Deutschland 1945-1999: Die politische Regulierung der öffentlichen Stellung von Religion und Religionsgemeinschaften. In *Demokratie und Politik in der Bundesrepublik 1949-1999*, ed. Ulrich Willems, 137–162. Opladen: Leske+Budrich.
———. 2003. Religion als Privatsache? Eine kritische Auseinandersetzung mit dem liberalen Prinzip einer strikten Trennung von Politik und Religion. In *Politik und Religion*, ed. Michael Minenberg and Ulrich Willems, 88–112. Wiesbaden: Westdeutscher Verlag.
———. 2004. Weltanschaulich neutraler Staat, christlich-abendländische Kultur und Laizismus: Zu Struktur und Konsequenzen aktueller religionspolitischer Konflikte in der Bundesrepublik. In *Religion und Politik: Zu Theorie und Praxis des theologisch-politischen Komplexes*, ed. Manfred Walther, 303–328. Baden-Baden: Nomos.
Williams, Rhys, and Nicholas J. Demerath III. 1991. Religion and political process in an American city. *American Sociological Review* 56 (4): 417–431.
Winandy, Julien. 2011. Religiöse und säkulare Argumente im Konflikt? Kritik einer irreführenden Dichotomie. In *Glaubensfragen in Europa*, ed. Elke Arien, Helmut König, and Manfred Sicking, 175–196. Bielefeld: Transcript.
Wolterstorff, Nicholas. 1997. Why we should reject what liberalism tells us about speaking and acting in public for religious reasons. In *Religion and contemporary liberalism*, ed. Paul J. Weithman, 162–181. Notre Dame: Notre Dame Press.
———. 2003. An engagement with Rorty. *Journal of Religious Ethics* 31 (1): 129–139.
Wolterstorff, Nicholas, and Terence Cuneo. 2016. *Understanding liberal democracy: essays in political philosophy*. Oxford: Oxford University Press.
Yamane, David. 2000. Naked public square or scrambling wall of separation? Evidence from legislative hearing in Wisconsin. *Review of Religious Research* 42 (2): 175–192.
Ziebertz, Hans Georg, and Ernst Hirsch Ballin. 2016. *Freedom of religion in the 21st century: a human rights perspective on the relation between politics and religion*. Leiden: Brill.

Part II
Catholic-Latin States with Low Level or Little Religious Pluralism

Chapter 2
Discursive Strategies of Catholic Churches in Assisted Reproduction Technology Regulation: Poland and Spain in Comparison

Anja Hennig

Abstract Although the Catholic Churches in Poland and Spain opposed the legalization of artificial reproduction techniques (ART), Spain enacted since the 1980s very permissive laws and Poland in 2015 an intermediate law. This chapter analyzes whether also the discursive strategies the Catholic Churches applied during the policy process differ. Following the debate about religious reasoning in post-secular societies, the concept of discursive strategies assumes that religious actors strategically decide how to insert which arguments with which semantics. The analysis reveals that whereas frames, semantics, and sources of legitimation similarly referred to the uncompromising positions of the papal encyclicals, often applying a secular language, unlike the Spanish bishops, the Polish bishops intervened directly, uncompromisingly, and in an authoritative way.

1 Introduction

For the past decades, the spread of permissive regulations in the arena of reproduction and biomedicine has provoked harsh criticism from the Roman Catholic Church (Hennig 2012; Knill et al. 2014; Grzymała-Busse 2015). The priority of individual choice and to heal infertility as underlying ethical principles contradicts Catholic moral teaching, which considers saving the unborn life and preserving traditional family structures as its fundamental responsibilities (Kalbian 2005).

Comparative studies on morality politics in Catholic states reveal two somewhat contradictory findings: in almost every Catholic state, the Catholic Churches intervene in decision-making processes concerning the regulation of abortion, gay marriage, or artificial reproduction techniques (ART). They issue letters to politicians or mobilize public protests. In the long run, however, the national Church hierarchies only delayed but seldom hindered the implementation of more

A. Hennig (✉)
European University Viadrina, Frankfurt (Oder), Germany
e-mail: ahennig@europa-uni.de

M. Weiberg-Salzmann, U. Willems (eds.), *Religion and Biopolitics*,
https://doi.org/10.1007/978-3-030-14580-4_2

permissive regulation (Knill et al. 2014)—also in Spain and Poland. Despite vigorous protests from the Catholic Churches and related groups, Spain has enacted very permissive laws and in 2015 in Poland an intermediate law since the 1980s. The study takes these different outcomes as the point of departure to investigate whether the Catholic Churches also applied discursive strategies and, thus, different arguments and modes of public intervention.

The concept of discursive strategies serves as a heuristic tool for comparatively analyzing the documented contributions of religious actors within the course of moral-political decision-making processes. Reflecting upon the liberal debate on religious contributions to public discourse in liberal democracies, the concept departs from the premise that religious actors and the Catholic Church in particular decide intentionally how to insert which arguments into the public and political discourse. Distinguishing between the framing of a policy problem, the justification of these frames, the language used, and the modes of intervention into politics, this concept disentangles the content of arguments from the way it is disseminated.

To comparatively analyze the discursive strategies of the Polish and Spanish episcopates since the 1980s (Spain) and since the 1990s (Poland), the chapter structures and summarizes the main documents issued as a response to several stages of the decision-making process concerning ART regulation. One conclusion is that both national Catholic Churches framed and justified their public statements similarly. Differences concern rather the modes of intervention.

The first two parts introduce to the main ethical controversies around ART legislation and the different regulation in Poland and Spain. The third part presents the concept of discursive strategies. As Catholic moral teaching continues to be a central source for public utterance of Catholic representatives, the fourth part summarizes the main doctrinal arguments related to ART. The firth part traces the legislative process with a particular focus on the contributions of the national Church hierarchies. In the concluding comparative analysis of the discursive strategies, the contribution ends by arguing that, though relying on similar arguments, the Polish bishops intervened unlike the Spanish, also directly, uncompromisingly, and in an authoritative way.

2 Incompatible Views on ART Legislation

ART legislation evolved in reaction to biomedical development and is continuously in progress (Engeli and Allison 2013). Most risks and uncertainties are usually associated with cloning, embryonic stem cell research, and genetic manipulation. However, also comparatively low invasive assisted reproduction techniques are ethically contentious. The core ethical question concerns the beginning of life and the status of the embryo. From a life-ethics perspective, from the moment of "conception" an embryo is a person with intrinsic rights. From a pro-choice or pro-healing position (for religious or nonreligious reasons), the embryo is an early form of human life but not a human individual or a person. This view justifies using an embryo for the sake of providing infertile couples with a child (Banchoff 2011, 7).[1]

The main artificial reproduction techniques associated with the risk of destroying embryos concern the numbers of embryos created and implanted, the cryo-conservation of remaining embryos, and the pre-implantation genetic diagnosis or screening (PGD/PGS) of embryos. As PGD implies selecting only those embryos capable of life or without genetic defects, while permitting gender selection and, sooner or later, genetic manipulation (Brezina and Zhao 2011, 5), it also bears the notion of eugenics and is thus ethically problematic, not only from a pro-life perspective (Banchoff 2011, 4–5). Cryo-conservation, the freezing of fertilized eggs for additional cycles, can save embryos but does not solve the problem of the remaining embryos. To donate them to embryonic (stem cell) research is unacceptable from an ethics-of-life perspective (Brezina and Zhao 2011, 4).

Another controversy demanding political regulation of ART concerns the involvement of donators of gametes and surrogate motherhood, both particularly opposed by the Catholic Church because of the significant sacral dimension of human reproduction in marriage. If gamete donation is legal, it is the welfare of ART children and their right to know their biological parents which clashes with preserving the anonymity of donors (Mills and Tanturri 2015). Since granting anonymity is considered to facilitate donation, states fearing a shortage of donors may decide on anonymity. For similar reasons, regulation allows or prohibits compensation for any third persons involved.[2] Those against compensation fear the "trading of children" and exploiting of women (Spar 2006).

The third type of conflict concerns to what extent the eligibility for ART treatments should be restricted with regard to gender and age. Supporters of the traditional family concept as the Catholic Church disagrees with a liberal pro-choice position excepting the variety of gender and family relations and would allow access of ART to single women and homosexual couples. A related conflict concerns state coverage of ART. All EU countries with ART legislation now provide some sort of financial support, but the extent to which non-married and homosexual couples or single women are entitled to benefit from this varies. Most Catholic states only recognize heterosexual couples (see Hennig forthcoming; Berg Brigham et al. 2013, 674).

From a country-wide perspective, regulation of ART varies greatly on a continuum between permissive and restrictive laws. Permissive regulations tolerate most forms of assisted reproduction except commercial surrogacy, grant unlimited access, and allow research using supernumerary embryos (Storrow 2011, 725). In addition, they give priority to the anonymity of donors. Intermediate jurisdictions have strict rules in gamete donation and bans on surrogacy. They may allow PGD/PGS only in special cases and prohibit the creation of embryos for research purposes. They

[1]For the evolution of ethical controversies concerning embryology from a secular and religious perspective, see Banchoff (2011, 4–12).

[2]A case in point is the UK where the government decided in 2011 to provide compensation of £750 (Campbell 2011).

permit however research on remaining embryos. Access is restricted only to stable heterosexual couples. The restrictive type stands in contrast to the permissive approach by placing limits on embryo selection and stem cell research following PGD. Other restrictions bar participation by third-party gamete donors or surrogates (Storrow 2011, 735).

3 Varying ART Legislation and Church–State Relations in Poland and Spain

Comparative morality policy research reveals that in areas related to "life-and-death matters," the most restrictive laws and the prevailing intermediate laws exist in states with majoritarian Catholic Churches (Hennig forthcoming). In Spain, however, not only ART but also bioethics in general have been regulated very liberally over the last four decades. Poland's recent law "curing infertility" fits into the intermediate type. Employing a pro-life and traditional-family position, the law puts the destruction of embryos under penalty and restricts access to heterosexual couples only. At the same time, Polish legislation allows anonymous gamete donation, embryo selection after PGD, and stem cell research with restrictions (Reuters 2015; Strzałkowski 2015). In Spain, with the exception of human cloning and surrogacy, the laws from 2006 and 2007 permit all ART procedures (Ory and Devroey 2013).

Given the Vatican's unanimous opposition to a permissive ART regulation outlined in the next section and the attempts of the Spanish and the Polish Church hierarchies to oppose the legalization of ART, the different legislations imply that the Polish Church was more successful in framing the political discourse than the Spanish Church, despite the fact that public opinion on ART was in both cases overwhelmingly supportive.[3] Factors accounting for the different political influence of the Church, in the Polish case, include the proximity of the Polish Catholic Church to political power (Hennig 2012; Grzymała-Busse 2015). Unlike in Spain, the post-dictatorial history of Polish Church–party relations is marked rather by a culture of cooperation than of conflict.[4]

[3]In 2010 in Poland, 94.3% Christians and 97.8% Catholics were in favor, in Italy 83.3% and 97.5%, and in Spain 78.3% and 95.7%; see country data of Pew-Templeton Global Religious Futures Project, http://www.globalreligiousfutures.org. In Poland, 76% support artificial fertilization for married and 62% for unmarried couples; http://wyborcza.pl/1,75398,18295221,CBOS__76_proc__Polakow_akceptuje_in_vitro_dla_malzenstw.html

[4]Moreover, in the realm of biotechnology also the market dimension of scientific competition and cross-border reproductive care pushes toward more permissive regulation (see Hennig forthcoming).

4 Discursive Strategies of Religious Actors

The proposed concept of discursive strategies relies on the premise that churches not only spread the gospel to the faithful but also defend their vital interests in the public sphere. They can be considered as interest groups (Warner 2000), who calculate with whom to ally and under what conditions which arguments to use. The term "discursive strategies" is borrowed from Michelle Dillon (2010) and conceptualized in reflection upon the liberal discourse on religious arguments in the public sphere of post-secular societies (Habermas 2001, 2011; Loobuyck and Rummens 2011; Urbinati 2014). It is structured along four empirical questions.

4.1 Framing of the Problem

The first question follows the basic tenets of framing analysis, which aims at elaborating how social and political actors construe reality and act accordingly (Benford and Snow 2000). A frame can be understood as interpretation scheme that structures the meaning of reality (Verloo 2005, 19). In the context of competing interpretations of policy problems, framing implies that actors strategically offer their specific meaning of the disputed issue in order to legitimate their action and certain solutions, attempting to influence the framing of the debate (Matthes 2014). In this vein, the first question for the analysis of the Catholic statements on ethically sensitive policies of vital importance for the Catholic doctrine simply asks "How is the problem framed?"

4.2 Secular or Religious Language

The second question departs from Habermas' "institutional translation proviso" as central element of his concept of the post-secular and takes the shifts in Catholic argumentation in the abortion debate since the 1960s into consideration (Hennig 2015). Discussing the normative options for religious contributions in the public sphere of liberal democracies, similarly to Rawls, Habermas contends that if religious arguments enter the formal public sphere, they must be translated into a secular language accessible to all. It is a normative argument according to which only secular reasons "push for taking a mutual perspective so that different communities can develop a more inclusive perspective by transcending their own universe of discourse" (Habermas 2011, 66). This process of translation has to be completed before debate can enter the formalized institutions of democratic decision-making such as parliament and government (Habermas 2005, 125). As a way of such translation, Habermas offers the example of "in the image of God" which could be translated into "human dignity" (Habermas 2011, 64).

The translation proviso, which is more complex than presented here, is disputed for several reasons. It remains, for example, unclear, according to which criteria a religious argument can be identified (Hennig 2015). According to the liberal thinker Robert Audi, it is almost impossible to clearly identify a religious argument, since it is often expressed in secular terms (Audi 1993). In his typology of religious arguments, only the first type, an argument with a theistic content such as Habermas' example, is clearly identifiable as religious. The other three types—the epistemic, the motivational, and the historical type—also contain secular vocabulary. Most importantly, some arguments exchanged in a political debate may be religious by motivation, "but except if the person offering the argument acknowledges his religious motivation, we have no legitimate ground to identify a secular expressed religious argument as translated." Religious argument, in a historical sense, "genetically traces explicitly or implicitly, by some mainly cognitive chain, such as a chain of beliefs, back to one or more arguments that are religious in one of the above senses" (Audi 1993, 683).

The brief discussion of Habermas and Audi suggests, on the one hand, a normative pressure, which may push religious actors to translate their public statements. On the other hand, these arguments are often articulated in a secular language already and, thus, not easily identifiable. Taking this into consideration, the second question asks "How are the arguments supporting these frames formulated, in other words, which semantics are used?"

4.3 Secular or Religious Sources of Legitimation

To understand that not only language but also the "source of legitimations" as I call it, is applied intentionally, Ronald Dworkin's book *Life's Dominion* is instructive (Dworkin 1993). Dworkin elaborates two different faith-independent religious objections to abortion, euthanasia, and individual freedom: according to what he calls a "derivative objection," embryos are seen as "creatures with interests of their own right from the start," including an "interest in remaining alive." Therefore, "they have the rights that all human beings have to protect these basic interests (...)." Abortion is wrong in principle because it violates a person's right not to be killed. According to the "detached objection," abortion is wrong in principle because it disregards and insults the intrinsic value of any stage or form of human life. In this view, human life is sacred in itself (Dworkin 1993, 11) and, thus, its protection justified on religious grounds. The Catholic Church included the derivative argument at the end of the 1960s. This gave the Vatican "a considerable political advantage in its campaign against abortion and strengthened its political position" (Dworkin 1993, 45/46).

The modern Catholic interpretation of the idea of the natural law is another example of legitimizing secularly expressed arguments grounded in theology. In Catholic theology, natural law is the overarching basis for civil law. Emphasizing the dignity of the individual and people's social rights and obligations, natural law also informs the Church hierarchy's understanding of what is morally right or wrong (Dillon 1999, 55). Since about two decades, however, natural rights arguments are increasingly used as "immunities against the order of law" (Hittinger 1997, 10), in particular in the context of morality politics (Hennig 2015). In the light of the assumed strategic choice of sources of legitimization, the third question asks, "How are the frames legitimized, thus, with reference to which authority?"

4.4 *Modes of Intervention*

The last element of discursive strategies connects the analysis of the content of public contributions of religious actors with their modes of intervention and, thus, the question "Who is how addressed?" It departs from Dillon's observation that "the discursive strategy of the Catholic Church" against abortion since the 1960s was to address its normative views on the public at large and "not simply to fellow believers". It is, thus, interesting to understand who is addressed: the faithful only, also the nonreligious, or also politicians? Furthermore, from a comparative angle it reveals whether the Church hierarchies address politicians directly or rather indirectly through the mobilization of the faithful (Meyer-Resende and Hennig 2015).[5] Moreover, in the context of process tracing it may be of importance to know whether the positions remained consistent or changed over time.

5 Bioethics and Discursive Strategies of the Vatican

The documents of the Polish and Spanish Catholic Church hierarchies to be analyzed following the concept of discursive strategies must be seen in the context of the unique hierarchical structure of the Roman Catholic Church and its twofold authority in interpreting "God's will" and protecting its interpretation in the form of the Catholic moral doctrine (Kalbian 2005). Accordingly, all documents national episcopates publish in reaction to political attempts to permissibly regulate morally sensitive issues are linked to papal encyclicals or statements issued by the Vatican Congregation for the Doctrine of Faith. National episcopates cannot contradict the Catholic doctrine in principle. They enjoy, however, autonomy in how they publicly defend or interpret the moral teachings.

5.1 *Humanae Vitae (1968)*

Regarding the position of the national churches on bioethics and ART in particular, four Vatican documents are fundamental. The first is the encyclical letter *Humanae*

[5]For a more elaborated model of direct and indirect intervention, see Hennig (2017).

Vitae (On Human Life), which Pope John Paul VI published in 1968 when ART was not yet a central issue. It reaffirms the Catholic moral teaching on the sacramental value of marriage as natural union of procreation and of the respect for life as God's gift. Considering any form of birth control as illicit, indirectly the encyclical opposes also ART (Pope John Paul IV, 25 July 1968).

5.2 *Donum Vitae*

The first document on bioethics is the "Instruction on respect for the human life in its origin and on the dignity of procreation replies to certain questions of the day, *Donum Vitae* (The Gift of Life)." Composed in 1987 by Joseph Cardinal Ratzinger, it confirms the doctrinal pro-life tenets and the relevance of marriage for natural procreation. While endorsing scientific progress in principle, the document warns science to serve the human and not to put its life at risk (Donum Vitae, 1, 2). Any scientific vocabulary that implies a "pre-life" stage of development by distinguishing between embryo and "pre-embryo" or "zygote" is rejected (Foreword *Donum Vitae*).

The general position on fetal life set out in the *Donum Vitae* marks the previously mentioned argumentative shift by adding to the detached view also the "derivative objection" (Dworkin 1993). In this vein, the document claims that "the innocent human being's right to life is inviolable from conception until death" (I.4.) and considers human life as "sacred because it involves the creative action of God and it remains forever in a special relationship with the Creator, who is its sole end" (I.5.). Based on this premise, embryos must be respected, just like any other human person and cannot be manipulated, destroyed, or frozen (Dworkin 1993, 39).

Another pro-life argument against in vitro expresses the fear that these techniques can "open the way to other forms of biological and genetic manipulation of human embryos," a practice framed as "abortion mentality" (II), which would already exist (I.6.). Consequently, "the law cannot tolerate that human beings, even in the embryonic stage, should be treated as objects of experimentation (. . .)" (III). ART, thus, is framed as a "threat to life."

The second fundamental frame construes ART as violation of the sacramental nature of marriage, which is in "its indissoluble unity" the only setting worthy of "truly responsible procreation." Whereas one could argue that from a religious point of view every human being is considered a gift of God, according to the Catholic doctrine life must be "the fruit and sign of the mutual self-giving of the spouses" and in collaboration with "the Creator." In vitro, thus, poses the problem of separating reproduction from the conjugal act between the spouses. Two arguments concerning rights are added without any further justification: "The child has the right to be conceived, brought into the world and brought up within marriage" (III.A). "Marriage," however, "does not confer upon spouses the right to have a child (. . .)." Since "a child is not an object" but "the supreme gift," sterile spouses are encouraged to adopt or educate children.

The document ends with implications the moral tenets would have for civil legislation and lawmakers alike. This "secular" section argues on behalf of the right to life as well as of the rights of the family and the institution of marriage as "fundamental moral values" and constitutive elements of civil society. In that view, political authorities are obliged to ensure that the civil law in the field of biomedicine "is regulated according to the fundamental norms of the moral law in matters concerning human rights, human life and the institution of family" (III)[6] (Congregation for the Doctrine of Faith 1987).

5.3 *Evangelium Vitae (1995)*

The encyclical *Evangelium Vitae* (The Gospel of Life) by Pope John Paul II develops a merely biblical justification of the doctrinal pro-life position in order to respond to the perceived increasing threat to life through contraception, abortion, in vitro fertilization, and assisted dying. Elaborating on the Cain and Abel story of brother's murder and the divine commandment "You shall not kill," the document deepens the arguments fleshed out in *Donum Vitae*. Arguing on behalf of a detached *and* derivated objection, the document now links up to the (from a Catholic view somewhat contradictory) human rights discourse: "The theory of human-rights is based precisely on the affirmation that the human person, unlike animals and things, cannot be subjected to domination by others." In reality, the "right to life" would be "trampled on" (16).

In vitro, in this context, is seen as a procedure "to exploit living human embryos and fetuses" by using it as "biological material" or as providers of organs or tissue for transplants in the treatment of certain diseases. Also, this document ends with a secularly formulated appeal to political leaders not to pass laws, which disregard the right to life and the dignity of the person (p. 85). The last paragraph emphasizes the perceived universality of "the Gospel of life": "(...) Life certainly has a sacred and religious value, but in no way is that value a concern only of believers" (Pope John Paul II, March 25, 1995, 94).

5.4 *Dignitas Personae (2008)*

The latest "Instruction *Dignitas Personae* on certain bioethical questions" published in 2008 under the papacy of Benedict XVI discusses the abovementioned documents in the light of the meanwhile regular practice of ART and genetic engineering. Going

[6]Pope Benedict XVI. "Donum Vitae (The Gift of Life)." Georgetown University, Berkley Center for Religion, Peace & World Affairs (Febrary 22, 1987). https://berkleycenter.georgetown.edu/publications/donum-vitae-the-gift-of-life, June 7, 2016.

through all existing ART procedures, it reaffirms the central argument on the defense of life and calls in religious terms for "courageous opposition to all those practices which result in grave and unjust discrimination against unborn human beings, who have the dignity of a person, created like others in the image of God" (Pope Benedict XVI, December 22, 2008). Addressed are "the faithful and all who seek the truth" (2).

Some arguments supporting the pro-life frame are particular. Two relate to the Pope's interpretation of "the natural" and, thus, of the natural law. According to the first, "(T)he embryonic body develops progressively according to a well-defined program with its proper finality (...)." In this vein, any intervention into this process is illicit (I.4.). Secondly, natural law is seen as a "source that inspires the relationship between the spouses in their responsibility for begetting children. The transmission of life is inscribed in nature and its laws stands as an unwritten norm to which all (sic!) must refer" (I.6.).

Two other arguments reflect more on socioeconomic and cultural contexts. In the case of freezing of oocytes, the document contends, "Researchers would sometimes (...) surrender the logic of objective desires and to economic pressures. And the procedure of pre-implantation diagnosis is considered the expression of a "eugenic mentality" that accepts selective abortion (...)" (II.22.). Furthermore, gene therapy, genetic engineering, and human cloning are construed as attempts to create "a new type of human being" and in religious terms a way "in which man tries to take the place of his Creator." While the statements concerning therapeutic cloning and stem cell research remain vague, the document develops a drastic future vision of human reproductive cloning as "biological slavery" (II.27.).

5.5 Discursive Strategies of the Vatican

In the light of these four Vatican documents related to ART and bioethics, one can summarize its discursive strategy as follows: whereas science is generally seen as positive if serving the human being, two doctrinal principles let frame ART as threats: First, the unconditional defense of life from conception to natural death. This notion is justified by the detached and the derivative argumentation. To the first, life as a gift from God is sacred in itself, and to the latter, every human person bears from its conception dignity and the right to life. The first refers to the authority of God and is semantically religious; the second is semantically secular but reflects the Catholic notion of natural law. Besides other arguments supporting the pro-life frame, the ART practice, and PGD in particular, is sometimes drastically framed as eugenic mentality, abortive mentality, or culture of death.

Secondly, given the sacramental character of heterosexual marriage understood as the only place for procreation, ART is framed as a threat to the family and society at large. Arguments supporting this frame relate to the rights of children to be born within marriage and the nonexistence of a right to a child.

The documents address the faithful in general, politically responsible Catholics in particular, and "truth-seeking" people. Semantically it is apparent that only sections addressing lawmaking apply a secular rights language. Against this background, the next part analyzes the discursive strategies of the Spanish and the Polish Catholic Churches in the context of ART decision-making.

6 ART Regulation and Catholic Discursive Strategies in Spain and Poland

6.1 Spain (1980s–2013): Contesting Spain's First ART Regulation in the 1980s

In the 1980s, the enthusiastic response to new reproductive technologies diluted the interpretation of artificial fertilization as morally wrong and shameful. Physicians took the lead in defining the terms of the debate, aiming to convince the public "that infertility is a disease that could be cured." They were eager and initially successful in aligning with politicians who were examining the issue around 1986 (Storrow 2011, 755). This clearly limited the room to maneuver for the Catholic Church. The first of two waves of moral-political liberalization began with the electoral victory of the PSOE in 1982. Abortion became legal but remained restricted, whereas ART regulation embraced biotechnological progress. Politics and society saw it as new solution to the "old problem of infertility that affected about 15% of the Spanish population" (Chaqués Bonafont and Palau Roqué 2012, 76).

The Catholic Church, though appreciating scientific progress in general and underlining the importance of regulating the biomedical field, opposed the planned laws (Spanish Bishops' Conference, March 22, 1988). Its first statement "Around the legal propositions concerning 'artificial reproduction technology and utilization of embryos and human fetuses or cells, tissue and organs'" from March 23, 1988, critically discusses every issue to be regulated. The main argument is centered on the defense of life, beginning with a religious justification. The Spanish episcopate demands the Christian community, political instances, and all citizens alike "to defend the supreme gift of life which we have received from God (...)" as well as "to protect marriage and family as fundamental societal institutions." Accordingly, the bishops demand a law that recognizes the right to life, the rights of family, and that of the child.

The second justification seems a tribute to the Spanish post-dictatorial situation. It refers to democratic principles and the relevance to accommodate "conscience and moral principles." Thus, "(i)n a democracy, the ultimate criteria of legislation cannot be interest of groups or political parties but the service to individual and inalienable fundamental rights" with its "moral imperatives." Legislators should recognize the doctrinal rights arguments and "have, thus, the obligation to be restrictive." The law, instead, being "inspired by an immoral perception of science and technique" would follow pragmatism (p. 2). The third justification is a universalistic approach, arguing that the defense of human life is nothing exclusively religious but exists also in other countries and even among agnostic scientists.

The fourth source of legitimizing a pro-life position is the reference to human rights, which the civil code would have to respect. Considering an embryo a "new individual with its own genetic code and inalienable dignity," the Church bemoans that the proposed law would depart from a false and scientifically disputed assumption of human life as beginning only with the 14th day after conception. Cryo-conservation, in addition, would "brutally stop the process every living human being has a right to." In this vein, the bills would reflect "a mentality of disregarding human dignity by even recognizing manipulation, direct abortion, direct euthanasia, and—this is a new aspect—the commercialization of the human being" (p. 3).

Furthermore, the document criticizes the wording in public discourse as shifting from "the dignity of human procreation" to "biological reproduction with technical assistance (. . .)." It goes on criticizing ART by applying the "no-right-to-child argument," problematizing that the planned law would allow in vitro to single mothers and lesbian couples. This "contradicts the right of the child to be born as a real son or daughter of a married couple, and, thus a known father and a mother." Not only is the perceived threat to the traditional family and procreation within marriage addressed but also the problem of a child's identity if the sperm donator is unknown (p. 5).

In the epilogue, the bishops assure that they consider their contribution as a critical voice among others from science and media and reaffirm their intention to engage in a responsible dialogue, portraying "the fundamental moral positions of the Catholic Church as shared by many who do not identify with it." The last sentence turns to the pro-life argument and links it to the democracy discourse: "If the state does not protect the rights of all its citizens" (this apparently includes the embryo), "it breaks with its own principles as a state of law" (p. 5).

Despite this criticism, in 1988 two different laws (42/1988 regulating research activities and 35/1988 about assisted reproduction techniques) were passed. According to these permissive laws, any woman could access any allowed ART. All basic techniques such as artificial insemination, in vitro fertilization, and egg and embryo donation were allowed—with the exception of cloning and surrogacy (Chaqués Bonafont and Palau Roqué 2012, 77). The medical community successfully adopted and dominated the "healing and progress" frame, while avoiding any conflict with human rights or the dignity of the human being. Political parties played a key role in channeling these demands and ways of thinking into political action. The Catholic Church and other civil societal actors were not involved (Chaqués Bonafont and Palau Roqué 2012, 77). Consequently, the permissive law provided that Spain, during the 1990s, became the core of "IVF tourism," thus profiting from different jurisdictions in Europe (Pavone and Arias 2012, 9–15).

6.2 *Mid 1990s–Mid 2000: In Defense of Adult Stem Cell Research*

In 1996, the conservative PP won the election with José Maria Aznar. During the following 8 years, the reactions to ART regulation and practice became more critical and the Church more active. At the end of the 1990s more than 40,000 frozen embryos were stored. This provoked a new controversy about the use of embryos for research and development activities (Chaqués Bonafont and Palau Roqué 2012, 78). In 2002, a campaign by a network of nondenominational as well as Catholic pro-life groups with links to *Opus Dei* around scientists from the Catholic University of Navarra aimed to restrict the creation of embryos to three and implanting all of them, in order to avoid cryo-conservation as a method that paves the way for stem cell research (Méndez 2002).

Following up the protests, in December 2002, the Episcopal Family Sub-commission of the Spanish Bishops' Conference issued an unusual statement. Its "Note about the use of human embryos for the investigation on stem cell research" argues in favor of adult stem cell research as a scientifically successful and ethically acceptable method (Spanish Bishops' Conference, December 19, 2002). The document explains scientifically how embryonic and adults stem cells serve as possible source for curing diseases: the first type at the cost of the embryo and the second extracted from a particular adult's tissue. They blamed the biotechnological industry and "opinion groups" to "confuse the public opinion" by campaigning in favor of embryonic stem cell research and "so-called" therapeutic cloning. They claimed they created false expectations about healing successes and were guided by economic and ideological interests. Quoting Pope John Paul II, programs investigating adult stem cell research would not get the necessary funding and support. Demanding "unconditional respect for the human embryo" and invoking the "embryo-is-a-person" argument, the document postulates an ethic-based science to promote human dignity.

6.3 *2003: Conservative Government and Policy Change in Accordance with the Church*

Embryonic research and therapeutic cloning dominated the bioethical debate until 2003, according to the Church a debate that lacked the involvement of the public. Based on available information about the planned reduction of the maximal number of embryos to be implanted, the episcopate issued the comment "reform for the better but very insufficient" (Spanish Bishops' Conference, July 23, 2003). Most points repeated the known arguments against ART practices, more emotionally speaking about the right to life of the "sons and daughters" which the current law would treat as objects for production. Also "the destiny of so-called surplus embryos would be uncertain." The Catholic moral doctrine and natural law are applied as sources of legitimation for demanding a fundamental reform of the "unjust law." Also, they criticized the term "pre-embryo" as a "linguistic fiction" suggesting the existence of a "pre-human reality" which does not deserve the respect as a human being.

The most remarkable part of this note also covered by the media concerned cryo-conservation. The bishops suggested that embryos, which "died via de-conservation" could be considered as "donors of their cells" and used under strict control for research (. . .). However, "for continuing the reform of the law, it would be indispensable to deepen the knowledge about the anthropological and moral principles the Church proposes (. . .)."

In November 2003, the PP government decided indeed to limit the number of embryos that could be created and implanted to three and, thus, responded to the earlier demand of the Catholic Church. However, while justifying the prohibition of the use of embryos for research purposes, this reform allowed the manipulation of cryo-conservated embryos created before 2004 and pre-implantation diagnosis with many restrictions (Chaqués Bonafont and Palau Roqué 2012, 79), thus paving the way for embryonic stem cell research (Europa Press 2003).

6.4 Mobilizing the Faithful Under the New Left Government

When in 2004 the PP government was replaced by the PSOE, the Catholic Church lost its few supporters in government. For the new premier, Rodriguez Zapatero, moral-political liberalization including the reform of the bioethics laws from 1988 was of major importance. The Catholic Church reacted to this "second wave of liberalization" with more public campaigns. Though embracing technological progress, the Church hierarchy now clearly opposed stem cell research with de-frozen embryos.

The government, however, proceeded with its plan to liberalize ART and bioethics. Therefore, the Spanish Bishops' Conference issued already in October a new "Note before the approval of the Law that amends the Law of Assisted Reproduction" (Spanish Bishops' Conference, October 29, 2004). It responds to an approved decree, which would "permit the production of embryos without restriction and opens the door for research with human beings in an embryonic stage." The document samples fundamental statements and doctrinal arguments on the role of science and the right to life published since 2002.

In reaction to the announced formulation of a completely new and all-encompassing bioethics law, the Church launched a third campaign against the social reforms of the government led by Rodrigo Zapatero.[7] At the end of March 2005, the Bishops' Conference presented its campaign "We were all embryos." Aiming at mobilizing against embryonic stem cell research, the argumentation printed in the secular mass media revealed clear religious references and semantics. As the date for the campaign, the bishops had chosen the tenth anniversary of *Evangelium Vitae*.

The central argument concerned the defense of life. Instead of referring to human rights, this time in the press conference the bishops used the detached justification stating that "life as a precious gift from God of Life is sacred and inviolable," reminding with reference to *Evangelium Vitae* the Bible's commandment "You shall not kill."

[7]The other two addressed assisted dying and the legalization of same-sex partnerships.

Thus, under the PSOE government, the Catholic Church turned its attention more to the mobilization of the faithful, a shift that includes the use of religious language and neglects any willingness to compromise. The ultimate aim, however, was to hinder the legalization of stem cell research with remaining frozen embryos. Contrary to its compromising posture from July 2003 regarding the use of de-frozen embryos as "donors of their cells," they now opposed the use of embryos in principle. The bishops justified this shift with the planned reforms of the PSOE government, which "does not depart from the respect for the human person from its first hour and days of existence (...)" (El País, March 30, 2005).

Only 2 months later, the episcopate published its paper "For science that serves the human life." The main themes concerned the misuse of science and in particular the planned unlimited production of embryos (Spanish Bishops' Conference, May 25, 2005). Whereas even genetics would have opened "unexpected horizons for prevention and curing," some medics and laws would treat human beings as if they were "things or animals, which you can produce, manipulate or merchandize." With reference to the arguments on human dignity, the embryo as a person and the embryo's right to be conceived raise the problem of the question of the remaining embryos. The use of adult stem cells is promoted as a promising example of science in the service of human life and as an authentic, because harmless, therapy (Part 1).

As in 2003, the bishops demanded to restrict the number of embryos produced with a clear statement against stem cell research: "Thawing frozen embryos, reanimating them and then destroying them to use as stem cells" cannot be justified not even for any supposed therapeutic end (3.3, 3.4). The priority of the pro-life position is mirrored also in the voiced critique against any type of cloning: "There is no justification to eliminate cloned embryos for the sake of experimentation". And sensitive to the power of words, the bishops bemoan that "nuclovos" or "nuclear transference" is used in politics to disguise reality and to avoid justified alarmism and aversion in society. For the first time, the statement ends on a semantically religious note: "We plead to God of Life to illuminate those who are responsible for these delicate matters." (...) "Let's jointly search for real progress which does not risk the loss of the sacred right to life and adequate family conditions".

6.5 *Political Arguments*

The political scene, instead, adhered in general to the "healing frame." A leaflet informing the public about the planned new law on bioethics and presenting the party positions reveals that the PP advocates the positions of the Catholic Church only to a minor extent, when criticizing the lack of alternative methods to stem cell research as, for instance, with the placenta. The other critique concerns hormonal stimuli the law would allow, which would "worsen the health situation of women." The PSOE clearly adhered to the "healing frame," arguing that the law "facilitates best the situation for infertile couples to get children. The application of techniques will prevent and cure diseases." In a similar vein, the CiU also emphasized the potential to improve the conditions of life for certain persons and, in some cases, also to cure life-threatening diseases.[8]

6.6 2006: Toward the Permissive Law 14/2006

A month before the permissive law came into force, the episcopate published a critical analysis of the legal project (Spanish Bishops' Conference, June 30, 2006). The comprehensive document addresses Catholics as well as those familiar with humanist ethics and aims at giving information on the prospective changes. Beginning each of the six sections with a question, it has the character of an information sheet. These questions, however, already bear the Catholic "threat-to-life" interpretation of the most important paragraphs: firstly asking: "Why is the production of human beings in labs against human dignity?" and "Why do you have to treat an embryo from the first moment on with the respect that a human being deserves?" With reference also to *Evangelium Vitae*, the answers refer to the rights and dignity of an embryo as a person.

The third question underlines the problem that the new law would now clearly define "an embryo created in-vitro as a group of cells resulting from progressive division of the ovocite since 14 days or more after its fecundation" and asks "Why is the definition "pre-embryo" unfounded and false (. . .)?" "A pre-embryo", the answer starts "is an embryo less than 14 days old. Any other definition would, according to the bishops, "lack a scientific or philosophical basis." The fourth question asks whether the new law authorizes the production of embryos also for research and industry. According to the bishops, and they seem right, the concept of "pre-embryo" would serve as a legal point of departure also for the production of embryos for research. The fifth question asks "whether PGD is a curative or a eugenic technique and what happens with the so called baby-to-cure?" The document considers both as "eugenic practice," arguing that "you have to cure the sick but not at the cost of the healthy one."

The last question is simple but generates a complex answer with links to the "threat to family frame": "Does the law allow cloning?" It begins with Dolly as "a sheep without parents," deducing that the disassociation from parental structures is already accepted, thus "How much is missing to produce a child without father and mother at all?" The document acknowledges that the law forbids cloning. However, the allowed practice of reproductive cloning would signal that other variations of cloning are possible and normal. To that end, the law would avoid the "nasty term by replacing it for instance with nuclear transfer." In the conclusion, the Church presents itself as legitimate actor in this discourse, insisting in the relevance of its teaching, despite a "certain unpopularity." Nonetheless, the document ends with a

[8]The source, a public information leaflet comparing the different positions, unfortunately does not exist anymore.

reference to God and the Bible: "The unborn belong to God's loving plan, which does not forget any of His creatures."

The thus criticized law on "techniques of assisted human reproduction" was passed in May 2006 and included the "born-to-cure" practice. Moreover, Law 14/2006 eliminates most of the restrictions on the use of frozen embryos and the number of embryos created, as introduced by the PP. Despite rather limited knowledge concerning the complex issue of ART, and stem cell research in particular, public opinion was overwhelmingly supportive (Chaqués Bonafont and Palau Roqué 2012, 81).

6.7 2007: The Permissive Bioethics Law

The last legal reform concerned the law on Biomedical Research (Ley de Investigación Biomédica) and, thus, also on stem cell research and therapeutic cloning. The episcopate did not cease to warn against the planned regulations. The bishops also raised the problem of an increasing request for egg cells. Furthermore, they criticized again the absence of the attempt to regulate adult stem cell research. This time they explicitly addressed Catholic legislators who would not be able "to vote for the law without objectively disagreeing with the Catholic doctrine." Turning to a religious language, the note ends with reference to God enhancing the Catholic communities to spread the Gospel of Life (Spanish Bishops' Conference, October 19, 2006). Thus, the political arguments remained similar. The center-left applied the "healing frame" and underlined that cloning of human being remains forbidden. The PP blamed the law for ignoring the new possibility to reprogram adult skin cells. In June 2007, the congress also passed the law on "biomedical investigation" which regulates therapeutic cloning and forbids reproductive cloning (Europa Press 2007).

When in 2011 the PP came to power again, it made no changes to the law. In 2013, however, the Spanish health minister announced the exclusion of lesbian couples from coverage of fertility treatments, arguing, "The lack of a man is not a disease." Given the fact that 80% of treatments are performed in the private sector, this initiative clearly has a symbolic character (Prats 2013), one reflecting the Catholic critique of the liberal Spanish gay-marriage law.

6.8 Poland (1990s–2015): The 1990s—The Catholic Fight Against Abortion

In Poland, ethical questions concerning reproductive medicine were discussed from the mid-1980s onward, while artificial fertilization became regular practice. When, after 1989, the first democratic and morally conservative government stopped the public funding of research centers, infertility treatments continued in the public-health sector (Szamantowicz 2005, 51). As the Catholic Church, in an alliance with Catholic politicians, did all it could to restrict permissive abortion laws after 1956, in the 1990s bioethics and existing ART practice were not public issues. Instead, in Poland, the medical community allowed various infertility treatments, despite protests from the Association of Catholic Medics who supported the pro-life stance of the Polish Catholic Church (Chańska 2005, 40).

After years of partisan conflict on the part of the Catholic Church, in 1997 the newly elected, morally conservative post-Solidarnośc government enacted a restrictive abortion law, which is still in force (Szawarski 2005, 107). Nonetheless, in the aftermath of the pacified abortion conflict, both political camps avoided politicizing the pending regulation of ART (Kulawik 2011, 75). Accordingly, ART was, with the exception of PGD and surrogacy, practiced liberally and gained in economic relevance (Szamantowicz 2005, 53).

6.9 2007: In Vitro on the Agenda: The Church in Opposition

The silence only broke in 2007 when the new center-right government run by the Civic Platform (PO) proposed a bill. The Church hierarchy immediately launched an anti-in vitro campaign in the media. It made clear that "the ultimate goal is to persuade the government to ban ART in Poland." In an open letter the bishops address the deputies and senators directly (Polish Bishops' Conference, December 15, 2007) pleading to share their "worries about marriage and family" given the recent political statements about "in vitro," in Poland a synonym for ART. A source of legitimation is Pope John Paul II, a moral authority for many Poles. He is also the reference for most of their arguments against in vitro, which the document considered "sophisticated abortion." They apply the known-rights arguments but use a more emotional language, when for instance denying parents the right to a child, "especially as this "right" is always related to the death of his brothers and sisters."

As Christmas was approaching, the bishops' direct appeal to political decision-makers mirrors their self-image as representing a moral authority in a Catholic nation: "We trust that for the honored deputies and senators the question of family and defense of life of every child (. . .) are of the utmost importance." "We wish that truth always guided us" (. . .). "Let the born-again Christ bless us in the shelter of our beloved fatherland."

With the intention of holding discussions with the government, in mid-2008 the bishops established a Bioethics Expert Council (Ekai, October 22, 2010). In his first statement, its head, Cardinal Henryk Hoser, strove for urgent regulation, emphasizing that no regulation should involve the everyday destruction of many human embryos, which he considered also trading objects. The in vitro method would thus be a kind of "selective abortion." A few months later, the Bioethics Council reinforced its position in drastic words: a legal project is acceptable only if it completely bans the murder and the freezing of human existences. Accordingly, all parliamentarians who care about the protection of human rights should vote for the complete ban of in vitro, a position which the conservative PiS appropriated. The appeal ends with motivating politicians to keep on "fighting about the content of the law" and reminding them that they have an "ethical duty to engage in the whole legislative process to best limit its harmful elements," a position that can be traced back to Cardinal Ratzinger's doctrinal note from 2003 about the political engagement of Catholics (Ekai, October 22, 2010).

At the end of 2008, eventually Premier Donald Tusk installed a bioethics expert group, led by the conservative Catholic Jarosław Gowin, to draft a comprehensive law in line with the requirements of the EU bioethics convention not then ratified. Gowin's position closely resembles the view of the Catholic Church: access only to married couples, unconditional protection of the embryo as a person, and no gamete donation. In addition, embryo selection should be prohibited. While Tusk promoted his plan to provide state funding for fertility treatments, the Church tried to convince Catholic policymakers to support a prohibitive law (Rzeczpospolita 2008).

A month later, in an interview Bishop Tadeusz Pieronek compared in vitro as a method of creating human beings "against nature" and "similar to the literary idea of Frankenstein." In substance, however, he criticized infertile couples using gamete donation: "Parents will chose sex, eye-color, hair, genial genes" (...). Their children "will be as made by Frankenstein" (...). In the end, Pieronek developed this "horrible perspective," a classical dam break argument, to "make clear that infertile married couples exist, but not the right to a child." He justifies his position mainly religiously, arguing that children are a gift from God, resulting from a natural process directed by God's will and codified by the Natural Law, which cannot be broken. Asked how he would respond as a confessor to a married couple who underwent an ART treatment, he dramatically repeated the unconditional pro-life position of the Catholic Church: "I would tell them the same as what I would tell a person who had had an abortion or committed murder" (Harpula 2009).

6.10 Since 2009: In Vitro Dividing the Center-Right

Between 2009 and 2012, several bills covering the whole ideological spectrum were discussed.[9] The PO itself was divided along liberal-conservative lines, proposing a prohibitive and a more permissive approach (Zaczek 2010). In autumn 2012, Tusk presented a program financed by the health ministry that covered infertility treatments for heterosexual couples from 2013 to 2016. Although promising "maximal protection of life" (Polskie Radio, October 22, 2012), the conservative wing of the PO and the Church hierarchy criticized the program (The Economist, July 23, 2012).

In a press conference the bishops justified their political involvement arguing that the role of the Catholic Church is not to predetermine how politicians shall vote but on which values they should base their vote. In a letter to the President, to the Head of Sejm and Senate, Jozef Michalik and Henryk Hoser, the former head of the Polish

[9]For a detailed analysis in English of these and previous bills, see Gozdecka (2012).

Bishops' Conference, the latter head of the Church's Bioethic Committee, and the Family Council, however, went a step further claiming that Catholic parliamentarians supporting the funding of IVF would run the risk of being excommunicated, given that "for the birth of one child several human beings have to die (. . .)." "In vitro seems to be the younger sister of eugenic" (Ekai, October 22, 2010).

6.11 Uncompromising Episcopate

The most comprehensive position paper "About bioethical challenges confronting the contemporary human being" was published in March 2013. It clearly draws on John Paul II *Evangelium Vitae*, as the subtitle of the first part, "the human being's goodwill with regard to the civilization of death and the culture of life," suggests. The position paper "aimed at creating a Christian's consciousness in accordance with the Church's teaching" and a "qualified knowledge concerning the beginning of life." Part two, departing from "God as donator of life," addresses the "basic destruction of human life in the pre-natal phase." Part three addresses arguments concerning "the truth about human sexuality and responsibility for the child." Part four clarifies the position of the bishops. Part five puts particular emphasis on the rights and duties of Catholic politicians and displays an uncompromising position. Referring to John Paul II, a political compromise would be possible only if it helps achieve the greater good and not just solve ethical conflicts or offer criteria for the good. However, "in the case of in vitro, anti-conception, gender ideology or the institutionalization of same-sex partnerships, such a compromise remains impossible" (Polish Bishops' Conference, March 5, 2013a).

Tusk consistently adhered to the planned regulation despite criticism from the conservative wing of his party, to which Gowin also belonged until its demission at the end of April 2013 (Gazeta Wyborzca, April 29, 2013). In June 2013, the Bishops' Conference issued a rather unusual communiqué in which the expert group on bioethics provided information about "the manipulation of scientific information concerning in vitro" (Polish Bishops' Conference, June 24, 2013b): The experts would observe the distribution of tendentious information, which would intentionally avoid providing information about scientific results proving genetic and health-related effects of the in vitro method. Not disseminating this information implies not telling the truth. Also included were scientific articles, which "prove the risks for children conceived by in vitro fertilization."

6.12 2015: Direct Church Intervention in the Context of ART Decision-Making

In spring 2015, the new Prime Minister Eva Kopacz (PO) proposed a rather permissive bill, but one penalizing the destruction of embryos (including cryopreserved ones) and limiting access to heterosexual couples only, a formulation to which the liberal and conservative wings of the PO finally agreed (Reuters 2015). The Catholic Church hierarchy immediately increased its pressure on Catholic politicians. Within the following 4 months of parliamentary discussion, the episcopate issued three quite different documents: in March 2015 an appeal of the presidium of the episcopate regarding the legislative process; in June 2015 a declaration of the Bioethical Expert group concerning the presentation of the Polish law "healing infertility" from June 25 (Polish Bishops' Conference, June 26, 2015c); and in July 2015 a letter to the Polish president just before the expected signature of the bill.

The first statement is less concerned with the bill itself but with Catholic politicians involved in the decision-making process, especially those who take part in the Eucharist and receive Holy Sacraments. The document presents the criteria upon which a politician should base the vote, also meant as "predisposition to receive the Holy Communion." The media considered this as putting pressure on Catholic politicians. Left-wing politicians claimed that "deputies, who follow it, ought to lose their position" (natemat.pl 2015).

As main "criteria for orientation" the bishops expose two obligations for Catholics: The first is the "faithfulness in Christ" and "the teaching of the Catholic Church" which considers a compromise in moral issues impossible. Politicians shall remain "in a comprehensible way unambiguous." The second obligation is "to wisely engage in public life and to remain truth-worthy at the same time."

What follows is to show that the law contradicts Catholic moral teaching, in particular because it "objectivizes the child" and lacks "esteem for life." The following arguments resemble the Catholic doctrine in a quite emotional language (objectify human embryos by calling them "group of cells," massive destruction of conceived human beings, destroying the fundaments for the development of a child). As a source of legitimation, they refer to Pope Francis with a quote in which he referred to in vitro as "non-ecological and destroying the basis of intergenerational justice." As a secular source of legitimation, the document refers to the public opinion center of the *Sejm* and the Highest Court, which would reveal that every deputy was aware of how to vote. "A Catholic," the bishops conclude, "cannot responsibly vote for the law in its current version."

Whereas, so far, the documents had addressed Catholic politicians in general, on July 17, 2015, Stanisław Gądecki directs a letter directly to President Komorowski, written with its "religious identity as president of the episcopate" and with its secular one as "citizen of the Polish Republic" (Polish Bishops' Conference, July17, 2015a). Cordially asking Komorowski to pass the law back to the *Sejm* or to send it to the Constitutional Court, the document refers merely to secular or secularly expressed sources of legitimization. Given its "incompatibility with the Polish constitution," it would be a "dictate to review the law, which is obliged to unconditionally protect human life." It would be his duty to underline the fact that in the controversy regarding in vitro, not only are opponents and supporters involved, but also the "unborn children, a third party, which one tries to completely omit and to transform into objects" (...). At this point, Gądecki refers to the Catholic Church as a "pupil of Christ and defender of this marginalized third party." He ends with a surprising appeal "for a law for the globalization of solidarity and against the globalization of the exclusion of the unprotected human beings already before their birth."

Despite these direct interventions, on July 21, 2015, president Komorowski passed the bill into law, justifying his decision with the words: "I am not the president of conscience. Many Poles awaited this law with hope. That is why I signed it" (Woźnicki 2016). The right accused him of defending (a reference to Pope John Paul II) a "civilization of death." The law "Healing Infertility" came into force on November 2015.

Just the day after, the bishops issued a communiqué now addressing Catholics in general (Polish Bishops' Conference, July 22, 2015b). Expressing their "deepest pain," they plead to "remember the fundamental good within the teaching of the Catholic Church," and, thus, the known two main doctrinal principles: "the right to life and physical integrity from conception to natural death, the marital unity with the respect of the right to become father or mother of their own children" and, with reference to *Dignitas Personae*, that "human life results as the fruit of a proper marital act, an act of love between wife and husband." The document ends reminding "with responsibility before God" that "Catholics cannot support in vitro, while married couples desiring offspring, must choose morally acceptable methods to heal infertility."

In October 2015, just before the election which brought the right-wing Law and Justice Party (PiS) into power, the health minister signed a second edict supporting public funding. Right after the election, the new church-friendly government abolished the program and used the "limited funds to heal sick children." While the existing law remained untouched (money.pl 2015), it made ART procedures unaffordable (Hennig 2016).

7 Discursive Strategies in Comparison

Despite the different evolution of the laws—in Spain since the late 1980s covering ART, embryonic research, and cloning; in Poland in the past decade focusing merely on ART procedures—the discursive strategies of the Spanish and the Polish Catholic Churches reveal interesting similarities. The concluding section structures the outcome of the comparative analysis following the four questions generated in the first part.

1. How is the problem framed?

As to the first questions, it does not come as a surprise that, given the consistent line of argumentation within the Vatican documents, both the Polish and the Spanish episcopates framed biomedical and artificial reproduction techniques predominantly as a "threat to life" and as a "threat to the family." However, whereas the Polish bishops continuously adhered to an uncompromising position centered on these frames or the drastic variation of Pope John Paul II's "culture of death" frame, the Spanish bishops took a more compromising stance. Despite their doctrinal opposition, they partly adopted the "healing frame" dominant in political discourse and promoted their idea of adult stem cell research as an embryo-saving substitute to embryonic stem cell research, an argument the Vatican appropriated only 7 years later.[10] Furthermore, in 2003 they shifted their pro-life position toward accepting the use of de-frozen embryos for research, which, however, was readopted by the left-wing government.

2. How are the arguments supporting these frames formulated; which semantics are used?
3. And how are they legitimized, with reference to which authority?

To justify the mentioned "threat frames," both episcopates referred in their documents to religious *as well as* to secular sources of legitimization. The first were expressed both religiously ("God of Life," children as God's gift, God's will (not) to become parents, sacramental nature of marriage, procreation only within marriage "in cooperation" with God) and secularly (embryo as a person with its own rights, natural law, no intervention into natural process, right to life and dignity, children's right to have parents, etc.). In line with Audi, however, most of these secularly expressed arguments can be traced back to the Catholic "derivative objection" and to the reinterpretation of the natural law. The only "real" secular sources of legitimation are "human rights," human dignity protection in particular.

A different justification of life protection is mirrored in the first document the Spanish episcopate had issued in the 1980s. Its reference to democratic values links apparently to the context of Spain's transition to democracy. Another difference concerns the specific attention only the Spanish bishops paid to the wording of the debate, aiming at defending the "pro-life frame" by deconstructing scientific terms such as "nuclear transfer" and "pre-embryo" used in legal documents.

4. Modes of intervention: Who is addressed how?

Concerning the question who was addressed, one can conclude that both Church hierarchies spoke to the faithful, political decision-makers and Catholic citizens alike. Only the Polish bishops, however, sent letters (and not only notes) appealing directly to Catholic politicians and even threatening with excommunication. Furthermore, the reference to an explicit religious source of legitimation and the use of religious language varied depending on the stage of the legal process and the audience addressed. The most striking coincidence between a particular type of language, source of legitimation, and audience addressed were the sequence of documents addressing politicians in the wake of the finalization of the Polish bill. Whereas the letter to the Polish president almost lacked any religious language, it prevailed in the documents issued after the approval of the law, when the politically

---

[10]The Vatican started promoting adult stem cells in 2011, when its Pontifical Council for Culture launched a collaboration with US bio-pharmaceutical company NeoStem; see Speciale, Alessandro (Apr 9, 2013) Vatican gets behind adult stem cell research. National Catholic Reporter. https://www.ncronline.org/news/vatican/vatican-gets-behind-adult-stem-cell-research

defeated Catholic Church spoke no longer as a political actor but as a societal moral authority.

The Spanish bishops, who also tend to respond immediately to every new stage in the legal process only once, addressed political decision-makers directly, using a mix of secular and religious language to justify their arguments. Instead, they were keen on appealing also to agnostics or persons of other faith. When with the coming to power of the PSOE, the Church lacked a traditional ally in government and concentrated more on mobilizing the faithful for public campaigns, the bishops turned to a more religious language.

With regard to changes over time, both cases exhibit a merely continuous and in principle similar way of argumentation. Given the priority of the "threat to life" and "threat to family" frames, the objections against artificial fertilization and procedures involving genetic engineering also did not change. Exceptions are the mentioned shift in argumentation of the Spanish episcopate concerning the use of de-frozen embryos and the increased frequency of Polish documents directed at political decision-makers before the approval of the law.

8 Conclusions

Coming to the initial question to what extent the discursive strategies in the process of ART decision-making also differ, two conclusions can be drawn: First, whereas in both cases frames, semantics and sources of legitimation were with some exceptions similar, mainly the modes of intervention differed. In the Polish case, a direct, uncompromising, and authoritative way of intervening corresponded with an intermediate law. In Spain, instead, the very permissive law corresponded with a more modest discursive strategy, including the aim to substantially contribute to the public discourse. Thus, although the Spanish bishops clearly expressed their intention to be a partner in dialogue, the medical and scientific community dominated the framing of the debate.

The Polish bishops being merely concerned with defending doctrinal principles and monitoring the sacramental duties of Catholic politicians nonetheless successfully disseminated the doctrinal pro-life argument. On their side were the unabated moral authority of the Polish Pope John Paul II and a substantial number of Church supportive parliamentarians. At the same time, however, the Church was not able to hinder the complete ban of ART.

In more general terms, one can conclude that, whereas Catholic Church hierarchies apply similar frames and semantics, their modes of intervention differ depending on the political support of the respective churches over time. Moreover, analyzing the public contributions of religious actors along the proposed concept of discursive strategies means disentangling semantics from the substance of an argument and including a strategic momentum of framing and modes of intervention. Finally, whereas different discursive strategies cannot account for different moral-political legislation, they can shed light on the discursive involvement and impact of

Catholic Churches as strategic moral-political actors determined by national patterns of relations between Church and conservative parties.

References

Audi, Robert. 1993. The place of religious argument in a free and democratic society. *San Diego Law Review* 30 (4): 677.

Banchoff, Thomas. 2011. *Embryo politics: ethics and policy in Atlantic democracies*. Ithaca, NY: Cornell University Press.

Benford, Robert D., and David A. Snow. 2000. Framing processes and social movements: an overview and assessment. *Annual Review of Sociology* 26: 611–639.

Berg Brigham, K., B. Cadier, and K. Chevreul. 2013. The diversity of regulation and public financing of IVF in Europe and its impact on utilization. *Human Reproduction* 28 (3): 666–675.

Brezina, Paul R., and Yulian Zhao. 2011. The ethical, legal, and social issues impacted by modern assisted reproduction technologies. *Obstetrics and Gynecology International* 2012: 686253.

Campbell, Denis. 19-10-2011. Egg and sperm donors to be paid more compensation. *The guardian*, http://www.theguardian.com/society/2011/oct/19/egg-sperm-donors-more-compensation

Chańska, Weronika. 2005. Reproduktionstechnologien: Stand der Debatte. In *Biopolitik grenzenlos: Stimmen aus Polen*, ed. Heidi Hoffmann, 19–49. Herbolzheim: Centaurus.

Congreation for the Doctrine of Faith. 1987. Instruction on respect for human life in its origin and on the dignity of procreation replies to certain questions of the day. http://www.vatican.va/roman_curia/congregations/cfaith/documents/rc_con_cfaith_doc_19870222_respect-for-human-life_en.html

Congregation for the Doctrine of Faith. 20-06-2008. Instruction “dignitas personae” on certain bioethical questions. *Congregation for the doctrine of faith*, http://www.vatican.va/roman_curia/congregations/cfaith/documents/rc_con_cfaith_doc_20081208_dignitas-personae_en.html

Dillon, Michelle. 1999. *Catholic identity: balancing reason, faith, and power*. Cambridge: Cambridge University Press.

Dillon, Michelle. 2010. Can post-secular society tolerate religious differences? *Sociology of Religion* 71: 139–156.

Dworkin, Ronald. 1993. *Life's dominion: an argument about abortion, euthanasia, and individual freedom*. New York: Vintage Books.

Ekai. 22-10-2010. W Sejmie dziś debata nad projektami ustaw bioetycznych. *Ekai*, https://ekai.pl/w-sejmie-dzis-debata-nad-projektami-ustaw-bioetycznych/

El País. 30-03-2005. La Iglesia lanza una campaña contra la investigación con células madre embrionarias. *El País*, https://elpais.com/sociedad/2005/03/30/actualidad/1112133604_850215.html

Engeli, Isabelle, and Christine Rothmair Allison. 2013. Diverging against all odds? Regulatory paths in embryonic stem cell research across Western Europe. *Journal of European Public Policy* 20 (3): 407–424.

Europa Press. 13-11-2003. El PP rechaza todas las enmiendas a la Ley de Reproducción Asistida. *El Mundo*, http://www.elmundo.es/elmundosalud/2003/11/13/biociencia/1068732656.html

———. 14-06-2007. El Congreso aprueba la Ley de Investigación Biomédica, que regula la clonación terapéutica. *El Mundo*, http://www.elmundo.es/elmundosalud/2007/06/14/biociencia/1181832966.html

Gawlik, Paweł. 29-04-2013. Jarosław Gowin odwołany. Nowym ministrem sprawiedliwości ma być poseł PO Marek Biernacki. *Gazeta Wyborcza*, http://wyborcza.pl/1,76842,13825101,Jaroslaw_Gowin_odwolany__Nowym_ministrem_sprawiedliwosci.html

Gozdecka, Dorota A. 2012. The Polish Catholic Church and the regulation of IVF in Poland: Polarised political discourses and the battle over "proper reproduction". *feminists@law* 2 (1): 1–40. http://journals.kent.ac.uk/index.php/feministsatlaw/article/view/26/112

Grzymała-Busse, Anna. 2015. *Nations under god: how churches use moral authority to influence policy*. Princeton, NJ: Princeton University Press.

Habermas, Jürgen. 15-10-2001. Glaube, Wissen – Öffnung. Zum Friedenspreis des Deutschen Buchhandels. Eine Dankrede. *Börsenverein des deutschen Buchhandels*, 9–15. https://www.friedenspreis-des-deutschen-buchhandels.de/sixcms/media.php/1290/2001_habermas.pdf

———. 2005. *Naturalismus und religion*. Frankfurt/Main: Suhrkamp.

———. 2011. Dialogue. Jürgen Habermas and Charles Taylor. In *The power of religion in the public sphere*, ed. Judith Butler, Eduardo Mendieta, and Jonathan VanAntwerpen, 60–69. New York, NY: Columbia University Press.

Harpula, Wojciech. 16-01-2009. Pierwowzorem in vitro jest Frankenstein (Interview with Bishop Tadeusz Pieronek). *onet Wiadomosci*, https://wiadomosci.onet.pl/pierwowzorem-in-vitro-jest-frankenstein/7exbs

Hennig, Anja. 2012. *Moralpolitik und Religion: Bedingungen politisch-religiöser Kooperation in Polen, Italien und Spanien*. Würzburg: Ergon-Verlag.

———. 2015. Habermas's translation proviso and conservative religious actors in the public sphere. In *Religious pluralism: a resource book*, ed. Aurelia Bardon, Maria Birnbaum, Lois Lee, and Kristina Stoeckl, 95–102. Fiesole: European University Institute.

———. 18-10-2016. Hofiert und ideologisch gespalten. Die katholische Kirche und die PiS-Regierung. Polen-Analysen, 189. http://www.bpb.de/internationales/europa/polen/235700/analyse-hofiert-und-ideologisch-gespaltendie-katholische-kirche-und-die-pis-regierung

———. 2017. Macht, Moralpolitik und Katholizismus: strategien und Erfolgsbedingungen katholischer Kirchen im Vergleich. In *Die Europäische Integration und die Kirchen Bd. III. Personen und Kontexte*, ed. Ines Dingel and Jan Kusper, 99–120. Mainz: Beihefte der Veröffentlichungen des Institutes für Europäische Geschichte.

———. forthcoming. Religion vs. economy? The regulation of artificial reproduction techniques in Italy, Poland, and Spain. In *The boundaries of life*, ed. Tilo Beckers and Pascal Siegers. Berlin: Springer.

Hittinger, Russel. 1997. Natural law and Catholic moral theology. In *A preserving grace: Protestant, Catholics, and natural law*, ed. Michael Cromartie, 1–30. Grand Rapids, MI: William B Eerdmans.

Kalbian, Aline H. 2005. *Sexing the church: gender, power, and ethics in contemporary Catholicism*. Bloomington: Indiana University Press.

Knill, Christoph, Caroline Preidel, and Kerstin Nebel. 2014. Brake rather than barrier: the impact of the Catholic Church on morality policies in Western Europe. *West European Politics* 37 (5): 845–866.

Kulawik, Teresa. 2011. The politics of human embryo research in Poland. In *Exploring central and Eastern Europe's biotechnological landscape: transforming the life science*, ed. Peter T. Robbins and Farah Huzair, 45–78. Wiesbaden: Springer.

Laura, Chaqués Bonafont, and Anna M. Palau Roqué. 2012. From prohibition to permissiveness: a two-wave change on morality issues in Spain. In *Morality politics in Western Europe: parties, agendas and policy choices*, ed. Isabelle Engeli, Christopher Green-Pederson, and Lars Thorup Larsen, 62–87. London: Palgrave.

Loobuyck, Patrick, and Stefan Rummens. 2011. Religious arguments in the public sphere: comparing Habermas with Rawls. *Ars disputandi Supplement Series* 5: 237–249. https://dspace.library.uu.nl/handle/1874/294494

Matthes, Jörg. 2014. Zum Gehalt der Framing-Forschung: Eine kritische Bestandsaufnahme. In *Framing als politischer Prozess: Beiträge zum Deutungskampf in der politischen Kommunikation*, ed. Frank Marcinkowski, 17–28. Baden-Baden: Nomos.

Méndez, Rafael. 30-09-2002. Grupos católicos y conservadores maniobran para recortar la Ley de Reproducción Asistida. *El País*. https://elpais.com/diario/2002/09/30/sociedad/1033336801_850215.html

Meyer-Resende, Madalena, and Anja Hennig. 2015. Shunning direct intervention: explaining the exceptional behaviour of the Portuguese Church hierarchy in morality politics. *New Diversities* 17 (1): 145–160.

Mills, Melinda, and Maria Letizia Tanturri. 25-08-2015. Assisted reproduction, late fertility and childlessness. *Interview with population Europe*, http://www.population-europe.eu/statement/assisted-reproduction-late-fertility-and-childlessness

Money.Pl. 02-12-2015. Program in vitro już nie z budżetu państwa. Nie stać nas na jego finansowanie. *Money.Pl.* http://www.money.pl/gospodarka/wiadomosci/artykul/program-in-vitro-juz-nie-z-budzetu-panstwa,7,0,1969671.html

natemat.pl. 31-03-2015. Episkopat grozi posłom wykluczeniem z Kościoła za in vitro. Senyszyn: Kto posłucha, powinien stracić mandat. *natemat.pl.* http://natemat.pl/138477,episkopat-grozi-poslom-wykluczeniem-z-kosciola-za-in-vitro-senyszyn-kto-poslucha-powinien-stracic-mandat

Ory, Steven J., and Paul Devroey. 2013. *IFFS surveilliance 2013*. International Federation of Fertility Societies, https://c.ymcdn.com/sites/iffs.site-ym.com/resource/resmgr/iffs_surveillance_09-19-13.pdf

Pavone, Vincenzo, and Flor Arias. 2012. Pre-impantation genetic testing in Spain: beyond the geneticization thesis. *Working paper Intituto de Politicas y Bienes Públicos (IPP)*, no. 12.

Polish Bishops' Conference. 15-12-2007. In vitro. *Polish Bishops' conference, council for family affairs*, http://diecezja.waw.pl/973

———. 05-03-2013a. O wyzwaniach bioetycznych, przed którymi stoi współczesny człowiek. *Polish Bishops' conference*, http://www.opoka.org.pl/biblioteka/W/WE/kep/bioetyczny_05032013.html

———. 24-06-2013b. Komunikat Zespołu Ekspertów ds. Bioetycznych Konferencji Episkopatu Polski w sprawie manipulacji informacjami naukowymi dotyczącymi procedury in vitro. *Polish Bishops' conference*, http://episkopat.pl/komunikat-zespolu-ekspertow-ds-bioetycznych-konferencji-episkopatu-polski-w-sprawie-manipulacji-informacjami-naukowymi-dotyczacymi-procedury-in-vitro/

———. 17-07-2015a. List Przewodniczącego KEP do Prezydenta RP. *Polish Bishops' conference*, http://episkopat.pl/list-przewodniczacego-kep-do-prezydenta-rp-2/

———. 22-07-2015b. Prezydium KEP po podpisaniu przez Prezydenta RP ustawy o "in vitro". *Polish Bishops' conference*, https://ekai.pl/wydarzenia/temat_dnia/x90839/prezydium-kep-po-podpisaniu-przez-prezydenta-rp-ustawy-o-in-vitro/

———. 26-06-2015c. Oświadczenie Zespołu Ekspertów ds. Bioetycznych Konferencji Episkopatu Polski w sprawie uchwalenia dnia 25 czerwca 2015 roku przez Sejm Rzeczypospolitej Polskiej ustawy o leczeniu niepłodności. *Polish Bishops' conference, expert group on bioethics*, http://www.leczenie-nieplodnosci.pl/pl/aktualnosci/oswiadczenie-zespolu-ekspertow-ds-bioetycznych-konferencji-episkopatu-polski-w-sprawie-uchwalenia-dnia-25-czerwca-2015-roku-przez-sejm-rzeczypospolitej-polskiej-ustawy-o-leczeniu-nieplodnosci/

Pope Benedict XVI. 22-12-2008. Address of his holiness Benedict XVI to the members of the Roman Curia for the traditional exchange of Christmas greetings. http://w2.vatican.va/content/benedictxvi/en/speeches/2008/december/documents/hf_ben-xvi_spe_20081222_curia-romana.html.

Polskie Radio. 22-10-2012. Refundacja in vitro. Pomoże 15 tys. par w 3 lata. *Polskie radio*, http://www.polskieradio.pl/5/3/Artykul/709010,Refundacja-in-vitro-Pomoze-15-tys-par-w-3-lata

Pope John Paul II. 25-03-1995. Evangelium vitae. *Pope John Paul II*, http://w2.vatican.va/content/john-paul-ii/en/encyclicals/documents/hf_jp-ii_enc_25031995_evangelium-vitae.html

Pope John Paul VI. 25-07-1968. Humanae vitae. *Pope John Paul VI*, http://w2.vatican.va/content/paul-vi/en/encyclicals/documents/hf_p-vi_enc_25071968_humanae-vitae.html

Prats, Jaime. 23-07-2013. El 80% de la fecundación se realiza en el sector privado. *El País*, http://sociedad.elpais.com/sociedad/2013/07/23/actualidad/1374611147_796311.html

Reuters. 10-03-2015. Polish government backs bill to regulate IVF treatment. *Daily mail*, http://www.dailymail.co.uk/wires/reuters/article-2988390/Polish-government-backs-bill-regulate-IVF-treatment.html

Rzeczpospolita. 11-10-2008. Gowin: ustawą bioetyczną zajmie się nadzwyczajna komisja. *Rzeczpospolita*, http://www.rp.pl/article/20081211/KRAJ/312119824-ap-1

Spanish Bishops' Conference. 23-03-1988. Acerca de las proposiciones de la ley sobre técnicas de reprodución de embriones y de fetos humanos o de células, tejidos u organos. *Spanish Bishops' conference*, http://www.conferenciaepiscopal.es

———. 19-12-2002. Nota sobre la utilización de embriones humanos en la investigación sobre células madre. *Spanish Bishops' conference*, http://www.unav.es/cdb/ceescelulasmadre.html

———. 23-07-2003. Una reforma para mejor, pero muy insuficiente. *Spanish Bishops' conference*, http://www.conferenciaepiscopal.es/documentos/Conferencia/embriones_julio03.htm

———. 29-10-2004. Nota ante la aprobación des Decreto Ley que aplica la Ley de Reproducción Asistida. *Spanish Bishops' conference*, http://www.aebioetica.org/rtf/13BIOETICA-55.pdf

———. 25-05-2005. Por una ciencia al servicio de la vida humana. *Spanish Bishops' conference*, http://www.bioeticaweb.com/por-una-ciencia-al-servicio-de-la-vida-humana-25052005/

———. 30-06-2006. Algunas orientaciones sobre la ilicitud de la reproducción humana artificial y sobre las prácticas injustas autorizadas por la Ley que la regulará en España. *Spanish Bishops' conference*, http://www.conferenciaepiscopal.es/documentos/Conferencia/reproduccion.htm

Spanish Bishops' Conference (CEE). 19-10-2006. El proyecto de Ley de Investigación Biomédica no protege el derecho a la vida y permite la clonación de seres humanos. *Spanish Bishops' Conference (CEE)*, http://www.conferenciaepiscopal.es/el-proyecto-de-ley-de-investigacion-biomedica-no-protege-el-derecho-a-la-vida-y-permite-la-clonacion-de-seres-humanos-2/

Spar, Debora L. 2006. *The baby business: how money, science, and politics drive the commerce of conception*. Boston: Harvard Business School Press.

Storrow, Richard F. 2011. Religion, feminism and abortion: the regulation of assisted reproduction in two Catholic countries. *Rutgers Law Journal* 42: 725–762.

Strzałkowski, Patryk. 25-06-2015. Ustawa o in vitro: Co zakłada projekt uchwalony przez Sejm? *Gazeta Wyborcza*, http://wiadomosci.gazeta.pl/wiadomosci/1,114871,18246390,Ustawa_o_in_vitro__Co_zaklada_projekt_uchwalony_przez.html

Szamantowicz, Marian. 2005. Die Unfruchtbarkeitstherapie mit den Methoden der Assistierten Reproduktionstechnologie (ART) in Polen. In *Biopolitik grenzenlos. Stimmen aus Polen*, ed. Heidi Hoffmann, 51–54. Herbolzheim: Centaurus.

Szawarski, Zbigniew. 2005. Ethics and Prenatal Screening. In *Biopolitik grenzenlos. Stimmen aus Polen*, ed. Heidi Hoffmann, 107–122. Herbolzheim: Centaurus.

The Economist. 23-07-2012. Poland and IVF: deeply divisive. *The economist*, http://www.economist.com/blogs/easternapproaches/2012/07/poland-and-ivf

Urbinati, Nadia. 2014. The context of secularism: a critical appraisal of the post-secular argument. In *Constitutional secularism in an age of religious revival*, ed. Susanna Mancini and Michel Rosenfeld, 14–31. Oxford: Oxford University Press.

Verloo, Mike. 2005. Mainstreaming and gender equality in Europe: a critical frame analysis approach. *The Greek Review of Social Research* 117: 11–34.

Warner, Carolyn. 2000. *Confessions of an interest group: the Catholic Church and political parties in Europe*. Princeton: Princeton University Press.

Woźnicki, Łukasz. 22-07-2016. Prezydent Komorowski podpisał ustawę o in vitro. Jest jedno "ale". *Gazeta Wyborcza*, http://wyborcza.pl/1,75398,18403024,in-vitro-prezydent-komorowski-podpisal-ustawe-jest-jedno.html

Zaczek, Danuta. 29-10-2010. Pięć z sześciu projektów ustaw dot. in vitro do komisji. *Polskie Radio*, http://www.polskieradio.pl/13/641/Artykul/558353,Piec-z-szesciu-projektow-ustaw-dot-in-vitro-do-komisji

Chapter 3
The Role of Religion in Debates on Embryo Research and Surrogacy in France

Jennifer Merchant

Abstract This contribution explores an area where religion in France has always played a very important role in the various debates on bioethical issues. Among various religious groups, the Catholic Church has been the most active and outspoken presence in these debates. This study explores their implication—via several activist groups—and the influence and impact they have in two specific areas: research on embryos and the issue of surrogacy.

1 Introduction

Religion has always played a very important role in the various debates on bioethical issues in France. That being said, there is no doubt that the Catholic Church has been the most active and outspoken presence in these debates. This chapter explores its implication—via several activist groups—and the influence and impact the latter have in two specific areas: *research on embryos and the issue of surrogacy.*

2 Embryo Research

Given the lack of consensus among French scientists and lawmakers on the definition of the human embryo's status, embryo research in France has suffered compared to research being conducted elsewhere. Great Britain and even Belgium, for example, distinguish embryos for a parental project from those that possess a legal status for the purposes of research (Belgium defining the latter as "artifacts").

Despite this lack of consensus, embryo research is nonetheless being conducted in France, and the legal framework, like that in the UK and Belgium, also makes the

---

Translated from French by Elizabeth Hargrett, UC Berkeley.

J. Merchant (✉)
Université Paris II Panthéon-Assas, Paris, France
e-mail: jennifer.merchant@u-paris2.fr

M. Weiberg-Salzmann, U. Willems (eds.), *Religion and Biopolitics*,
https://doi.org/10.1007/978-3-030-14580-4_3

distinction between embryos destined for a parental project from those that have been donated for research. Several conditions are required for such research to be approved by health authorities: the scientific relevance of the project, a specific medical purpose, the impossibility of carrying out research without the use of donated embryos, and respect of ethical principles (e.g., no reproductive cloning) must all be demonstrated before a research protocol is approved.

Furthermore, only embryos that have been conceived through in vitro fertilization, and are no longer the subject of a parental project, can be donated to science. In other words, embryos cannot be created for the express purpose of conducting scientific research. In the majority of cases, parents choose to destroy "supernumerary" (i.e., surplus) embryos, while just under a third of parents decide to donate their embryo(s) to research. If the couple chooses to do so, their written consent must be obtained, and renewed after 3 months.

According to many French researchers, this framework is steeped in contradictions and governed by excessively rigid criteria. Moreover, once the French Biomedicine Agency (*Agence de la biomédecine*) approves a research protocol, the latter is vulnerable to attacks by interest groups that are against any form of embryo research. In a forum recently published in *Le Monde*, 146 French researchers denounced what they identified as a legal "guerilla warfare" being waged against embryo research authorizations by the Jérôme Lejeune Foundation, a Catholic anti-abortion association,[1] and implored public authorities to reconsider the organization's public interest status, which it obtained at the time of its founding in 1996. Among the signatories were approximately 30 scholars and professors of the *Collège de France*, as well as the presidents of several academic and medical organizations.[2]

Indeed, 30 lawsuits were filed against authorizations issued by the French Biomedicine Agency (ABM)—affecting nearly all of the current projects in the field.[3] Generally, protocols are attacked on the grounds of legal technicalities. For example,

[1]Jérôme Lejeune was a French geneticist who purportedly located the gene for Trisomy 21. https://www.fondationlejeune.org/. His foundation finances both research on chromosomes and anti-abortion campaigns. It was in fact Marthe Gautier who had first made this discovery and who was rightfully recognized as such by the Inserm Ethics Committee in September 2014: https://www.la-croix.com/Actualite/France/Decouverte-de-la-trisomie-21-le-role-de-Marthe-Gautier-reconnu-2014-09-24-1211275, https://www.inserm.fr/qu-est-ce-que-l-inserm/l-ethique-a-l-inserm/saisines-et-notes-du-comite-d-ethique

[2]"146 médecins et chercheurs accusent la Fondation Lejeune d'entraver la recherche sur l'embryon," *Le Monde*: http://www.lemonde.fr/sciences/article/2017/03/30/des-chercheurs-et-des-medecins-denoncent-des-entraves-a-la-recherche-sur-l-embryon_5102966_1650684.html

[3]On July 28, 2017, the French Council of State rejected five of these lawsuits. *Conseil d'État*, No 397413, ECLI:FR:CECHR:2017:397413.20170728, Inédit au recueil Lebon, 5ème – 4ème chambres réunies, M. Guillaume Leforestier, rapporteur, M. Nicolas Polge, rapporteur public SCP BORE, SALVE DE BRUNETON, MEGRET; SCP PIWNICA, MOLINIE, avocats: https://www.legifrance.gouv.fr/affichJuriAdmin.do?oldAction=rechJuriAdmin&idTexte=CETATEXT000035317236&fastReqId=900899153&fastPos=1. On June 21, 2017, the Administrative Tribunal of Montreuil rejected 19 more of these lawsuits; however, the Foundation Jérôme Lejeune is appealing: http://montreuil.tribunal-administratif.fr/Actualité/Actualités-Communiques/Communique-de-presse-du-21-juin-2017

one Lyon researcher's protocol on embryonic development, which received authorization from the ABM, was attacked because the embryo donation consent form that had been completed by the parents referred to the bioethics law of 2004, rather than the revised law of 2011. Parental consent has also been challenged at I-STEM, one of the largest French laboratories for research and development focused on human embryonic stem cells. In this case, donation consent was obtained abroad (as the laboratory uses imported human embryonic stem cells), and French law remains unclear on the appropriate consent forms for this kind of situation.

For researchers, this poses a major obstacle to French research, to the extent that some researchers have given up on their projects and taken them abroad. Those who remain express uncertainty regarding the legal framework for their research, as well as the attacks on the ABM that can follow protocol approval. Often, they refrain from continuing their research for fear of retaliation and point out that every day embryos that are no longer the purpose of parental projects are destroyed.

The President of the Jérôme Lejeune Foundation (FJL) defends the association's legal activism by referring to its will to "defend life"—including that of embryos even those that have only existed for a few days—in the name of human rights. For researchers, understanding this first stage of life is becoming more and more crucial when it comes to recognizing the conditions for the onset of certain diseases (such as rare genetic illnesses like infantile spinal muscular atrophy) and eventually developing treatments. According to many scientists, "these studies are only part of what is needed to develop quality medicine that meets people's needs. They are at the service of life."

3 Surrogacy

In France, the issue of surrogacy is plagued with conflict, controversy, and brazen displays of homophobia. Nothing better illustrates this climate than the campaign that religious groups, among others, organized and carried out to prevent the passage of France's "Same-Sex Marriage Law." Following a year of massive demonstrations organized by these groups (which were mostly made up of and financed by Catholic associations and conservative political groups that came together under the umbrella of a movement they called "La Manif pour tous"), the law was passed in May 2013.[4]

Here, we do not intend to trace the genealogy of the "La Manif pour tous" (MPT) movement (meaning "Demonstration For All," a recuperation of their adversaries' use of the expression "Marriage For All"), nor identify the movement's principal actors or understand all of its nuances. Several extremely well-documented investigative reports have already done so, beginning with journalist Samuel Laurent's in-depth investigative and interactive article entitled "Derrière la grande illusion de

[4]Government of France. "Le mariage pour tous: Faire progresser l'égalité des droits." 15 May 2017. http://www.gouvernement.fr/action/le-mariage-pour-tous

la 'Manif pour tous,'" which was published in *Le Monde*.[5] Other sources also have amply demonstrated that the movement is in fact constituted of a nebula of semi-official and often ghost groups and associations and has strong ties to various religious and political organizations that have interests and demands that are both personal and legal (Laurent, *Le Monde* 2013). The investigative report published in *Le Monde* shows that this movement, which paints itself as neutral and "unbiased," actually actively intervenes in various public domains: in law, politics, education, training, communication, culture, and the economy.[6] It is also important to note the often violent verbal tendencies of the movement's general discourse, which have been described by Beatrice Fracchiolla in her striking analysis, in which she reveals a "strategy of targeting and stigmatization that contributes to a reduction and fossilization of meaning" (Fracciolla 2015). Here, we aim to focus on the words the "Manif pour tous" as well as the falsehoods this movement disseminates on "gender theory," the "inevitable" legalization of surrogacy that gender theory will bring on, and ultimately the abolition of sex differences, and the collapse of democracy.

3.1 Elements of Discourse on the Subject

In an online report published in anticipation of the fourth annual Summer University Session of the "Manif pour tous" held in Etiolles on September 17 and 18, 2016, 40 proposals were listed, all of which had the primary objective of "respecting the best interests and basic needs of the child who is today threatened by societal reforms inspired by gender ideology" ("La Manif pour tous," *Politiaue de la Famille* 2016). In this document, which summarizes all of the movement's positions that were expressed during the parliamentary debate on same-sex marriage, the MPT systematically returns to what it considers to be its cornerstone, the notion that "the family is the intersection of the difference between the sexes and the difference between generations" (ibid.: 62). According to the MPT, only a nuclear family unit that includes a man/father and a woman/mother can meet the needs of a child at all unavoidable life stages: conception and birth; when a child begins to understand his or her personal identity and build relationships with others; when the child enters a social group, then the society at large, and finally the child-turned-adult's contribution to social and civil peace (ibid.: 62). According to the MPT, the differences between the sexes and generations, embodied in the nuclear family, are thus necessary conditions for the well-being of the child and the maintenance of "democracy."

[5]Samuel Laurent, "Derrière la grande illusion de la 'Manif pour tous,'" *Le Monde*. http://www.lemonde.fr/societe/article/2013/03/21/manif-pour-tous-la-grande-illusion_1850515_3224.html

[6]For an important discussion of the role of the Roman Catholic Church in the fabrication of the "concept" of "gender theory" and its appropriation by MPT, see Sophie van der Dussen, et al.,: *Habemus Gende! Déconstruction d'une riposte religieuse*, Brussels, Editions de l'Université de Bruxelles, 2015. See also Odile Fillod, "Le mariage raté du Vatican et de la science," April 30, 2013: http://allodoxia.blog.lemonde.fr/2013/04/30/mariage-genre-vatican-science/

As such, marriage and parenthood must fall in line with "the sexual reality of humanity," and as such this distinction guarantees "the complementary differences between the two sexes" (ibid.: 62).

In this document, as well as the militant speeches given at every MPT demonstration, another argument is systematically put forward: the idea that a child needs "plausible parenthood"—in other words, a *father and a mother* "of which the child cannot be deliberately deprived" (ibid.: 33). According to this line of thinking, "*plausible* parenthood" (author's emphasis) is the foundation upon which each child's identity is built, and is thus essential for a healthy upbringing (ibid.: 33). By "plausible parenthood," the MPT means that the child is conceived/born, in the best-case scenario, from a sexual relationship between a husband and his wife, and in the worst-case scenario, by medically assisted procreation—but never using sperm donation, a technique that should be abandoned as it "turns the father into a stranger."[7]

Given this argument, which explicitly names gamete donors as "parents," one might think that surrogacy would be acceptable according to the MPT. If a woman without a uterus and her husband, for example, went through the IVF process using their own gametes, and implanted the embryo into the uterus of another woman they know, and with whom they are in regular contact during the pregnancy, one could imagine that the MPT might look kindly upon this "gift of conception" (Théry 2010) as a supreme act of altruism with the aim of bringing a child into the world—the couple would be truthful to the child, and might explain that the surrogate was her "gestational mother," or perhaps use another gratifying term. With this scenario, however, the MPT is faced with a difficult question: who is the mother—the gestational mother, or the one who passes on her genetic heritage? Unable to explicitly escape taking a position, the MPT adds to the confusion by implying that the woman who gives birth is necessarily the mother of the child, and nothing more is said about the "genetic" maternity of the woman who supplies her eggs (ibid., p. 35). As a result of this thinking, the MPT strongly condemns surrogacy (as well as the transcription into the French Civil Registry of birth certificates that have been drawn up abroad), citing the violent "ripping of the child from their mother's womb." To this end, the MPT has proposed an amendment of the French law (which, in Article 16–7 of the Civil Code, already refuses to recognize any surrogacy contract, whether altruistic or commercial) to repeal the "Taubira Circular" (which directs family law judges to register the child), thus calling for the refusal of transcriptions of the birth certificates of children born abroad via surrogacy, which they fear would open the door to the legalization of surrogacy.[8,9]

---

[7]Indeed, the MPT's fifth proposal is to abolish anonymous gamete donation. Ibid., page 46.

[8]On this particular subject, see Merchant, Jennifer, and Karène Krief-Parizer. "A Baby's Citizenship and Kinship Ties After Surrogate Birth: The Case in France" in *Handbook of Gestational Surrogacy: International Clinical Practice & Policy Issues*, ed. E. Scott Sills, Cambridge: University Press, 2016, pp. 131–139.

[9]"The following are the amendments to the French Civil Code that have been proposed by the MPT." *Politique de la famille et intérêt de l'enfant: 40 propositions concrètes*, pages 49–50.

According to this logic, the best interests of the child depend on profound changes in the law, as well as the repeal of the "Taubira Circular," which would bar children born of surrogacy from being recognized as the children of their intended parents, even if the gametes came from the father and the intended mother, as in the case of a heterosexual couple. The best interest of the child, according to the MPT, also calls for a revision of the Constitution, namely the insertion of a principle of respect for the best interests of the child in accordance with international treaties France signed. The MPT also calls for the constitutionalization of the principle of the nonavailability of the human body (ibid.: 35), as well as the outright denunciation of the European Convention on Human Rights, which contradicts the prior requirement that France respect its international commitments concerning the best interests of the child (ibid.: 35).

Article 1

Rewrite article 227–12 of the penal code as follows:

1. In the first paragraph, replace "6 months" with "5 years," and replace the number "7500" with "150,000."

2. In the second paragraph, replace "1 year" with "5 years," and replace the number "15,000" with "150,000."

3. Replace the third paragraph with the following:

"The act of mediating between a person or couple wishing to have a child, and a woman who has agreed to carry a child for them, is punishable by 5 years of imprisonment and a fine of 150,000."

4. After the third paragraph, insert the following two sentences:

"The purchase or sale of children, whether in France or abroad, is forbidden."

"An attempt to commit one of the offenses enumerated in this article, regardless of whether or not the act is carried out, or the act of presenting one of these offenses in a favorable light, is punishable by 5 years' imprisonment and a fine of 100,000 euros."

5. After the fourth paragraph, insert the following sentence:

"If any of the offenses enumerated in this article are committed abroad by a French citizen or resident, French law is applicable as an exception to the second paragraph of Article 113–6, and the provisions of the second paragraph of Article 113–8 are not applicable."

Article 2

The Government is to submit a report to the Parliament on bills which could propose the adoption of a specific international convention on the prohibition of surrogacy or gestational surrogacy.

Article 3

Replace Article 16–7 of the Civil Code with the following sentence:

"The adoption by the parent of a child born as a result of surrogacy is prohibited, both in France and abroad."

Article 4

After Article 47 of the Civil Code, Article 47–1, insert the following:

"Art. 47–1—No official document drawn up abroad, regardless of its legal nature, that notes, attests to, recognizes or proves parentage resulting from surrogacy, including in cases where it concerns a parent or parents of French nationality, can be transcribed into the French Civil Registry."

"Nowhere in the French territory can a decision from any authority, or any official document (regardless of its legal nature), legally recognize surrogacy or gestational surrogacy, including in the case of foreign citizens, regardless of their nationality."

3.2 The "Dangers of Marriage for All": The Circular Reasoning of Surrogacy and the Abolition of the Sexes

According to the MPT, "gender theory" is an attack on democracy that seeks to abolish sex differences and replace them with "subjective identities" and threatens the best interests of the child, as well as his or her right to be raised in a way that conforms to his or her assigned sex/gender. On the MPT website, under the heading "Understand," one can find a subheading entitled "Democracy in the Face of Gender"[10] which directs the user, among other things, to reject the lowering of standards that "gender theory" brings on. According to the MPT, in its text "The Ideology of Gender,"[11] this supposedly monolithic and unified "gender theory" is a scourge that represents a palpable threat to democracy, French culture, and Western civilization. In the MPT's publicity and communication materials, slogans intentionally reproduce gender stereotypes (girls dressed as fairies, boys dressed as blue knights, and asexuation/hermaphroditism symbolized by a snail) and present them as values to be defended against "gender theory" and the manifold threats it poses to society.[12]

Another poster used by the MPT in their demonstrations reject "gender" as being a bona fide term, and insist on only using the term "sex." From that flows the importance they place on the role of the father in that it must be a man.[13] If this sexual distinction is not made—mother/woman, father/man—then peace for the entire society would be threatened and the population would be plunged into chaos and ruin.[14] Indeed, at the heart of the threat of "gender ideology" lies the rejection of "the rooting of the feminine or masculine condition in a naturally sexed body" ("La Manif pour tous," *L'idéologie du genre* 2016: 8), and the promotion of the idea that "sexual identity is only a social construct, independent of any biological reality such as the sexed body" (ibid.). One demonstration MPT poster brings that point to the fore, stating "There are no eggs in testicles."[15]

---

[10]Publication of "La Manif pour tous," *La démocratie face au genre*: http://www.lamanifpourtous.fr/comprendre/la-democratie-face-au-genre/

[11]Publication of (La Manif pour tous), *L'idéologie du genre*: http://www.lamanifpourtous.fr/wp-content/uploads/2016/07/LMPT-L-ideologie-du-genre.pdf

[12]"Don't Touch Our Gender Stereotypes." "The Theory of Gender at School": http://www.20minutes.fr/societe/1274497-20140115-20140115-affiches-rassemblements-manif-tous-esquisse-retour

[13]"We Want Sex, Not Gender." "Zero Dad, Not Possible": http://blog.francetvinfo.fr/mariage-adoption-gay/2013/01/14/manif-pour-tous-une-demonstration-de-force-trop-millimetree.html

[14]"Whoever Destroys the Family Threatens Civil Peace": http://www.brain-magazine.fr/article/page-president/17605-Les-15-pires-slogans-de-la-Manif-Pour-Tous

[15]"There Are No Eggs in Testicles": http://leplus.nouvelobs.com/contribution/768440-frigide-barjot-a-l-elysee-une-redoutable-femme-politique-a-la-com-bien-huilee.html

This deliberate misreading of the past 40 years of contributions to the study of the dimension of gender in the human and social sciences[16] brings the MPT to the very same kind of conclusion it denounces: an ideological affirmation without any empirical basis. As a result, one can find muddled and cobbled-together observations such as "gender ideology is destructive, obscurantist, anti-social, anti-popular [. . .], anti-natural [. . .], of totalitarian essence [. . .], increases child behavioral problems [. . .], participates in social impoverishment through the recognition of subjective identities," and so on ("La Manif pour tous," *Politique de la famille*: pp. 17–20). This insistence on "the natural order" and "the rooting of the feminine or masculine condition in a naturally sexed body" is both the basis for the MPT's arguments and the result of their opposition to surrogacy practices and donor-assisted reproduction. Thus, the difference between the male and the female genders fails to serve as an effective foundation for this edifice—already actively weakened by "gender theory," which has supposedly been "thriving" since the opening of parliamentary debates on the "marriage for all" bill.[17] Only if a child is conceived and born of procreation between a man and a woman, preferably "under the covers" and within the bounds of marriage (as divorce is considered another source of the erosion and trivialization of the institution), can a society protect its weakest members ("La Manif pour tous," *Politique de la famille*: 21).[18,19]

The MPT has also mobilized other arguments, some of which echo the concerns of many feminists and social scientists who denounce the exploitation of women's bodies, or the commodification of children through the practice of commercialized surrogacy. Although this point has been the subject of international intellectual debates fostered by academic research for more than 30 years, the MPT supports its position against surrogacy almost exclusively through the use of sensational headlines which, in reality, are dramatic and exceptional cases that all sides wish to avoid (with "pro-surrogacy" advocates pushing for regulation of the practice and those "against" surrogacy calling for its global prohibition). Regardless of one's stance, and contrary to these posters' provocative messages, recent research shows that intended parents do not usually hide their means of procreation from their surrogacy-born children (Golombok et al. 2013). At present, surrogacy contracts

[16]See, among others: Anne Fausto-Sterling, *Myths of Gender: Biological Theories About Women and Men*, New York, Basic Books, 1992 & *Sexing the Body: Gender Politics and the Construction of Sexuality*, New York, Basic Books, 2000; the website "Gendered Innovations" by Londa Schiebinger, http://genderedinnovations.stanford.edu/; Catherine Vidal, *Cerveau, Sexe et Pouvoir*, Paris, Belin, 2015 (first edition 2005), *Nos cerveaux, tous pareils tous différents*, Paris Belin, 2015 & "Des différences aux inégalités entre les femmes et les hommes," ENS Lyon 2015: http://cle.ens-lyon.fr/plurilangues/hommes-et-femmes-ont-ils-le-meme-cerveau-249676.kjsp?

[17]For an overview of the debates between the left and the right in the National Assembly on the subject, see the documentary *La sociologue et l'ourson* by Etienne Chailloux and Matthias Théry, Quark Productions, 2016: https://la-sociologue-et-l-ourson.com/

[18]"ART, Surrogacy. Make A U-Turn": http://jepreferelenoirleblogdejul.tumblr.com/post/45109314436/les-affiches-de-la-manif-pour-tous-d%C3%A9cryptage#.V-5eC1vIYeM

[19]"One Father + One Mother For Filiation": http://jepreferelenoirleblogdejul.tumblr.com/post/45109314436/les-affiches-de-la-manif-pour-tous-d%C3%A9cryptage#.V-5eC1vIYeM

that have been drawn up in a regulated context and which take into account the best interests of every actor involved (starting with the unborn child and the surrogate, and followed by the intended parents) are generally carried out without conflict, and the number of legal disputes brought before the courts remains extremely low (less than one per thousand) (Peng 2013). This important perspective on the reality of surrogates and the children to whom they give birth (Merchant 2012; Jadva et al. 2015; Jacobson 2016; Lance and Merchant 2016, 149–164) is denied by the MPT, which does not hesitate to extend their analogies to animal breeding and even science fiction.[20]

Thus, according to the rhetoric of "La Manif pour tous," sex differences are embodied in the nuclear family, which is therefore an absolute necessity in guaranteeing the well-being of the child, and by extension, social peace. These recommendations take aim at both the private lives of those involved—the intimacy of the couple, the development of the child—and their public lives, with marriage supposedly acting as a reflection of "human nature," and parenthood a way to demonstrate "the sexual reality of humanity," a necessary and sufficient condition to ensure "the difference and complementarity of the sexes" (ibid.: 62). This argument is based on the reasoning that A justifies B, which in turn requires the prior validity of A. In this case, it is argued that men and women are different, which is why they get married and start families, thus proving the difference and the complementarity of the sexes. In other words, the conclusion leads one back to the initial assumption, a phenomenon that logicians identify as a *circular argument*, or more generally, a *petition of principle*.

Beyond the sophistry evident in the presentation of the MPT's short and superficial arguments, the condition of the possibility of the existence of circular reasoning—and where its interest lies—is that it witnesses a logical link of *co-existence* and *co-relation* between propositions A and B.[21] As such, analysis of the construction of the MPT's discourse—not its content, but its rhetorical form, thus taking into consideration the logical implications of its discursive architecture—must focus its attention on the clarification and qualification of the link between the institution of marriage and the family, sex differences, and sexual reproduction. These very general questions reflect the ontological questions of the double "origin myth," which we can find both at the individual, even psychoanalytic level (where do I come from, and who made me?), and the level of the humanity (where do we come from, as humans?).

---

[20]"Farm-Raised Mothers": http://jepreferelenoirleblogdejul.tumblr.com/post/45109314436/les-affiches-de-la-manif-pour-tous-d%C3%A9cryptage#.V-5eC1vIYeM, "I Too Want to Know Where I Come From. All of Us Are Born From a Man and a Woman": http://jepreferelenoirleblogdejul.tumblr.com/post/45109314436/les-affiches-de-la-manif-pour-tous-d%C3%A9cryptage#.V-5eC1vIYeM. These images of "industrialized farm-raised surrogates" deliberately bring to mind dystopian fiction such as Aldous Huxley's *Brave New World* and Margaret Atwood's *The Handmaid's Tale*.

[21]See B. Fracciolla, op.cit.

4 Conclusion

Through these two examples, research on embryos which is strictly regulated in France and surrogacy which is entirely illegal, we nonetheless witness a decided attempt on the part of some religious groups to both set back the clock on research and reestablish a traditional family configuration despite notable evolution in many other modern democracies. The most active of these religious groups is the Catholic Church which works behind the scenes via associations and interest groups, some of whom (keep) only exist on paper. This is a particular characteristic of the French society, in comparison for example with the USA where religious groups assume their public positions and activism. Proof of this is the dismay and oftentimes shock that the French population felt when demonstrations launched by the MPT against the same-sex marriage law filled the streets on several occasions with hundreds of thousands, at one point almost a million persons. Sociologists, political scientists, and other academics were completely taken by surprise at the extent of mobilization and financial resources to garner them that was taking place before their eyes. The French public at large, the majority of whom are in favor of same-sex marriage, were also stunned. Once the law was passed, religious groups and the MPT did not rest. They continue to try to obstruct embryo research as well as hold frequent and large demonstrations against surrogacy, curious acts in light of the fact that surrogacy is once again prohibited in France.

References

Dupont, G. 2017. 146 médecins et chercheurs. http://www.lemonde.fr/sciences/article/2017/03/30/des-chercheurs-et-des-medecins-denoncent-des-entraves-a-la-recherche-sur-l-embryon_5102966_1650684.html

Fracciolla, B. 2015. Violence verbale dans le discours des mouvements antagonistes: le cas de 'Mariage pour tous' et 'Manif pour tous'. *Argumentation et analyse du discours* 14|2015. http://journals.openedition.org/aad/1940; https://doi.org/10.4000/aad.1940. Accessed 09 April 2015.

Golombok, S., L. Blake, P. Casey, G. Roman, and V. Jadva. 2013. Children born through reproductive donation: a longitudinal study of psychological adjustment. *Journal of Child Psychology and Psychiatry* 54 (67): 653–660. https://doi.org/10.1111/jcpp.12015

Jacobson, H. 2016. *Labor of love: gestational surrogacy and the work of making babies*. New Brunswick, NJ: Rutgers University Press.

Jadva, V., S. Imrie, and S. Golombok. 2015. Surrogate mothers 10 years on: a longitudinal study of psychological wellbeing and relationships with the parents and child. *Human Reproduction* 30 (2): 373–379. https://doi.org/10.1093/humrep/deu339

Lance, D., and J. Merchant. 2016. Surrogacy in context: Ukraine and the United States. In *Assisted reproductive technologies in the global South and North: issues, challenges and the future (Routledge studies in the sociology of health and illness)*, ed. Virginie Rozée and Sayeed Unisa, 149–165. London: Routledge.

Laurent, Samuel. 2013. Derrière la grande illusion de la 'Manif pour tous'. *Le Monde*, http://www.lemonde.fr/societe/article/2013/03/21/manif-pour-tous-la-grande-illusion_1850515_3224.html

Manif pour Tous online publications. 2016. *Comprendre l'essentiel*: http://www.lamanifpourtous.fr/comprendre/comprendre-lessentiel/; *Sexe ou genre*: http://www.lamanifpourtous.fr/sexe-ou-genre/; *L'Idéologie du genre*: http://www.lamanifpourtous.fr/wp-content/uploads/2016/07/LMPT-L-ideologie-du-genre.pdf; *Comprendre la démocratie face au genre*: http://www.lamanifpourtous.fr/comprendre/la-democratie-face-au-genre/; *Politique de la famille*: http://www.lamanifpourtous.fr/politique-de-famille/

Merchant, J. 2012. Une GPA 'éthique' est possible. *Travail, genre et sociétés* 28 (2): 183–189. https://www.cairn.info/revue-travail-genre-et-societes-2012-2-p-183.htm

Merchant, J., and K. Krief. 2016. A baby's citizenship and kinship ties after surrogate birth: the case in France. In *Handbook of gestational surrogacy: international clinical practice and policy issues*, ed. E. Scott Sills, 131–139. Cambridge: Cambridge University Press.

Peng, L. 2013. Surrogate mothers: an exploration of the empirical and the normative. *American University Washington College of Law Journal* 21 (3): 555–582. https://pdfs.semanticscholar.org/5c37/3b587c667ab575f063a656f27b43b149ea10.pdf

Théry, I. 2010. *Des humains comme des autres: Bioéthique, anonymat et genre du don*. Paris: Editions EHESS.

Chapter 4
Embryonic Silences: Human Life Between Biomedicine, Religion, and State Authorities in Austria

Ingrid Metzler and Anna Pichelstorfer

Abstract This contribution seeks to develop a better understanding of the role of the Catholic Church in Austrian biopolicy making. It argues that the ambiguous place of the Catholic Church in Austrian culture and politics helps us to make sense of the particular shape of biopolicy debates. The argument is based on an exploration of moments in the politics of assisted reproductive technologies (ART) and in vitro fertilization (IVF) embryo research over the past three decades, which are situated in the values, norms, and imaginaries of the political culture of Austria's postwar Second Republic. The study shows that biopolicy debates focused on the use of ART technologies in reproductive medicine and the legitimate shape of Austrian families. This focus went along with a blurred vision on IVF embryo research and constitutive silences on IVF embryos. The brief moments in which IVF embryos were given center stage revealed that unborn human life is entangled in competing visions on its moral status and disagreements about the appropriate place of the Catholic Church in Austria's public life. In line with the norms and imaginaries of Austrian political culture, such disputes were not publically debated but pragmatically circumvented and silenced.

1 Introduction

In January 2015, the Austrian Parliament amended the "Law on Reproductive Medicine" (*Fortplanzungsmedizingesetz*, FMedG), modifying it quite substantially 23 years after its initial formulation. The amendment involved attenuations of some of

The writing of this chapter has been enabled by a grant awarded by the Austrian Science Fund (FWF) as a Hertha Firnberg-Research Fellowship (Grant Number T 811-G27).

I. Metzler (✉)
Department of Science and Technology Studies, University of Vienna, Vienna, Austria
e-mail: ingrid.metzler@univie.at

A. Pichelstorfer
Department of Science and Technology Studies, University of Vienna, Vienna, Austria

Department of Political Science, University of Vienna, Vienna, Austria

M. Weiberg-Salzmann, U. Willems (eds.), *Religion and Biopolitics*,
https://doi.org/10.1007/978-3-030-14580-4_4

the law's previous most central prohibitions. First, access to assisted reproductive technologies (ART) was no longer confined to heterosexual couples. It was now extended to lesbian couples. (Gay couples and single women and men continue to be excluded.) Second, both sperm and egg donation, which had been limited to the use of sperm donations for in vivo inseminations, were now also allowed for in vitro fertilization (IVF). Last but not least, the revised law also allowed the deployment of genetic technologies to inform decisions on which embryos to transfer, prohibiting pre-implantation genetic diagnosis (PGD) in principle, yet allowing its use in well-defined exceptions. Other prohibitions, such as the ban on surrogate motherhood or the prohibition of embryo research, were not altered (Bundesgesetzblatt 2015).

The revisions were intriguing not only in terms of their content. They were also remarkable in procedural terms. They were the result of a long process in which experts, bureaucrats, and politicians deliberated on how to revise the FMedG in light of a number of biomedical and societal developments as well as several court rulings. However, most Austrian citizens might have simply missed these deliberations. They took place not on the visible "front stage," but on the more secluded "back stages" of Austrian policy making, such as in working group meetings of parties, in intergovernmental committees, or in expert hearings. Thus, these deliberations were not publicly observable.

Once deliberations came to the fore in the form of a bill introduced by the coalition government formed by the "Social Democratic Party" (*Sozial-Demokratische Partei Österreichs*, SPÖ) and the "People's Party" (*Österreichische Volkspartei*, ÖVP) in November 2014, much of the politics had already been accomplished. Even members of the liberal "NEOS" and the "Green Party," two of the four parties in opposition during the 25th legislative period, welcomed the revisions. On the parliamentary floor, only members of the "Freedom Party" (*Freiheitliche Partei Österreichs*, FPÖ) and the tiny "Team Stronach" criticized the law. Outside parliament, organizations of the Catholic lay movement deplored that the law had been revised substantially without the public taking note, mobilizing first-person accounts of persons afflicted by the negative side effects of ART and calling for public debates (Katholischer Familienverband Österreichs 2015; Lebenskonferenz 2015). However, it seems fair to claim that these calls went unnoticed outside of the "Catholic bubble."

Thus, Austria moved from a restrictive law on ART to a more permissive one without this transition involving vocal public debates. To readers unfamiliar with Austrian politics, this might come as a surprise. Austria is usually referred to as a Catholic country, and the relative strength of the Catholic Church in a country tends to be mobilized as a factor to explain why "moral issues" are politicized or biopolitical issues are "moralized," in particular in cross-country comparative work (Knill et al. 2015). However, readers more familiar with Austrian biopolicy making might not be taken by surprise. Indeed, the policy process preceding the amendment of the FMedG seemed to be a rehearsal of patterns previously described, and often also bluntly criticized, in the literature on biopolicy making in Austria.

These patterns include a bifurcation of all things "bio." Austria has witnessed intense debates about "green biotechnology" in the 1990s (Grabner 2003; Felt

2015). Yet, when compared to these vocal debates or the controversies on "red biotechnology" in other countries (Jasanoff 2005; Banchoff 2011), debates on "biomedical policies" in general or on ART and IVF embryos in particular are rather inaudible (Grießler 2010). Policy debates are not altogether absent, but they are shaped by a distinct "topology." Issues tend to meander through secluded back stages, "flaring up" only briefly on publicly visible stages, such as on the floors of the Austrian parliament (Grabner 2003; Gmeiner 2006; Bogner 2007; Prainsack and Gmeiner 2008; Grießler 2010; Zichy 2010). Moreover, they also tend to be performed by a fairly regular "cast" of experts, stakeholders, bureaucrats, and representatives of interest groups. While this puzzling of elites in seclusion is at times criticized by actors in the debate and observers of the debate, these critical voices are not sufficient to disrupt what might be a distinctively Austrian "political culture" (Jasanoff 2005)—a nation-specific way in which a polity "acquire[s] communal knowledge for purposes of collective action" (Jasanoff 2005, 9).

How can we make sense of this? Why is there such a lack of public debate or full-blown controversies in Austria? Or—put in positive terms—how have Austrian policy makers managed to keep certain issues of policy making out of the limelight, which in other countries received unruly attention? And what is the role of religion—or, more precisely, of the Catholic Church—in this silence? In this contribution, we try to provide answers to these questions.

To do so, we start by building on secondary literature on the role of the Catholic Church in Austrian biopolicy making. A consistent part of the scholarly work on Austrian biopolicy making describes the Catholic Church as an important actor and/or (cultural) factor that shapes debates and law in Austria, often, however, mainly alluding to the Catholic Church's salience (Grabner 2003; Gmeiner 2006; Prainsack and Gmeiner 2008; Grießler 2010; Griessler and Hager 2016). Two contributions sought to shed light on how the Catholic Church influences biopolicy making in more detail. Both Christopher Knill and colleagues and Micheal Zichy argue that the Catholic Church shapes policy making in Austria mainly through influencing the policy positions of the ÖVP, the conservative "Austrian People's Party," the historical representative of the Catholic electorate. Thus, the Catholic Church can delay but not stop permissive reforms in the field of "morality policies" (Knill et al. 2014). Zichy notes that while this influence is "not strong enough" to impose conservative policies, it is "evidentially strong enough to impede political debates on a number of ethical issues, such as stem cell research or PGD" (Zichy 2010, 153).[1]

Agreeing with this in principle, we wish to understand *how* these delays and impediments are achieved in practice. We ask: How do Catholic actors, such as clergymen or associations of the Catholic lay movement, and Catholic values, sensibilities, and moralities shape biopolicy debates and biolaw in Austria? How and why do Catholic values and sensibilities matter in practice? And what does this tell us about the power (Bachrach and Baratz 1962; Digeser 1992) of the Catholic Church? We will base our argument on a discussion of a selected number of

[1]Unless otherwise noted, all translations from German originals into English are ours.

moments in debates on ART and IVF embryo research over the past three decades. Comparing regularities and differences over a relatively long period of time helped us to distill patterned focal points and blind spots in these debates, admittedly at the price of more detailed explorations of the many ways in which such publicly observable effects were produced in practice. We will argue that policy makers drew pragmatic boundaries between ART practices involving having children and embryo research in laboratories. While the former were diligently defined by law, the latter were pragmatically sidestepped and silenced. We will interpret these blind spots and blurred visions in law and public debates not as acts of neglectful omissions but as an effect of a tacit agreement to sidestep matters of contention between Catholic voices and the presumably secular rest.

We base our argument, firstly, on an analysis of statements made in public, analyzing how biomedical technologies were framed and problematized at these moments and who had a say in these debates (Jasanoff 2005; Wagenaar 2011). We interpret these moments through "agnostic lenses." Thus, rather than contrasting biopolicy making in Austria with an understanding of how it ought to be made, we wish to describe and to better understand salient patterns in these debates in the first place, seeking to avoid using debates in other countries as the implicit normative baseline. Secondly, we also brought in "ethnographic sensibilities" (Prainsack and Wahlberg 2013) to Austrian biolaw, situating these in the culture that we both inhabit. In order to lessen our "native lenses," we implicitly compared Austrian debates with those in Germany and Italy. Thirdly, in order to understand the stability of the patterns, we adopted "historical sensibilities," situating the IVF embryo debates in the "political culture" (Jasanoff 2005) of postwar Austria with its values, norms, and "political imaginaries"—or "causative fictions," such as "metaphors, ideas, images, or conceptions that acquire the power to regulate and shape political behavior and institutions in a particular society" (Ezrahi 2012, 3). We will argue that IVF embryo politics in Austria was both shaped by actors performing major tenets of the imaginaries of Austrian political culture and spaces where major elements of these imaginaries were "rehearsed," disrupted for a while, and eventually reestablished and "stabilized" (Felt 2015, 117–119).

2 Background: The Catholic Church in the Austrian Polity and Its History

2.1 *The Position of the Catholic Church in Austria*

When it comes to the role that the Catholic Church is allowed and expected to play in Austrian culture and politics, experiential knowledge and official records are divided. Austria seems to be a deeply Catholic country. Roman Catholics constitute by far the largest group among the population, even if their numbers have fallen from 89% in 1970 to 75% in 2001 and an estimated 64% in 2016 (Knill et al. 2014, 856;

Goujon et al. 2017, 11).[2] Austrian culture is also permeated by Catholic rituals, or as Julia Permoser aptly notes, "in the national imaginary 'Austrian culture' is strongly associated with (largely desacralised) Catholic traditions" (Permoser 2014, 254). There are also many urban legends on the ways in which the Catholic Church continues to influence Austrian culture and public policies.

However, this experiential knowledge of the perceived salience of the Catholic Church in Austrian culture and politics is at odds with the position that the Catholic Church is officially permitted to hold in encoded law. Christopher Knill and colleagues note:

> (T)he Catholic Church has no special position within the Austrian state. The national constitution defines the principle of religious freedom; (Knill et al. 2014, 856).

Hence, there is a mismatch between written law, which prescribes the Catholic Church to be one of several religious denominations to which the state keeps an equal distance, and a much more ambiguous role that the Catholic Church is perceived to be tacitly allowed to hold in Austrian culture and politics in practice. In the following sections of this chapter, we will argue that this distance between constitutional law and a much more ambiguously perceived reality is helpful in understanding the dynamics of the politics of IVF and IVF embryo research in Austria. In this section, we start with brief excursion into the history of the two Austrian republics, and the ways in which parts of it have been selectively forgotten, in order to understand how and why this distance began to emerge.

2.2 The Position of the Catholic Church in the History of the Austrian Polity

In the years now recorded as *Erste Republik* or "First Republic" (1918–1934), religion was a major cleavage between competing societal and political groups—or *Lager*—in particular between (often Anti-Catholic) Social Democrats and the Catholic Christian Socialists (the antecedent of the ÖVP), as well as the predominantly Protestant "German nationalists." The Catholic Church was skeptical toward the new republican polity (Bruckmüller 2013, 48) and also became an active force in its daily politics (Pelinka and Rosenberger 2000; Prainsack 2005, 542). The conflict between the Social Democrats and Christian Socialists eventually evolved into a bloody civil war between armed Social Democrats and Christian Socialists. In the period of the Austro-fascist *Ständestaat* (1934–1938), the relationship between the state and the Catholic Church was redefined (Hanisch 2005; Bruckmüller 2013, 48). For instance, the preamble of the then newly encoded "May constitution" contained an explicit

[2]In terms of numbers, the group of Roman Catholics is followed by individuals without denomination (Statistik Austria 2007). These have grown from 4% in 1971 to 12% in 2001. In 2016, researchers have estimated the number of individuals without denomination to be 16% (Goujon et al. 2017, 11).

Table 4.1 Austrian history in the twentieth century

1919–1933	First Republic
1933–1938	"Ständestaat": Austro-Fascism
1938–1945	Austria is part of Nazi Germany
1945–	Second Republic

reference to God (Suppanz 2005, 58). In 1938, the period of Austro-Fascism was brought to an end with the *Anschluss* of Austria to Nazi Germany (Table 4.1).

When Austria was reimagined as a democratic nation after 1945, much of this "imagination work" was accomplished by members of the SPÖ and the ÖVP and hence by the successors of formerly bitter enemies (Pelinka and Rosenberger 2000). They reestablished the written text of the constitution of the First Republic (in its version of 1929). The difference between the bloody past of the First Republic and the then newly emerging New Republic was not made formally—through a new constitutional text—but informally, through new values, practices, and institutions and thus through a new political culture. Collaborative deliberation on the level of elites in the form of a new "consensus democracy" was to help channel conflicts from streets to the back stages of Austrian politics (Pelinka and Rosenberger 2000), helping to ensure the independence of the newly reestablished nation and the durability of its newly imagined democracy. In this new understanding of good democracy, elites were not only entitled to retreat and deliberate solutions on behalf of the groups and the people they represented; they were actually expected to do so. Channeling conflicts to roundtables, thus producing silence on the streets, was framed as a historical achievement.

In the context of elites seeking to set aside their differences, the adoption of previously encoded constitutional settlements on religion as well as silence on what these ought to mean in practice was not an act of neglectful omission. Not only from the vantage point of hindsight, the adoption of previously agreed upon laws and the commitment of the new polity to keep an equal distance to all denominations (must have) seemed, in the words of Ezrahi, an "empirically baseless fabrication" (2012, 3). However, the reinvention of this "political fiction" was "necessary" (Ezrahi 2012, 4) for those elites to overcome, forget, and silence what had previously been a matter of armed contention between them, namely the conflicting views of the social democrats and conservatives about the appropriate position that the Catholic Church should hold in the Austrian polity. Perhaps, indeed, elites *had* to silence the matter of contention of their "fratricidal war" of the First Republic in order to succeed in their efforts to reimagine the new nation (Bruckmüller 2013, 28). In the early years of the Second Republic, issues concerning the relationship between the state and religion eventually evolved into a "discursive taboo" (Suppanz 2005, 29).

We deem the major tenets of the "imagination work" in the postwar years (the imaginary that consensus ought to be achieved by elites in seclusion and the political fiction that the Catholic Church is neither allowed nor expected to have a special place in the Austrian polity) to be helpful to better understand the peculiar case of Austrian IVF embryo politics. More generally, the politics of unborn human life is both shaped by actors performing elements of this imaginary *and* a space where major tenets of this "imagination work" were rehearsed, disrupted, and eventually reestablished and institutionalized. As all successful political imaginaries, these were fictions with consequences, which were also "institutionalized and sustained" and eventually "naturalized" in Austrian political culture (Ezrahi 2012, 5). Over time, they gained "sufficient credibility and adherence to attain the status of performative imaginaries that produce behavior that, in turn, affirms them" (Ezrahi 2012, 3). Over the decades of the Second Republic, most policy makers seemed to be reluctant to engage with issues on which the Catholic Church has or might have a say, forcing them to position themselves on the appropriate space of the Catholic Church in Austrian public life and unpack the political fiction that was necessary for the SPÖ and the ÖVP—the two parties whose interactions shaped the Second Republic—to be able to collaborate. This also seems to be the lesson drawn on later from the abortion debates in the 1970s, in which conflicts disrupted consensus and the relationship between the Catholic Church and the Austrian state was for a while exposed.

2.3 A Moment of Disruption: The Abortion Debates of the 1970s

In January 1974, Austria passed a reform of the Criminal Code, liberalizing the provisions on interruptions of pregnancies or abortions. The reform stipulated three exceptions in which these were exempted from punishment: first, when conducted within the first 3 months after conception and after a consultation with a physician; second, when performed so as to avoid a serious threat to life or the health of a pregnant woman; and, third, in case of a risk that the child might be seriously disabled (Grießler 2010).

The reform of the abortion law was one of the rare contentious moments in the Second Republic. It was passed 3 years after the SPÖ managed to win an absolute majority of votes in the general elections of 1971, forming a single-party government. It was preceded by debates within the SPÖ, whose mainstream was then at pains to enact a new culture of cooperation with the ÖVP and the Catholic Church, and therefore reluctant to endorse the women movement's call for a liberalization of abortion legislation (Mesner 1999, 190). Once consensus had matured within the SPÖ, the planned reform triggered criticism outside of it, eventually culminating in an open conflict between the SPÖ government and Catholic and conservative forces. Inside parliament, both parties in opposition, the ÖVP and the FPÖ, contested the planned reform. Opposition toward the reform also spilled over from parliament to other areas. In 1971, groups of the Catholic lay movement established *Aktion Leben* ("Action [for] Life)" and collected almost 900,000 signatures for a popular petition. The contestation of the reasoning of the SPÖ government reached parishes across the country, while the Archbishop of Vienna also participated in a demonstration in front of the St. Stephan's Cathedral in Vienna (Mesner 1999; Gmeiner 2006, 24; Grießler 2010, 160).

The conflict involved different visions of how the state ought to take care of human life (Lehner 2011). Opponents of the planned reform focused on the ontological and moral status of unborn human life, framing embryos and fetuses as embodiments of human life that begins with the fusion of gametes. They deduced a moral duty of the state to protect this early human life, assisting women to carry pregnancies to term (Prainsack and Gmeiner 2008, 384; Lehner 2011, 40). Proponents of the reform, on the other hand, did not venture into deconstructing this understanding. They sidestepped it, "avoid[ing] visibly evaluating the moral status of [the unborn.]" (Grießler 2010, 157–158; see also Lehner 2011). Social democrats framed the issue at hand differently—not as taking an ontological position on what unborn human life was or ought to be, but, in more pragmatic terms, as a matter of how the state ought to take care of (pregnant) women (Dohnal 2013, 51). Enshrining a difference between abortions in the first trimester (when these were allowed in general) and the second and the third term (when these were allowed for exceptional reasons), the eventually enacted law made an implicit difference between unborn human life at an early stage (in the first trimester) and a more advanced stage (after the first trimester); however, this difference was also rooted in the understanding that abortions were riskier when these were performed at a later stage. Ultimately, the law subordinated the question of how it ought to take care of human life to the question of how the state ought to deal with (pregnant) women, remaining constitutively silent on the moral status of unborn human life.

2.4 *Resealing the Black Box of Unborn Human Life*

Forty years after its formulation, the abortion law seems to be set in stone. However, the words in the book have not been translated into a public infrastructure of care. While abortions are allowed, access to them is unevenly distributed in Austria and often conducted in private clinics. At the same time, Catholic associations assist pregnant women in precarious situations. Hence, competing understandings on how the state ought to take care of unborn human life have not altogether disappeared. Instead, they were relocated from the political limelight to more dispersed sub-political spaces (Mesner 1999; Gmeiner 2006, 24; Grießler 2010).

The persistence of these contrasting visions can also be related to how the abortion law reform was achieved. This was not enacted by a coalition government but by a single party (the SPÖ) (Table 4.2). Involving substantial mobilization from "below," the abortion debates were one of the rare moments in which issues left the contained spaces of usual Austrian policy making (Mesner 1999, 200–202), disrupting Austrian political culture.

The bitter conflicts of the 1970s over how the Austrian Republic ought to take care of women and/or unborn human life have not worked their way into the collective memory (many of our fellow citizens are either no longer or simply not at all aware of them), but they have made their way into the memory and tacit practices of social democratic and conservative elites. Members of the SPÖ and the

Table 4.2 Governments of Austria's second republic

Begin	Composed by ÖVP and SPÖ or ÖVP and SPÖ ("grand coalition")	Single-party governments or other coalitions	Biolaw
1945	ÖVP, SPÖ, KPÖ		
1949	ÖVP, SPÖ		
1952	ÖVP, SPÖ		
1953	ÖVP, SPÖ		
1956	ÖVP, SPÖ		
1959	ÖVP, SPÖ		
1960	ÖVP, SPÖ		
1961	ÖVP, SPÖ		
1963	ÖVP, SPÖ		
1964	ÖVP, SPÖ		
1966		ÖVP	
1970		SPÖ	
1971		SPÖ	Reform of the abortion law
1975		SPÖ	
1979		SPÖ	
1983		SPÖ, FPÖ	
1986		SPÖ, FPÖ	
1987	SPÖ, ÖVP		
1990	SPÖ, ÖVP		FMedG
1994	SPÖ, ÖVP		
1996	SPÖ, ÖVP		
1997	SPÖ, ÖVP		
2000		ÖVP, FPÖ	
2003		ÖVP, FPÖ	
2007	SPÖ, ÖVP		
2008	SPÖ, ÖVP		
2013	SPÖ, ÖVP		Reform of the FMedG
2016	SPÖ, ÖVP		

ÖVP seem to have overcome the conflict on abortion by repairing the political imaginaries of postwar political culture, as well as by mobilizing the constitutive silence on unborn human life of the abortion law. Erich Grießler convincingly argues that the debate about interruptions of pregnancies of the 1970s constitutes an "'original conflict' ('*Urkonflikt*') of bioethical debates" (Grießler 2010, 144), noting that the memories of these moments continue to shape practices of contemporary policy makers, who do their best to avoid issues involving unborn human life. Similarly, Barbara Prainsack and Robert Gmeiner argue that the abortion conflicts have rendered discussions "of the *status* of the embryo (...) impossible" (Prainsack and Gmeiner 2008, 384). We agree with Prainsack and Gmeiner that "the embryo [is] powerful in structuring the discursive space" (Prainsack and Gmeiner 2008,

384). However, we do not interpret this as the effect of a shared understanding of the moral status of unborn life. On the contrary, we understand unborn human life to be so performative in Austria, because policy makers agree to disagree about its status and prefer to silence this matter of contention, sidestepping it pragmatically or leaving it altogether unaddressed.

As we will show in the next section, this pattern was rehearsed when the SPÖ joined forces with the ÖVP in a coalition government, preparing a government bill on ART. While it enshrined wordy (pro-)visions on the appropriate shape of Austrian families—and hence on what policy makers could agree upon—the law remained fairly monosyllabic on the status of unborn human life in petri dishes and freezers. Thus, it put unborn human life back into a black box, therewith implicitly allowing the Catholic Church to be seated on top of it.

3 The Reproductive Medicine Law

In 1992, the Austrian Parliament enacted the "Law on Reproductive Medicine" (*Fortplanzungsmedizingesetz*) (FMedG). The law defined the permissible uses of ART and the legitimate shape of Austrian families. It restricted access to ART to heterosexual couples and banned surrogate motherhood, embryo research, and gamete donation, making an exception for the use of donor sperm in in vivo inseminations.

The law was based on a bill prepared by a coalition government formed by the SPÖ and the ÖVP, and hence by the two parties that formed most of the coalition governments in the Second Republic, yet had been on the opposing sides in the abortion debates in the 1970s. They problematized ART in different ways. The ÖVP was mainly concerned with protecting the proper shape of families, which in their view should continue to consist of heterosexual couples and children and the integrity of human life. Thus, the ÖVP shared some sensibilities with *Donum Vitae*, the document in which the "Congregation for the Doctrine of the Faith" gave answers to questions raised by ART and IVF embryo research in 1987 (Congregation for the Doctrine of the Faith 1987). Remarkably, *Donum Vitae* was also included among the expert sources in the explanatory material of the eventually agreed upon government bill (Parlament 1991b). The SPÖ emphasized the need to provide legal certainty for biomedical professionals, intended parents, and children and to protect women and their bodies from being exploited by the destructive forces of reproductive technologies (Grießler and Hadolt 2006). Hence, similar to the abortion debates in the 1970s, the two parties problematized human reproduction and life differently. However, while the two governing parties diverged in the ways in which they problematized ART, they converged in their understanding that it was their responsibility to make the common ground for a law that could direct ART practices toward socially desirable ends (Hadolt 2007).

Visions of "nature" and "natural reproduction" (Grabner 2003, 206) allowed the two parties to bridge their different moralities—the "conservative defense of a traditional family model (ÖVP) and the protection of women against the exploitation and commercialization of their bodies (SPÖ)" (Griessler and Hager 2016, 74). Similar to the "Law Forty," which was enacted in Catholic Italy in 2004, the Austrian FMedG inscribed into law visions of "nature" and "naturalness," allowing only those artificial reproductive technologies in vitro that mimicked what already existed in vivo (Metzler 2007; Testa 2011; Kopetzki 2012). Ranked according to their level of "technicity" or distance from natural processes, ART was permitted only as "ultima ratio" (Parlament 1991b, 11) for couples whose ailing bodies needed a helping hand to reach a socially permissible end: to normalize their heterosexual couplehood with children.

Thus, "nature" allowed policy makers to build a common ground on how ART ought to be used in reproductive medicine. When the law was finally approved, it was about the proper shape of Austrian families and socially permissible reproductive therapies. However, the law was not a law on IVF embryos or experimental research conducted in and restrained to laboratories, which was apparently considered morally troublesome and, in any way, scientifically far-fetched; policy makers did not deem science in laboratories in need of their attention. A report by a parliamentary committee explained:

> The creation of a legal framework for specific techniques of medical reproduction shall not be intermingled with questions that concern a potential abuse of insights from biology and genetics. The frequently mentioned possibilities of cloning, the creation of chimaera or interspecies entities (...) are not related at all to medical assistance in the fulfillment of reproductive wishes (Parlament 1991a, 2).

With the deliberate exclusion of laboratory practices, the issue of the status of IVF embryos, of what they were ontologically and what they ought to be morally, which was central in the reasoning of the Roman Catholic Church and also a major concern in the laws in Austrian neighboring countries Germany and, later, Italy, was also pragmatically sidestepped in Austrian law. Tellingly, the FMedG did not even use the term "embryos," apparently choosing to bypass known matters of contention, referring to them with ambiguous terms. A note explained that the term "embryo" was explicitly avoided in the law "because neither scientific terminology nor everyday use—given different ideologies—is unambiguous or consistent" (Parlament 1991b, 15). Seeking to keep such ambiguities at bay, the law used the term "cells capable of development" (*entwicklungsfähige Zellen*) instead, importing it from the German Embryo Protection Act (Metzler 2016). However, while in German law *Entwicklungsfähigkeit* was used to diligently define those human lives whose dignity the state had to protect, Austrian policy makers reappropriated this term in order to produce ambiguity, bypassing debates on the status of unborn human life and leaving unpacked the questions that the then governing social democratic SPÖ and the more Catholic ÖVP had previously proven to disagree about in the abortion debates.

Visions on what "cells capable of development" were or could be were not discussed explicitly but implicitly inscribed in between the lines of the FMedG. Paragraph 9 stipulated that "cells capable of development must not be used for anything other than reproductive purposes," thus prohibiting embryo research.

Paragraph 10 also invited biomedical professionals to fertilize no more egg cells than needed for a pregnancy. What this meant in practice was, however, left open to the interpretation of biomedical professionals. Hence, while the generation of "surplus" human life in reproductive medicine was a central concern in the official moral doctrine of the Roman Catholic Church and also a prominent issue in German and in Italian law (Jasanoff 2005; Metzler 2007; Testa 2011), Austrian law admitted silently and pragmatically that some embryos might be lost in reproductive treatments. Stipulating that these must not be used for anything other than reproductive purposes, the law restricted the possible trajectories of IVF embryos to developing into children of their parents. It remained silent on embryos—or "cells capable of development"—outside of this parental custodianship. These were nonentities in legal terms. As we will see in the next two sections, IVF embryos reemerged in the following two decades—often brought to the public attention by Catholic actors. They proved to be the main spokespersons of IVF embryos in particular and unborn human life in general.

4 Human Embryonic Stem Cells in Austria

In the early years of the twenty-first century, policy makers in many countries puzzled over how to accommodate the hopes and anxieties of human embryonic stem cell (hESC) research. Scientifically classified as bearing the potential to form all kinds of cells and tissues from adult human beings, the cells crafted from human IVF embryos were socially imbued with promises of therapies for incurable conditions (Waldby 2002). In many countries across the globe, they became the matter of contentious politics in which the biological potentials and social promises clashed with understandings on the state's mandate to protect the lives and dignity of IVF embryos (Jasanoff 2005; Banchoff 2011). When compared to the vibrant public lives dwelt elsewhere, hESCs bore a fairly quiet life in Austria.

In legal terms, hESC research was indirectly governed by paragraph 9 of the FMedG, which prohibited the use of "cells capable of development" for anything other than reproductive purposes. Legal experts agreed that this implied that Austrian scientists were not allowed to dismantle an IVF embryo for the sake of producing hESC lines. However, existing hESC lines, which were imported from abroad, were not classified as "cells capable of development," and thus deemed outside of the area of application of the FMedG (Bioethikkommission beim Bundeskanzleramt 2002, 1; Kopetzki 2008).

Hence, when hESCs began to emerge in the late 1990s, the legal status quo in Austria was comparably undefined as in Germany (with the Embryo Protection Act) and, later, in Italy (with "Law Forty"). However, the politics that emerged from this indefiniteness was different. Neither did biomedical professionals, legal experts, bioethicists, and politicians join forces to collectively agree on regulations to publicly define under which circumstances hESCs might be legally imported as happened in Germany (Sperling 2004). Nor, on the other hand, did patient organizations and biomedical professionals contest the state's prohibition to use (Italian) IVF embryos, as in Italy (Metzler 2007; Prainsack and Gmeiner 2008). In Austria, hESCs remained the "esoteric stuff" that only few biomedical professionals were reported to be involved with (Schönthaler and Wagner 2008), rather than growing into the stuff of contentious politics, many Austrian citizens were purportedly "ignorant about" hESC (Freistetter 2015), and a small group of legal experts, bioethicists, and clergymen proved to disagree about it (Körtner 2008; Virt 2008). Overall, Austrian policy makers and stakeholders seemed to prefer the combination of legal restrictions (on the use of Austrian IVF embryos) and ambiguity (on funding for hESCs imported from above) over contestations in public.

The public silence on hESC was interrupted only twice. A first dispute evolved in the early noughties around the question whether hESCs should be included among the matters eligible for being funded in the then upcoming sixth framework program of the European Union (Kneucker 2008; Prainsack and Gmeiner 2008; Grießler 2010). In December 2001, the Austrian government, then formed by an (unusual) coalition between the ÖVP and the FPÖ, articulated opposition toward this prospect (Gmeiner 2006, 30; Körtner 2008, 2). After this announcement, the responsible minister Elisabeth Gehrer (ÖVP) asked the newly established "Bioethics Commission at the Federal Chancellery" for an opinion. The commission published its statement in May 2002 (Körtner 2008, 1). As in many opinions and statements that were to follow, the commission was split in a permissive and a restrictive part (Bioethikkommission beim Bundeskanzleramt 2002). In the short "resolution" on hESC research of 2002, however, both sites made visible efforts to search for a common ground. In this statement of six pages, all members agreed that research on "'surplus' embryos" and on "somatic cell nuclear transfer ('therapeutic cloning')" should not be eligible for funding (Bioethikkommission beim Bundeskanzleramt 2002, 2). However, members disagreed on the eligibility for funding of "already existing" hESCs. A tight majority of 11 members deemed that this should "not be rejected in principle" (Bioethikkommission beim Bundeskanzleramt 2002, 3–4). A minority of eight members noted that the therapeutic aims of hESC research did "not amount to an ethical justification of the destruction of embryos" (Bioethikkommission beim Bundeskanzleramt 2002, 5). Subsequently, minister Gehrer made the *minority* position into her own and continued to work against European funding for any kind of hESC research (Kneucker 2008; Prainsack and Gmeiner 2008). This did not stop hESC research from eventually becoming eligible for European funding (Salter 2007). Yet, it did have lasting effects on the (absence of) debates on hESC research in Austria.

Journalists, commentators, and scholars converged in classifying Gehrer's actions as being "sensibly" motivated by Catholic moralities (Kneucker 2008, 67), at times also bluntly dismissing them as an inappropriate effort to "appease Roman-Catholic and conservative circles" (Grabner 2003, 207). This interpretation does not seem entirely ungrounded. Globally known as a fervent opponent of the instrumentalization of the lives of IVF embryos for the sake of promises that were also deemed far-fetched, the Roman Catholic Church was among the few actors in Austria that spoke out against the then globally emerging hESC research in public. For instance, in November 2001, the Austrian Bishops Conference issued a press release, asking for "a general prohibition on cloning and all 'exploitation' of embryos or foetuses" (quoted in Gmeiner 2006, 28), and inviting the Austrian government to follow "its own ethically responsible path" (quoted in Gmeiner 2006, 28; see also Zichy 2010, 150). However, the reluctance of the minister to explicitly draw upon Catholic values to justify her decision in public and the predominantly dismissive and insinuating terms in which her decision was commented on by others showed, firstly, that "Catholic values" were not deemed to belong to the repertoire of arguments from which Austrian politicians were allowed to draw when envisioning socially desirable ends or reasoning in public. Moreover, secondly, the short presence of hESCs on Austria's political stage also helped to frame them as potential matters of contention between some actors (purportedly) misguided by "Catholic moralities" and a secular rest. They disagreed on both—an appropriate Austrian approach toward hESC research and the appropriate place of Catholic morality in co-defining it.

The latter framing of hESC research was rehearsed in 2009, when the Bioethics Commission issued a longer report on hESC research. This consisted of two diverging positions: a permissive position A supported by a majority of 17 members and a restrictive position B supported by 5 members (Bioethikkommission beim Bundeskanzleramt 2009). The two sites were sensibly more polarized than in the commission's previous statement. The report also provided ample space for what the previous commission had reportedly deliberatively sidestepped: the issue of the status of IVF embryos (Körtner 2008; Zichy 2010).

Position B noted that "considerations of the problem" of hESC research could not "avoid a discussion on the status of the human embryo" (Bioethikkommission beim Bundeskanzleramt 2009, 71).[3] It argued that human embryos should not be treated as a generic form of "human life," which might also be "present in living human cells," but as "*someone's* life" (Bioethikkommission beim Bundeskanzleramt 2009, 71). Stating that while it might be impossible to come up with a "clear definition of the beginning of someone's life," position B drew on this uncertainty to make a case for precaution and against the instrumentalization of human embryos (Bioethikkommission beim Bundeskanzleramt 2009, 72). The minority wrote:

> Someone who cannot know whether something is the case (for example, that a human being has started to exist), may not act as if he knew with certainty that this was not the case. There is, therefore, due to rational reasons, an ethical imperative (not based, for example, on 'Weltanschauung' or religious prejudice) to choose the safer path and to accept the resulting practical consequences (Bioethikkommission beim Bundeskanzleramt 2009, 72).

Following position B, it was in light of rational reasoning that the state should take legal precaution, protecting unborn human life from infringements. Hence, even known Catholic theologians seemed to find it inappropriate to use religious terms, when reasoning in public.

The majority position A was substantially longer, much more permissive, effectively taking the opposite stance. It also provided ample space to the status question,

---

[3]The English translation is taken from the original. This statement of the Bioethics Commission is available in both German and English.

qualifying position B as only just one among several possible positions on the status of IVF embryos. This, position A underlined, could not provide a common ground in a "pluralistic society" (Bioethikkommission beim Bundeskanzleramt 2009, 22), nor could the state privilege the moralities and sensibilities of one group. While position B mobilized the uncertainty on the status of IVF embryos for a call for comprehensive legal restrictions, position A translated this uncertainty in an invitation to a much lighter touch.

Thus, the report illustrated that in the early years of its existence the Austrian Bioethics Commission constituted a semi-secluded micro-cosmos in which the 1970s disputes about what unborn human life was and what it ought to be were played out intensively. It further showed that these positions were discursively structured along a religious line: the reasoning of a (then exclusively male) Catholic minority was opposed by the reasoning of a more heterogeneous majority. Moreover, specific arguments on the ontological and moral status of IVF embryos were also implicitly tied to debates about the appropriate place of Catholic moral reasoning in Austria's public life. Thus, the Austrian Bioethics Commission was one space where the "political fiction" of the state's equal distance to all denominations was opened up.

The short moments in which hESCs surfaced on the Austrian political stage were long enough to underline that the pluripotent hESCs also bore the potential of contentious politics between Catholic sensibilities and a seemingly secular rest. At the same time, their presence was too short to make these cells signify beyond a small group of scientists and bioethicists. As entities that were neither morally innocuous enough to be collectively embraced nor monstrous enough to be in need of being "kept out" (Felt 2015) of Austrian laboratories, hESCs were allowed to keep their morally undefined and legally ambiguous status. At a time in which the coalition government had returned to the normalcy of the Second Republic with an SPÖ-ÖVP government, hESC merely had to be kept away from the political stage—as they effectively were.

As we will show in the next two sections, similar conflict lines also shaped the debates on another biomedical technology involving IVF embryos: pre-implantation genetic diagnosis (PGD). Just as hESC research, PGD "flared up" on the public stage only occasionally, without engendering soul-searching in public, as in Germany (Herrmann 2009), or triggering contestations of the state's laws by couples and patients as in Italy (Metzler 2012). However, as we will show in the next two sections, unlike hESC research, PGD began to gain significance beyond a small circle of bioethicists and scientists, eventually growing into an issue that policy makers took care of.

5 Pre-implantation Genetic Diagnosis

At the end of the 1980s, while Austrian policy makers were still debating how to order ART, a team of British biomedical professionals demonstrated that genetic analysis conducted on a cell removed from an early stage IVF embryo could make informed decisions on which IVF embryo to transfer (Handyside et al. 1990). Therewith, they proved that what is by now known as "pre-implantation genetic diagnosis" (PGD) could help carriers of genetic conditions avoid the genetic lottery of nature. Despite the temporal overlap between scientific demonstrations in the United Kingdom and legal sense-making in Austria, the FMedG did not mention PGD. However, when debates on PGD began to emerge one decade later, the dominant (but not unanimous) interpretation was that PGD was at odds with two of the law's provisions: first, the law's prohibition to use "cells capable of development" for anything other than reproductive purposes (paragraph 9, article 1) and, second, the restriction of access to ART to infertile couples (paragraph 2) and thus the exclusion of (fertile) carriers of genetic conditions (Kopetzki 2012).

Debates on PGD in Austria began to take shape in 2004, when the Austrian Bioethics Commission issued a report on it. As in the case of the two hESC reports, the Austrian commission's members were split into two parts. A majority of 11 members voted for a limited permission of PGD. A minority of six members voted against a liberalization of PGD. The reasoning of the two sides showed the same patterns discussed in the previous section on the hESC debates: the commission was divided along religious lines, with different views on what human IVF embryos were and how the law should protect them (Bioethikkommission beim Bundeskanzleramt 2004; Bogner 2007).

Despite the common beginnings of the hESC and PGD debates in Austria, the meaning-making around PGD subsequently developed differently. On the one hand, skepticism toward the "selection" of IVF embryos was not only voiced by members of the Catholic Church but also articulated by more secular associations. They problematized PGD as a medicalization of women's bodies and value judgments on human life with disabilities and invited the state to take measures to protect this life from being discriminated against. Both also warned that permitting PGD would amount to a return of the eugenic past of Nazi Austria. On the other hand, proponents of PGD were not limited to biomedical professionals, but also included actors concerned about the health and reproductive rights and freedoms of women. They underlined that the deemed ban on PGD was at odds with the state's permission to interrupt a pregnancy after prenatal diagnosis. Thus, PGD in Austria began to gain importance beyond a small circle of actors. Framed not as basic science (as hESCs) but as an issue of public health, it became a matter that the state's law had to address.

In July 2005, the Ministry of Health, then overseen by Maria Rauch-Kallath (ÖVP), took a first initiative. It circulated a draft bill for an amendment of the *Gentechnikgesetz*, the "Gene Tech[nology] Act" of 1994 (Bogner 2007, 324). This draft also included provisions that should allow PGD to test the "viability of embryos," which will die "before, during, or shortly after birth" (Parlament 2005, 16–17). The provisions were dropped after a storm of protest from disability rights associations (derStandard.at 2005), Catholic organizations (Österreichische Bischofskonferenz 2005), and the Minister of Justice, then overseen by Karin Gastinger, a former member of the FPÖ (Bundesministerium für Justiz 2005).

Between 2005 and 2015, the provisions of the FMedG were criticized several times, in particular by legal scholars, biomedical professionals, and the majority of the Bioethics Commission (Bioethikkommission beim Bundeskanzleramt 2012). However, in the absence of a common ground that could bridge the differences between the SPÖ and the ÖVP, the two parties that had returned to constitute the Austrian government after a brief interlude of the ÖVP with the FPÖ, PGD remained an issue of consensus seeking in seclusion and not a matter of limelight politics (Bogner 2007).

Nonetheless, the FMedG was eventually changed. As we will show in the next section, appeals by Italian and Austrian citizens to judges in courtrooms helped to make a difference. In 2011, the European Court of Human Rights (ECHR) issued a judgment on the case *SH and others* vs. *Austria* brought to court by two Austrian couples claiming access to IVF using sperm and ovum donation. The ECHR confirmed the FMedG as constitutional, implicitly advising the Austrian government to review it (European Court of Human Rights 2011). In 2012, the ECHR deemed Italian law inconsistent in *Costa and Pavan* vs. *Italy*, which—similar to Austrian law—allowed carriers of genetic conditions access to prenatal diagnostics and abortions, yet not to PGD (Hanafin 2013). Finally, in 2013, a lesbian couple successfully challenged the FMedG's restriction of the access to ART to heterosexual couples, when the Constitutional Court of Austria declared large parts of the FMedG unconstitutional, providing the government with a 1 year period to revise the law (Verfassungsgerichtshof 2013). Thus, courts "forced" the SPÖ-ÖVP government to take actions, while also providing them with a depoliticized common ground. As we will show in the next section, this common ground no longer consisted of "natural facts," as in the past, but of "legal facts" and scientific evidence.

6 Revisions of the FMedG

In 2015, the Austrian parliament revised the FMedG fairly substantially. The amended law diversified the desirable shape of Austrian families, allowing access to ART to lesbian couples and permitting gamete donation. Reproductive rights continued to be tied to couples, on the ground that single parenthood should not be intentionally produced. Thus, "natural families" consisting of couples and children remained the norm. However, "natural families" were no longer defined in biological terms alone but also included social ties. In the words of the Minister of Justice Wolfgang Brandstetter (without affiliation to a party), these were "more important for the well-being of the child than biological parenthood" (Parlament 2015, 89).

Next to extending the permissible shape of families, the revised law also allowed some form of PGD, banning it in principle but making a number of exceptions: in the case of repeated failures of IVF, repeated miscarriages or stillbirths, or if one of the intended parents was a carrier of a severe genetic condition. Brandstetter explained that in light of the reasoning of the ECHR in *Costa and Pavan* vs *Italy*, there was a need to "adjust the imbalance between the protection of the embryo in vitro and in vivo, in order to spare women and couples with a desire for a child a tentative pregnancy" (Parlament 2015, 88). Thus, the new provisions translated the reasoning of the ECHR into Austrian law, giving women with troubled reproductive histories "a chance, to perhaps have a child" (Parlament 2015, 89) while sparing them repeated experiences of "pregnancies on trial," as Minister of Health Sabine Oberhauser (SPÖ) underlined (SPÖ-Bundesorganisation 2015). The provisions were restrictive, so as to ensure that human life would not be "selected," and that PGD would not be used to make differences between "worthy and unworthy life," or between "perfect and non perfect [life]" (Parlament 2015, 104).

The amendments of the FMedG were facilitated by a depoliticized common ground that courts had begun to make. Brandstetter framed the law as being "stipulated by constitutional and fundamental rights' requirements," facts which were "established by high courts and which do not get to be debated" (Parlament 2015, 87–88). In addition to the "legal facts," advancements made in reproductive medicine were discursively used to argue for a revision of the law. Proponents of a more permissive regulation, among them biomedical professionals, LGBT organizations, and the majority of the Bioethics Commission, highlighted the safety of ART, countering the arguments of opponents, who propounded health risks for children and women. "Societal transformations" and "lived realities" were brought to bear on the debates as well. For instance, members of the Green Party argued that the Austrian population had now accepted different forms of families. These "lived realities of families and citizens" should be recognized by Austrian biolaw (Parlament 2015, 79).

A variety of actors, such as lay Catholic organizations and a minority of the Bioethics Commission, opposed these arguments. They emphasized the importance of the well-being of children born through ART. They criticized the lack or poor quality of scientific data, underlying that the side effects of ART were still uncertain. Thus, even Catholic organizations followed the new rules of this "fact-based policy making," basing their arguments on scientific evidence or the lack thereof. They questioned and deconstructed the purported legal necessity of the reform (Österreichische Bischofskonferenz 2014) and highlighted the necessity of scientific expertise to provide more "thorough assessments" (Aktion Leben 2011). Furthermore, they used a procedural argument often made in the literature on biopolicy making in Austria, mobilizing the "lack of public debates" around ART in Austria to criticize the law and called it a "mockery of parliamentarianism" (IMABE 2014). Despite this opposition, the revisions of the FMedG did not result in public debates.

Thus, legal facts and scientific evidence provided a depoliticized common ground that enabled the government to propose an amendment to the original FMedG. While even Catholic organizations outside of parliament used this language in their contestation of the bill, inside parliament there were only two parties that disagreed. Members of the FPÖ contested the evidence assembled on parliamentary floors, given that there were "studies for everything!" as deputy Herbert Kickl (FPÖ) deplored. Accusing the ÖVP of betraying its "own values and its own pretension, to be a purportedly conservative party on a Christian fundament" (Parlament 2015, 76), Kickl dismissed the bill as an effort to "turn creation on its head" (Parlament 2015, 77). Thus, he indicated that while even Catholic organizations had left the religious language behind, the FPÖ was ready to revitalize it, now claiming to be the true representatives of Catholics in Austria. In January 2015, it was overruled.

7 Conclusions

In this chapter, we have followed IVF embryos in Austrian politics to develop a better understanding of the many ways in which Catholic actors as well as Catholic values, sensibilities, and moralities shape Austrian biopolicy debates. We have argued that policy debates on IVF technologies in Austria have focused on the use of these technologies in reproductive medicine. This focus on reproductive medicine went along with a blurred vision of IVF embryo research in law and in policy debates, resulting in legal and discursive silences of embryos outside of parental custodianship.

Revisiting the past abortion debates as well as brief moments in which IVF embryos *were* put center stage, such as in the debate on hESC research, helped us to show that blind spots in debates and pragmatic silences in law were not just unintended omission. These were the effect of a tacit consensus between the SPÖ and the ÖVP to "black box" known matters of contention. In the rare moments in which IVF embryos were disentangled from this relationship of care with their intended parents, disputes on the ontological, moral, and legal status of the IVF embryo were performed along a cleavage between Catholic actors (or arguments labeled as Catholic ones by others) and more heterogeneous groups of legal experts and biomedical professionals. Thus, debates on the proper place of IVF embryo research in Austria were also tied to debates on the appropriate place of Catholic moralities in Austrian public policies. The one—the status of IVF embryos—could not be settled without having a say about the other—the appropriate place of Catholic sensibilities. Rehearsing the norms, values, and imaginaries of Austrian political culture, these matters of contention were not wrestled through in public. Instead, they were either altogether sidestepped and avoided (as in the hESC debates), or deliberated on the back stages of policy making until evidence and facts, elaborated in working groups or imported from courtrooms, provided a depoliticized common ground. Ultimately, both the status of unborn human life and the appropriate place of the Catholic Church in the Austrian polity remained undefined and unsettled.

These elements aggregate to a fairly ambiguous mosaic on the many ways in which Catholic actors and Catholic values shape IVF embryo politics in Austria. First, Catholic values and sensibilities shaped Austrian policies in a direct way, contributing to the content of the original version of the FMedG. The very prominence of "the family" and the initially narrow vision on its appropriate shape are an indicator of the salience of Catholic values in Austrian law. Importantly, however, these values could be brought to bear in Austrian law, as these were no longer deemed partisan religious ones but cultural or even natural ones. In more abstract terms, Catholic values and moralities matter most when they resonate with more secular sensibilities. Catholic values and moralities need to be translated into (Austrian) cultural ones or (universal) natural ones to become legitimate in public reasoning in Austria. This must be seen in light of the Second Republic's political fiction on the equal distance of the state to all recognized religions, which is constantly rehearsed, institutionalized, and also naturalized through this translation work. However, this influence comes at a price: once Catholic sensibilities are translated into cultural ones and get salience in public debates, these are no longer owned by the Catholic Church. For instance, the redefinition of legitimate forms of families in the revision of the FMedG in 2015 shows that what an appropriate Austrian family is is no longer amenable to be controlled by the Catholic Church alone.

Second, the Catholic Church also shaped Austrian biopolicy making in an indirect way, by hampering others from pursuing desired actions or—paraphrasing Bachrach and Baratz' famous definition of the "second face of power"—by "creating or reinforcing social and political values and institutional practices that limit the scope of the political process" (Bachrach and Baratz 1962, 948). Some matters, such as hESC research, are kept at a distance from the political stage and silenced instead. However, as we have shown in this chapter, these silences in public debates or blind spots in law are not produced by the Catholic Church alone. They are a joint product of conservative and social democratic policy makers, who prefer to sidestep and silence matters of contention, thus rehearsing elements of political imaginaries of Austrian political culture.

The silences of matters of contention produced the effect that unborn human life remains unnamed and undefined in Austria. What unborn human life is and what it ought to be has never been defined nor secularized in Austrian biolaw. It remained in a "black box" of unaddressed issues, on whose top the Catholic Church is allowed to sit as the main spokesperson of unborn human life (Callon and Latour 1981, 285). Intriguingly, however, the Catholic Church is seated on top of this black box not just because it has not been disenchanted by state policies; its position is also created by others, for instance, when concerns for the dignity of unborn human life are dismissed as mere Catholic sensibilities; or when others—including social scientists—allude to a somehow pervasive power of the Catholic Church in Austria.

Hence, third, an additional dimension of the power of the Catholic Church consists in the power invoked by others. This power is not exercised in practice but mainly attributed in theory. Nonetheless, it is performative too: it leads to the avoidance of potential conflicts and thus to public silences, producing a circular effect. It is because the Catholic Church is deemed powerful that it becomes powerful as a consequence. That is why we wish to conclude with the suggestion that the power of the Catholic Church is not only a reason for the peculiar shape of biopolicy making and its "embryonic silences" in Austria; the power of the Catholic Church is also an effect of a mode of biopolicy making that favors consensus, ambiguity, and silences over opening black boxes.

References

Aktion Leben. 2011. Dokumentation der Studientagung: Die Eizellspende. Schafft sie mehr Probleme als sie löst? Aktion Leben.

Bachrach, Peter, and Morton S. Baratz. 1962. Two faces of power. *The American Political Science Review* 56: 947–952.

Banchoff, Thomas. 2011. *Embryo politics: ethics and policy in Atlantic democracies*. Ithaca, NY: Cornell University Press.

Bioethikkommission beim Bundeskanzleramt. 2002. Stammzellenforschung im Kontext des 6. EU-Rahmenprogrammes. Beschluss der Bioethikkommission vom 3. April und 8. Mai 2002. http://archiv.bundeskanzleramt.at/DocView.axd?CobId=1115. Accessed 19 Dec 2017.

———. 2004. Präimplantationsdiagnostik. Bericht der Bioethikkommission beim Bundeskanzleramt. July. http://archiv.bundeskanzleramt.at/DocView.axd?CobId=6415. Accessed 19 Dec 2017.

———. 2009. Forschung an Humanen Embryonalen Stammzellen. Stellungnahme der Bioethikkommission vom 16. März 2009. http://archiv.bka.gv.at/DocView.axd?CobId=34240. Accessed 19 Dec 2017.

———. 2012. Reform des Fortpflanzungsmedizinrechts. Stellungnahme der Bioethikkommission beim Bundeskanzleramt. July 2. http://archiv.bundeskanzleramt.at/DocView.axd?CobId=48791. Accessed 19 Dec 2017.

Bogner, Alexander. 2007. Was heißt "Politisierung von Expertise?". *Österreichische Zeitschrift für Politikwissenschaft* 36: 319–335.

Bruckmüller, Ernst. 2013. Katholisches an und in Österreich. In *Öffentliche Religionen in Österreich. Politikverständnis und zivilgesellschaftliches Engagement*, ed. Jürgen Nautz, Kristina Stöckl, and Roman Siebenrock, 17–54. Innsbruck: Innsbruck University Press.

Bundesgesetzblatt. 2015. 35 Bundesgesetz, mit dem das Fortpflanzungsmedizingesetz, das Allgemeine bürgerliche Gesetzbuch, das Gentechnikgesetz und das IVF-Fonds-Gesetz geändert werden (Fortpflanzungsmedizinrechts-Änderungsgesetz 2015 – FMedRÄG 2015). Bundesgesetzblatt für die Republik Österreich. February 23.

Bundesministerium für Justiz. 2005. 15/SN-327/ME XXII. GP – Stellungnahme zum Entwurf elektronisch übermittelt. *Parlament*. August 19. https://www.parlament.gv.at/PAKT/VHG/XXII/ME/ME_00327_15/imfname_047940.pdf. Accessed 19 Dec 2017.

Callon, Michel, and Bruno Latour. 1981. Unscrewing the big leviathan: how actors macro-structure reality and how sociologists help them to do so. In *Advances in social theory and methodology*, ed. Karin Knorr-Cetina and Aaron V. Cicourel, 277–303. London: Routledge and Kegan Paul.

Congregation for the Doctrine of the Faith. 1987. Donum vitae. Instruction on respect for human life in its origin and on the dignity of procreation. Replies to certain questions of the day.

derStandard.at. 2005. "Aktion Leben"-Protest gegen Präimplantationsdiagnose. derStandard.at. August 6.

Digeser, Peter. 1992. The fourth face of power. *The Journal of Politics* 54: 977–1007.

Dohnal, Johanna. 2013. Über die politische Durchsetzbarkeit von liberalen Modellen. Zur Geburtenregelung in einer konservativen Öffentlichkeit am Beispiel Österreich (Wien, 18. September 1991). In *Johanna Dohnal. Ein politisches Lesebuch*, ed. Maria Mesner and Heidi Niederkofler, 51–57. Wien: Mandelbaum Verlag.

European Court of Human Rights. 2011. Information Note on the Court's cases-law No. 146, S.H. and Others vs. Austria [GC] – 57813/00. November 3. http://www.echr.coe.int/Documents/CLIN_2011_11_146_ENG_897517.pdf. Accessed 19 Dec 2017.

Ezrahi, Yaron. 2012. *Imagined democracies: necessary political fictions*. Cambridge: Cambridge University Press.

Felt, Ulrike. 2015. Keeping technologies out: sociotechnical imaginaries and the formation of Austria's technopolitical identity. In *Dreamscapes of modernity: sociotechnical imaginaries and the fabrication of power*, ed. Sheila Jasanoff and Sang-Hyun Kim, 104–125. Chicago: Chicago University Press.

Freistetter, Florian. 2015. Sehr peinlich. *FALTER HEUREKA*. April 2.

Gmeiner, Robert. 2006. Spotlight on the bioethical and biopolitical debate in Austria. *Acta Medica Lituanica* 13: 23–33.
Goujon, Anne, Sandra Jurasszovich, and Micheala Potancoková. 2017. *Demographie und Religion in Österreich. Szenarien 2016 bis 2046*. Wien: ÖIF-Forschungsbericht.
Grabner, Petra. 2003. Schlaglichter auf die österreichische biopolitische Debatte. *Österreichische Zeitschrift für Politikwissenschaft* 32: 201–211.
Grießler, Erich. 2010. "Weil es so ein heißes Thema ist, berühren wir es nicht an." Zur Regulierung kontroverser biomedizinischer Forschung in Österreich. In *Steuerung von Wissenschaft? Die Governance des österreichischen Innovationssystems*, ed. Peter Biegelbauer, 143–185. Innsbruck: StudienVerlag.
Grießler, Erich, and Bernhard Hadolt. 2006. Policy learning in policy domains with value conflicts: the Austrian cases of abortion and assisted reproductive technologies. *German Policy Studies* 3 (4): 693–742.
Griessler, Erich, and Mariella Hager. 2016. Changing direction: the struggle of regulating assisted reproductive technology in Austria. *Reproductive biomedicine & society online* 3. Brocher symposium: perspectives on access to reproductive healthcare: 68–76.
Hadolt, Bernhard. 2007. Die Genese der Reproduktionsechnologiepolitik in Österreich: Überlegungen zum Politiklernen in neuen Politikfeldern. *Österreichische Zeitschrift für Politikwissenschaft* 36: 285–302.
Hanafin, Patrick. 2013. Law, biopolitics and reproductive citizenship. The case of assisted reproduction in Italy. *Tecnoscienza* 4: 45–67.
Handyside, A.H., E.H. Kontogianni, K. Hardy, and R.M.L. Winston. 1990. Pregnancies from biopsied human preimplantation embryos sexed by Y-specific DNA amplification. *Nature* 344: 768–770.
Hanisch, Ernst. 2005. Der Politische Katholizismus als ideologischer Träger des "Austrofaschismus". In *Austrofaschismus. Politik – Ökonomie – Kultur. 1933-1938*, ed. Emmerich Tálos and Wolfgang Neugebauer, 5th ed., 68–87. Wien: LIT Verlag.
Herrmann, Svea Luise. 2009. Policy debates on reprogenetics. In *The problematisation of new research in Great Britain and Germany*. Frankfurt a. M.: Campus.
IMABE. 2014. Stellungnahme zur Novellierung des Fortpflanzungsmedizingesetzes. https://www.parlament.gv.at/PAKT/VHG/XXV/SNME/SNME_02282/imfname_375428.pdf. Accessed 19 Dec 2017.
Jasanoff, Sheila. 2005. *Designs on nature: science and democracy in Europe and the United States*. Princeton, NJ: Princeton University Press.
Katholischer Familienverband Österreichs. 2015. Über die Zukunft des Kinderbekommens soll am 21. Jänner im Parlament entschieden werden. http://www.kinderbekommen.at. Accessed 19 Dec 2017.
Kneucker, Raoul. 2008. Humane Embryonale Stammzellenforschung – Stand der Forschung. In *Stammzellenforschung. Ethische und rechtliche Aspekte*, ed. Ulrich H.J. Körtner and Christian Kopetzki, vol. Band 2, 62–76. Vienna: Springer. Schriftenreihe Ethik Und Recht in Der Medizin.
Knill, Christoph, Caroline Preidel, and Kerstin Nebel. 2014. Brake rather than barrier: the impact of the Catholic Church on morality policies in Western Europe. *West European Politics* 37: 845–866.
Knill, Christoph, Christian Adam, and Steffen Hurka, eds. 2015. *On the road to permissiveness? Change and convergence of moral regulation in Europe*. Oxford: Oxford University Press.
Kopetzki, Christian. 2008. Humane Embryonale Stammzellenforschung – Stand der Forschung. In *Stammzellenforschung. Ethische und rechtliche Aspekte*, ed. Ulrich H.J. Körtner and Christian Kopetzki, vol. Band 2, 62–76. Vienna: Springer. Schriftenreihe Ethik und Recht in der Medizin.
———. 2012. Altes und Neues zur Präimplantationsdiagnostik. *Journal für Rechtspolitik* 20: 317–337.
Körtner, Ulrich H.J. 2008. Stammzellforschung: Der bisherige Diskurs in der Bioethikkommission beim Bundeskanzleramt. In *Stammzellenforschung. Ethische und rechtliche Aspekte*, ed. Ulrich H.J. Körtner and Christian Kopetzki, vol. Band 2, 1–18. Vienna: Springer. Schriftenreihe Ethik und Recht in der Medizin.

Lebenskonferenz. 2015. Reproduktionsmedizin. Betroffene erzählen ihre Erfahrungen. http://repmedizin.info. Accessed 19 Dec 2017.

Lehner, Daniel. 2011. *Die österreichische Verhanlung des Schwangerschaftsabbruchs (1973–2009). Zur Veränderung der diskursiven Formation im politischen Feld. Masterarbeit am Institut für Soziologie*. Vienna: Universität Wien.

Mesner, Maria. 1999. Political culture and the abortion conflict. In *From world war to Waldheim. Culture and politics in Austria and the United States*, ed. David F. Good and Ruth Wodak, 186–209. New York: Berghahn.

Metzler, Ingrid. 2007. "Nationalizing embryos": the politics of human embryonic stem cell research in Italy. *BioSocieties* 2: 413–427.

———. 2012. On why states still matter: in vitro fertilization embryos between laboratories and state authorities in Italy. In *Bio-objects: life in the 21st century*, ed. Niki Vermeulen, Sakari Tamminen, and Andrew Webster, 1st ed. Burlington, VT: Ashgate.

———. 2016. Human life between biology and law in Germany. In *The Routledge handbook of biopolitics*, ed. Sergei Prozorov and Simona Rentea, 295–312. London: Routledge.

Österreichische Bischofskonferenz. 2005. 8/SN-327/ME XXII. GP – Stellungnahme zum Entwurf gescannt. *Parlament*. August 17. https://www.parlament.gv.at/PAKT/VHG/XXII/ME/ME_00327_08/imfname_047522.pdf. Accessed 19 Dec 2017.

———. 2014. Stellungnahme der Bischofskonferenz zur Novelle des Fortpflanzungsmedizingesetzes. https://www.parlament.gv.at/PAKT/VHG/XXV/SNME/SNME_02394/imfname_376174.pdf. Accessed 19 Dec 2017.

Parlament. 1991a. Bericht des Justizausschusses über die Regierungsvorlage (216 der Beilagen): Bundesgesetz, mit dem Regelungen über die medizinisch unterstützte Fortpflanzung getroffen (Fortpflanzungsmedizingesetz – FMedG) sowie das allgemeine bürgerliche Gesetzbuch, das Ehegesetz und die Jurisdiktionsnorm geändert werden. *Parlament*. https://www.parlament.gv.at/PAKT/VHG/XVIII/I/I_00490/imfname_261307.pdf. Accessed 19 Dec 2017.

———. 1991b. Regierungsvorlage. Bundesgesetz, mit dem Regelungen über die medizinisch unterstützte Fortpflanzung getroffen (Fortpflanzungsmedizingesetz – FMedG) sowie das allgemeine bürgerliche Gesetzbuch, das Ehegesetz und die Juris- diktionsnorm geändert werden. *Parlament*. July 30. https://www.parlament.gv.at/PAKT/VHG/XVIII/I/I_00216/imfname_260596.pdf. Accessed 19 Dec 2017.

———. 2005. 327/E XXII. GP – Entwurf elektronisch [ü]bermittelt. *Parlament*. https://www.parlament.gv.at/PAKT/VHG/XXII/ME/ME_00327/imfname_047049.pdf. Accessed 19 Dec 2017.

———. 2015. Stenographisches Protokoll. 59. Sitzung des Nationalrates der Republik Österreich, XXV. Gesetzesgebungsperiode, Mittwoch, 21. Jänner 2015. *Republik Österreich, Parlament*. https://www.parlament.gv.at/PAKT/VHG/XXV/NRSITZ/NRSITZ_00059/fname_419486.pdf. Accessed 19 Dec 2017.

Pelinka, Anton, and Sieglinde Rosenberger. 2000. *Österreichische Politik: Grundlagen – Strukturen – Trends*. Vienna: WUV Universitätsverlag.

Permoser, Julia Mourão. 2014. Austrian MEPs: Between privatisation and politicisation of religion. *Religion, State and Society* 42: 251–265.

Prainsack, Barbara. 2005. Religion und Politik. In *Politik in Österreich: Das Handbuch*, ed. Herbert Dachs, Peter Gerlich, Herbert Gottweis, Helmut Kramer, Volkmar Lauber, Wolfgang C. Müller, and Emmerich Tálos, 538–549. Vienna: Manz.

Prainsack, Barbara, and Robert Gmeiner. 2008. Clean soil and common ground: the biopolitics of human embryonic stem cell research in Austria. *Science as Culture* 17: 377–395.

Prainsack, Barbara, and Ayo Wahlberg. 2013. Situated bio-regulation: ethnographic sensibility at the interface of STS, policy studies and the social studies of medicine. *BioSocieties* 8: 336–359.

Salter, Brian. 2007. Bioethics, politics and the moral economy of human embryonic stem cell science: the case of the European Union's Sixth Framework Programme. *New Genetics and Society* 26: 269–288.

Schönthaler, Helia B., and Erwin F. Wagner. 2008. Stammzellenforschung. Ethische und rechtliche Aspekte. In *Stammzellenforschung. Ethische und rechtliche Aspekte*, ed. Ulrich H.J. Körtner and Christian Kopetzki, vol. Band 2, 12–32. Vienna: Springer. Schriftenreihe Ethik und Recht in der Medizin.

Sperling, Stefan. 2004. Managing potential selves: Stem cells, immigrants, and German identity. *Science and Public Policy* 31: 139–149.

SPÖ-Bundesorganisation. 2015. Oberhauser: Mit modernem Gesetz zur Fortpflanzungsmedizin schließt Österreich an internationale Standards an. *OTS.at*. January 21.

Statistik Austria. 2007. Bevölkerung nach dem Religionsbekenntnis und Bundesländern 1951 bis 2001. https://www.statistik.at/web_de/statistiken/menschen_und_gesellschaft/bevoelkerung/volkszaehlungen_registerzaehlungen_abgestimmte_erwerbsstatistik/bevoelkerung_nach_demographischen_merkmalen/022885.html. Accessed 19 Dec 2017.

Suppanz, Werner. 2005. Das Kreuz mit den Präambeln. Die österreichische Debatten über den Religionsbezug im europäischen Verfassungsvertrag. In *Die Werte Europas: Verfassungspatriotismus und Wertegemeinschaft in der EU?* ed. Helmut Heit, 57–69. Münster: LIT Verlag.

Testa, Giuseppe. 2011. More than just a nucleus: cloning and the alignment of scientific and political rationalities. In *Reframing rights: bioconstitutionalism in the genetic age*, ed. Sheila Jasanoff, 86–104. Cambridge, MA: MIT Press.

Verfassungsgerichtshof. 2013. G 16/2013-16, G 44/2013-14. December 10.

Virt, Günter. 2008. Stammzellforschung: Der bisherige Diskurs in der Bioethikkommission beim Bundeskanzleramt. In *Stammzellenforschung. Ethische und rechtliche Aspekte*, ed. Ulrich H.J. Körtner and Christian Kopetzki, vol. Band 2, 7–11. Vienna: Springer. Schriftenreihe Ethik und Recht in der Medizin.

Wagenaar, Hendrik. 2011. *Meaning in action: interpretation and dialogue in policy analysis*. Armonk: Sharp.

Waldby, Catherine. 2002. Stem cells, tissue cultures and the production of biovalue. *Health* 6: 305–323.

Zichy, Michael. 2010. Religion in den bioethischen Diskursen Österreichs. In *Religion in bioethischen Diskursen: Interdisziplinäre, internationale und interreligiöse Perspektiven*, ed. Friedemann Voigt, 133–158. Berlin: De Gruyter Verlag.

Chapter 5
Moral Politics in Ireland: From Religious Domination to Political Indifference

Mirjam Weiberg-Salzmann

Abstract Religion has played a significant role in Ireland in the formation of the state and in shaping national identity. It is anchored both institutionally and in the collective memory. Although Catholicism as a source of national identity has been weakened in recent decades, many Irish people retain a sense of pride in the century's long unifying band of Catholicism. However, the position of the church is today viewed much more critically. With regard to biopolitical topics, it is obvious that the dismissive attitude of the Catholic Church and the pro-lifers had led to an avoidance of a public debate among politicians, journalists, doctors, legal scholars and even economic actors in Ireland. Legal regulations exist for none of the new biotechnologies, not least because the right to a legal, unpunished abortion remains highly contested. The entire debate on biopolitics has taken place in a narrow institutional framework with limited participants and contents. The church operates like an institutional power player and would prefer to limit the discussion rather than to spread it. Neither the legal framework nor the social and ethical effects of the technologies are sufficiently opened up for discussion. The weakness of the political elite, the historical position of the Catholic Church as moral authority and societal power and the divide within the Irish population along a conservative-rural (religious) and modern-urban (secular) fault line are responsible for the lack of public (and political) discussions and the very slow institutionalisation of bioethics.

1 Religious Actors and Arguments in Liberal Democracies

This contribution originates with the continuing debate on separating church and state in liberal democracies and the legitimacy and functionality of religious interventions into the "secular" state. The separation of church and state and the creation of a secular state (in terms of neutrality towards religious communities and convictions) are seen as essential, though not sufficient, preconditions of stable democracies. Recent partial revisions of this strict separation can essentially be attributed to

M. Weiberg-Salzmann (✉)
University of Münster, Münster, Germany
e-mail: mirjam.weiberg-salzmann@uni-muenster.de

M. Weiberg-Salzmann, U. Willems (eds.), *Religion and Biopolitics*,
https://doi.org/10.1007/978-3-030-14580-4_5

growing inconsistencies between empiricism and theory. Interestingly enough, the debate on the role of religion in modern democracies focuses mainly on questions of the influence religious actors may have on politics ("what influence exists, if any?") and its structural and institutional prerequisites ("what elements of the system enable/ encourage any such influence?"). Questions concerning the characteristics of patterns of (religious) argumentation, how the content of arguments is structured and the discursive environment have essentially been marginalised (Banchoff 2008; Rowe 2012; Fox 2012; Heit and Pfleiderer 2012; Anderson 2006).

At least in the USA, the theoretical discussion about the use of religious arguments in public debates has never completely come to a standstill. Though religion was accorded a positive and practicable potential in political opinion and decision-making processes, this demonstrated most benefits when religion remained within the spheres of private life or civil society. The legitimacy of political decisions was highest if based on a generally intelligible, secular argumentation, thus accessible to all citizens. Religious arguments were meant to be broadly excluded from the political decision-making process to ensure equality of inclusion into civil society for all members of a community. The majority of theorists, who used to argue in a strictly secular-liberal way, have now moved towards arguments limiting religious arguments in public dialogue instead of excluding them. Hence there are only a few among the pluralist critics who would support an unconditional admission of religious arguments into public dialogue, least of all support exclusively religious reasons for a decision. The theoretical discussion has been characterised by two positions: on the one hand, there are the representatives of a liberal concept of democracy (e.g. Rawls, Habermas or Audi), who are willing to concede no space or merely a subordinate significance for religious arguments in the public sphere and who favour a strict separation between secular-public and private-religious spheres: only a religiously neutral state would be able to meet the demands of citizens' best interests and mediate conflicts in a pluralist society. Religious arguments, however, would be too particular and contain potential conflicts. On the other hand, there are the representatives of a pluralistic concept of democracy (e.g. Wolterstorff, Quinn and Bader), who consider a complete separation of the spheres both impossible and unnecessary. According to them, the neutrality of liberalism sustains an epistemological limitation of, inter alia, religious arguments and contains the unjustified insinuation that religious arguments are not generally intelligible and particularly conflict prone (Audi 2000; Rawls 1993, 1999; Habermas 2008; Wolterstorff 1997; Quinn 1997; McConnell 1999; Taylor 1998).

1.1 The Public Politics of Life Sciences

My analysis will focus in particular on the discursive forms and reasons underlying religious arguments in political debates. Based on a current controversial issue—the possibilities and limits of the life sciences—there is a discussion whether the democratic state should renounce all religious substantiation of its political decisions. Biotechnologies like pre-implantation diagnosis, stem cell research and cloning are a suitable object of study for several reasons: progress in the life sciences has made the human nature accessible to technical intervention to a hitherto unknown degree, thus putting more and more biopolitical topics on the political agenda. Furthermore, there is a deep-rooted moral and ethical disagreement on whether such technologies ought to be allowed or prohibited. Religious organisations are among the social groups decisively shaping these debates. The questions asked are: Who are the actors and agents who participate in the negotiation and decision-making processes on biotechnologies and what constitutes their arguments? Are religious actors likely to dispose of specific moral resources in answering these particular questions or does precisely this cultural and ideological pluralism of society prevent drawing on religious traditions? These questions arise all the more urgently since the modern state needs to decide more and more frequently on conflicts that need to be regulated by law but do not have fixed moral and ethical standards. While the influence of religion on politics has been intensely studied in recent years, less is known about the discursive strategies employed by religious actors in political debates. I begin with the assumption that the structure of a religion's arguments within these discursive processes depends less on a general institutional arrangement than on the specific manifestations of religion and the sociocultural context.

1.2 The Definition of Religious Arguments

Besides the sociocultural, e.g. the historical, relationship between politics and religion, the identification of religious arguments in these debates forms an essential part of my analysis. But what is a religious argument? In general, religious arguments are understood as all arguments which refer to God, theological positions or religious authorities and institutions. Although this definition is reasonably useful for research, there remains the question of whether it encompasses all the categories of religious arguments. Audi relies on four criteria by which he distinguishes religious from secular arguments. First, an argument is essentially religious if its content refers to a divine or religious authority/rule, a religious source/scripture or a religious leader. Second, an argument is religious if its premises and/or its conclusion cannot be understood or accepted independently of religious considerations (scripture, authority, etc.). Third, an argument is religious, if its proponent displays a pronounced motivation to enforce religious goals with a religious justification regardless of whether the content of the argument is overtly religious or not. The fourth criterion relates to the historical genealogy of the argument; hence, an argument is religious if one is able to trace it back to at least one religious proposition. The argument may be persuasive on its own or through its reference to a religious source, but the power of persuasion ultimately depends on the addressee. In nearly every case, the boundaries between religious and non-religious arguments are fuzzy or heavily dependent on the assessment of the observer. As Audi himself admits, without a thorough knowledge of the religion in question it is impossible to classify an argument as "religious" (Audi 2000).

For the purposes of my analysis, I found two levels of classification of religious arguments crucial: the parameters of the first level are content, source, intention, belief, convention and validity. They shall determine when an argument can be called religious. The labels "content" and "source" are relatively easy to identify as they point to obvious markers in the text. The two criteria of "convention" and "validity" relate to the reception of the argument and the social identity of the actor and can be traced by exploring their background within the religious tradition and society. The two criteria of "intention" and "belief" are difficult to operationalise. The religious significance can only be fully understood by reconstructing the articulation process and its genesis.[1] Whether the religious argument is deemed permissible in the wider public discourse primarily depends on the second level. This is the level of gradation: an argument can be purely religious, not clearly religious with a facultative component, non-religious used in a religious context or a secularised adaptation.[2] Each of these factors determines the admissibility of the religious arguments in public discourse. Except for the case of secular transfer, all of these gradations are considered problematic in liberal democratic theory. To analyse the role of religious actors and argument within the public debate, I will first outline the historical relationship between religion and politics in Ireland. The following chapter examines the biopolitical debate from the 1980s until today. I will then identify and classify the arguments used by different religious actors. The last chapter offers an explanation of the influence of religious agents on the public debate and of the specific role and application of religious arguments within the discussion of the life sciences in Ireland.

[1]First level: Religious arguments (a) refer to religious content, e.g. (quoted or paraphrased) statements from a religious text or a divine commandment of a given religious tradition (content); (b) call upon (quoted or paraphrased) statements of religious authorities or institutions (source) to justify their opinions; (c) are motivated in the sense that their reference to the content/their connection to the content or the concepts of a religion is not coincidental or unintentional (intention of the speaker/conscious use); (d) are motivated in the sense that their reasoning about the content of or ideas about a religion comes from a (internal or external/instrumental) conviction (conviction/motivation); (e) must be recognised, in how they are received by society, as belonging to a religious tradition (conventionality); and (f) must be (recognised and) accepted as statements and reasoning of a religious tradition by the members of that tradition (validity).

[2]Second level: The religious content of religious arguments can be defined as (a) purely religious arguments (quotation and paraphrase of "content" and "source"); (b) not clearly religious arguments with facultative components (ad hoc created arguments by single speakers as religious reinterpretation or pseudo-religious use); (c) not religious arguments, which are used in religious context (applicable religious transformation, e.g. natural law); and (d) secularised adaptation/transfer (absorption into general terminology, e.g. human rights).

2 Challenging the Catholic Heritage: The Case of Ireland

2.1 State and Church in Ireland

Religion has played a significant role in Ireland in the formation of the state and in shaping national identity. It is anchored both institutionally and in the collective memory. The long history of Catholicism in Ireland (fifth century: St. Patrick) and the struggle against British rule have shaped the cultural and political identity of the country up until today.[3] During the Reformation, the conflicts between Ireland and the United Kingdom were deepened. From the Irish side, a feeling emerged that colonisation and takeover could only be effectively resisted if the primacy of Catholicism were preserved. The Catholic religion became a symbol of resistance against a foreign power and the unifying element in national struggle for independence. At the end of the nineteenth century, Catholicism was an integral, perhaps even the most defining, element of Irish national identity. Being Catholic and Irish is (even today) inseparably connected within the collective national identity. Moreover, the Catholic Church and Catholic belief has been the anchor of the oppressed classes in Ireland. It offered moral and spiritual direction, as well as a bit of help for survival. The opinions and values of the Catholic Church influenced all levels of everyday life.[4]

Formally, church and state had already been separated in the nineteenth century and the Irish constitution at no point explicitly references Catholicism. While it does begin with a reference to religion ("In the Name of the Most Holy Trinity, from whom is all authority and to whom, as our final end, all actions both of men and States must be referred") and religious belief is taken into consideration by the state ("The State acknowledges that the homage of public worship is due to Almighty God"), it never explicitly names Catholicism. No religion is given a preference over another. In fact, however, the Catholic Church possessed an accentuated position in the Irish state during the twentieth century in education and in the welfare system, as well as in matters of public and private morality. Until 1922, Ireland remained in the hands of Great Britain and the influence of this union can still be recognised within many political institutions. The first constitution of the Irish Free State of 1922 incorporated the British liberal democratic tradition as well as that of Catholic social morality. However, the passive role of the church during the uprising and its opposition to a violent revolution minimised its direct influence on politics. Because nationalists nevertheless continued to invoke the Catholic heritage of Ireland, in the coming years more and more predominance over public morality and over some parts of social politics was entrusted to the Catholic hierarchy. While the new

[3]The church and monasteries were already considered by the ninth century to be the protectors and keepers of the Irish culture, identity and civilisation (see the Viking attacks) (Bartlett 2010, Chap. 1).

[4]The association of conservative-rural values with the Irish nation also began in the nineteenth century (Connolly 2007).

constitution of 1937 defined the state as sovereign, independent and democratic, it was also pointedly Catholic.[5]

Until the 1960s, personal relationships and friendships between politicians and the clergy remained the subject of divisive political debate. Church leaders were and are regularly consulted about new laws. Consultations between the hierarchy and politicians have, since the foundation of the state, formed a central element in the church–state relationship. Further, there exists a sort of moral self-censorship of politicians, whose proposals often incorporate in advance the position of the Catholic Church (Keogh 1986; Daly 2000). Generally, the church has not needed to exercise much public influence on politics. Religious questions were seldom controversial—exceptions were the debates about divorce and abortion—and the government often avoided an open confrontation with the church.[6] Moreover, the Catholic Church had great influence on society through its embeddedness in the welfare system. Schools, hospitals, orphanages, women's shelters as well as services for the poor and unemployed were and remain in their hands or are at least joint ventures of the church and the state. Since nearly all politicians are Catholic, they accept the Catholic Church as the sole representative in moral, ethical and religious questions, because they themselves share these convictions, because they grew up within institutional networks of the church or out of regard for their (Catholic) constituencies. Giving the church control over these social services was also simpler for politicians than building a functioning welfare state out of limited state resources (Fahey 1998).[7]

Until the 1970s, the constitution remained largely unchanged. This had much to do with the conceptions of Irish constitutional lawyers, who saw the constitution as nearly sacrosanct. In the 1960s, an understanding slowly arose according to which the written constitutional text could, if needed, be altered to better fit with changed societal realities (Elvert 1999, pp. 256, 260, 276 and 280). Significant among other things was the Supreme Court's understanding in 1963 that within the rights named

[5]This constitution, which is still in use today, consists of a preamble and 50 articles. In the preamble, as well as in the articles on family, on the position of women and on divorce, the great influence of the Catholic Church is reflected (Casey 1995; Coakley 1993, p. 31).

[6]Abortion was not a political issue between 1922 and 1959 because the Irish state took over the restrictive British laws. In questions of divorce and contraception, the British laws were too liberal: contraceptives were banned in 1935 and divorce in 1937. The church had made their approval of the new constitution dependent upon these bans. With the exception of a few examples (e.g. the rejection of a welfare programme for mothers and children in 1951), there were no open conflicts between church and state. It was not until the trend towards secularisation began that the Catholic Church had to publicly confront the government. Often the opinion of the public (population) was on the side of the Catholic Church. Surveys showed even in 1990 that the majority of the population considered it appropriate for the Catholic Church to express itself in questions such as that of abortion (81%), euthanasia (77%), non-marital partnerships (71%) and homosexuality (59%). At the same time, surveys show that—with respect to voting—these themes did not carry much weight (O'Halpin 1993, p. 201; Coakley 1993, p. 31; Laver and Marsh 1993, p. 117).

[7]Moreover, the majority of the Irish population tended to support a neoliberal individualistic welfare state (Millar 2008).

in the constitution (Art. 40–44), further rights were included, despite not being explicitly named. Out of these rights were inferred for example, in 1973, the right to "marital privacy" and the import ban on contraceptives was lifted six years later. In 1985, contraception was made even more accessible. In 1992, the court reinterpreted the abortion ban of 1983 and allowed an abortion in a case where the life of the mother was endangered by continuing the pregnancy.[8] Starting in the 1970s, about a dozen additional constitutional changes were put through by plebiscite. This led to a conservative backlash in the 1980s: it became clear that the Irish population continued to remain conservative/Catholic (or at best divided) in questions of social morality. In 1986 and 1995, the question of divorce stood for debate and, in 1983 and 1992, the right to abortion. In the countryside, the population voted almost unanimously against liberalisation (or abstained); in the cities in 1995, there was a razor-thin majority for a change in the law. In each case, the church had massively mobilised against the liberalisation of the law.[9] Surveys show that ca. 30% of the (mostly young urban) population wish for a modernisation of the constitution, 30% (of the older countryside) strictly reject this and a further 30% are indifferent.

The Catholic penetration of the Irish state can also be explained by the social structure of the country: together with independence came the division of Ireland; the more rural South and West became the Republic of Ireland, and Northern Ireland remained a province of Great Britain. The social stratification of the republic was extremely homogeneous and consisted largely of a Catholic rural population and a small-town bourgeoisie. Correspondingly, the party system was not, like in other European countries, determined by certain social groups or social background. In Irish society, there were no major social cleavages, around which a party system might have organised. Differences between the largest remaining urban centres of Belfast (and Cork) and the rest of the country were initially overshadowed by the question of whether to accept the country's division or the struggle against it. The party system was divided along these lines with Fine Gael (FG), who were for the treaty and ran the first government, and on the other side Fianna Fail (FF), who spoke out against the treaty and were first able to win the election in 1932. Both parties support the same ideology—nationalism, centred upon territory, language (Irish) and religion (Catholicism). They are Christian Democratic in orientation and place much weight on social encyclicals and papal statements.[10] The voting system in Ireland—representative voting rights with single transferable vote—facilitates a close relationship between candidates and voters. Local support and the rootedness

[8]Ireland continues today to have one of the most restrictive laws in the Western world and the right to life of the mother is considered to be at the same level as the right to life of the foetus (Porter 1996).

[9]In most cases, the official influence of the church was through parish priests, but it also worked closely together with anti-abortion groups.

[10]In the decades following independence, FF was anxious to appear as the more Catholic of the two parties. Until the 1960s, it was then FG, who always placed itself on the side of the Catholic Church. Starting around 1960, deference and clericalism then decreased more quickly in the FG (Nitsche 2012, pp. 27 and 235).

of the candidates in the constituency are the most important criteria. The interests of local representatives are those of their constituents and less those of the national level or of higher interests (Sinnott 1993; Mair 1993; Gallagher and Komito 1993). The administration itself is considered conservative and limits itself to necessary state functions. Within this small administration, personal relationships are predominant and it is relatively easy to find unity on a given political issue (Budge 2002, pp. 18, 21, 29, 42 and 47).

Since the 1970s, the influence of the Catholic Church has been reduced due to social and economic changes: The Catholic Church no longer presents itself as a political actor but continues to intervene actively in certain public debates whenever central elements of its religious convictions are touched upon. The discussion is no longer carried out using theological and moralistic arguments, but instead often using secular ones (by giving sociological or economic reasons). Tendencies towards secularisation in society can be observed in the level of church attendance and membership, which are in comparison to other European countries still high but as a whole on the decline.[11] Both speak to a privatisation of Irish Catholicism. Moreover, the European Court of Human Rights has repeatedly questioned the authoritarian position of the Catholic Church as a giver of ethical and moral norms.[12] Secular parties or parties aimed at specific social groups, such as Labour, Progressive Democrats, Worker's Party or the Green Party, receive increasingly more votes and coalition governments are becoming more common. The election of the party-less candidate Mary Robinson in 1990 as the first female state president likewise shows that Ireland is in the process of reform. But it is a slow process: in 2011 only 23 members or 13.8% of the full complement of 166 were women deputies in lower house of national parliament (National Woman's Council of Ireland 2010; Coakley and Gallagher 2004). The long underdeveloped political participation of people in citizens and social movements has also considerably improved in the last decade, and more and more people disapprove of church influence in elections or political decisions. Whether these tendencies will continue or provoke a countermovement has currently not yet been decided. Today, one can legitimately speak of a divide in Irish society along a religious-secular fault line (Ruane 1995; Hemton 2006).[13]

---

[11]Church attendance is highest in the border regions and lowest in Dublin and the east midlands of Ireland. A majority did not consider belief in God to be a criterion for judging a politician's (dis) qualifications and believed that church dignitaries should not influence government decisions (O'Mahony 2010). Newer statistics show that 84.7% consider themselves part of the Roman Catholic Church, 2.7% Church of Ireland, 2.7% other Christian Churches, 1.1% Islam, 1.7% unspecified and 5.7% not belonging to any religious community. CIA World Factbook: Country Report Ireland 2015. Jürgen Elvert reports that as of 1999 the country was still 91.6% Catholic.

[12]For example in 1992, the European Court of Human Rights ruled that the ban on information on abortion was in contradiction to the European Convention on Human Rights.

[13]On the history of the women's movement in Ireland, see Taylor (1999) and White (2007).

2.2 The Bioethical Debate in Ireland

As in many countries, the question of embryo research is tightly linked with the question of abortion. This is especially true in Ireland, since the discussion is still unresolved. Abortion was already forbidden before independence (since 1861) and remained criminalised within the Irish Republic. In 1929, a law additionally banned all information/literature on the subject; in 1935, the import and purchase of contraceptives was likewise banned. Up until the early 1980s, no parliamentarian supported abortion under any circumstances.[14] While access to contraceptives has been legalised since the end of the 1970s, in 1983 the anti-abortion movement succeeded, together with the Catholic Church, in constitutionally anchoring the abortion ban.[15] After a massive campaign, a constitutional amendment was created emphasising the right to life of the unborn and banning information on abortion: "The state acknowledges the right to life of the unborn and, with due regard to the equal right of life of the mother, guarantees in its law to respect, and as far as practicable, by its law to defend and vindicate that right..." (Article 40(3)3). Although the argumentation of the Pro-Life Amendment Campaign was religiously neutral ("inalienable rights due to the intrinsic humanity of the foetus"), the claim that a foetus has individual rights from the moment of conception was plainly Catholic. The language used by Irish anti-abortion groups was similar to American anti-abortion/pro-life discourse: the foetus was construed as an autonomous entity/subject threatened from without as a vulnerable, marginalised unprotected person. The pregnant woman was reduced to the (hostile) maternal environment.[16]

In 1992, debates came to a head over the case "X" in which a 14-year-old rape victim was barred from leaving the country in order to get an abortion. The court justified the decision with the constitutional amendment of 1983. After a heated 3-week long public debate, the Supreme Court decided that an abortion could be legal under certain conditions (if the life of the mother is endangered). At the same time, the judge called for a new legal ruling. Since then abortion has been legal—although only under extreme circumstances. Abortion critics immediately called for a new referendum in order to roll back the court ruling. In a 1992 referendum, the ban on abortion with the exception of the endangerment of the mother's life was confirmed, but the right to information and freedom of movement was allowed. The Catholic Church had advised its believers to vote "no" or argued that citizens had to decide according to their conscience whether or not they detested abortion (Murphy-Lawless 1993).

---

[14]One exception was Jim Kemmy. He was vilified in a public campaign and voted out in 1982 (Girvin 1992, p. 203).

[15]All movements were Catholic either in name or ideology, e.g. the Pro-life Amendment Campaign, Society for the Protection of the Unborn Child, Irish Family League, the Council of Social Concern, the Responsible Society and Family Solidarity.

[16]Kathryn Conrad shows marked parallels between nationalist discourses on the subject of Ireland (as a nation-state threatened from without and the last bastion of Catholicism) and the language surrounding the threatened foetus (Conrad 2001, p. 157).

The Catholic hierarchy additionally openly criticised the government and accused it of not doing enough for the absolute protection of life. It warned that a loosening of the laws would bring with it a liberalisation of morality. After the referendums, there was nothing in law preventing a doctor from performing an abortion within the conditions established by the Supreme Court, but in practice no abortions are likely to have taken place. The Irish Medical Council, moreover, changed its ethical guidelines after the referendums in a way that would lead to a doctor's removal from the medical register if he or she actually performed an abortion. This was all the more crucial as most hospitals in Ireland are owned and controlled by the Catholic Church. The legislative proposal of the government on abortion (1995) was rejected by the hierarchy by making reference to the papal position (Encyclical Veritas Splendor 1993). The church accused the Supreme Court of adopting a relativistic approach in moral issues and claimed its decision on abortion information was a "corruption of law" and therefore "unjust". The Catholic hierarchy was criticising the court, the parliament and the electorate for not adopting a natural law position (Girvin 1996, pp. 604 and 607; Smyth 1996, p. 117). In 2002, the 25th Amendment was narrowly rejected. It attempted to amend the Constitution of Ireland to tighten the constitutional ban on abortion and would have removed the threat of suicide as a ground for legal abortion in the state, while also introducing new penalties for anyone performing an abortion.[17] Reaching a more permissive legal regulation in Ireland was difficult, not least because state politics in this area were dominated by traditional Catholic social teachings. Using the concept of natural law and its "neutral" discourse, the Catholic social teachings have been translated into state law since the foundation of the state.[18] Unlike in France, for example, the Catholicism of Ireland is characterised by its tight connection to the Roman Curia and papal doctrine. Personal decisions of conscience and morality are subordinated to the Catholic Church (concerning the priority of rules, duties and obligations) (Dillon 2002).

When newer biopolitical issues are discussed in Ireland, this mostly takes place in the context of the discussion on "assisted human reproduction" (AHR). Since the debate lags behind that in other European countries by about 10 years and the controversy surrounding abortion is still a current issue, in vitro fertilisation (IVF) and artificial insemination constitute the primary themes of discussion. IVF has been practised in Ireland since 1987. Donor insemination and egg donation are available.

---

[17]It was introduced by the Fianna Fáil–Progressive Democrats coalition government and opposed by Fine Gael, the Labour Party, the Green Party and Sinn Féin.

[18]Hanafin describes the Irish constitution as deontological and an expression of Thomist models of society as advocated by the hierarchical structure of the Catholic Church. Already in the preamble, with its reference to God and divine law, the importance of natural law and its primacy over positive law is shown. This view makes a liberalisation of law, for example in the area of abortion, almost impossible, since the right to life from this point of view is absolute even when it comes into conflict with the autonomy of the individual. Along with the pluralisation of Irish society since the end of the 1970s, the constitutional consensus of the early years disappeared and the constitution was interpreted in the direction of individual rights (which were not in accordance with Catholic teachings). The rights in the constitution are no longer absolute (Hanafin 1996).

Although AHR is carried out in Ireland, there are no legal guidelines surrounding it. The only general guideline is a publication on the topic offered by the Medical Council, the "Guide to Ethical Conduct and Behaviour" (Section F 24.1 and 24.5, 1998), which appears every five years. This is only binding for practising doctors, but is not a guide for biologists or other scientists.[19] Since all actors tried to avoid a new polarised discussion about the beginning of life, the discussion of IVF was postponed in the 1990s. What initially began as a potentially conflictual ethical debate became then a discussion on a valuable medical technique to handle infertility. This could also be seen in the senate debate on Dr. Mary Henry's defeated Private Member's Bill on the regulation of ART (Regulation of Assisted Human Reproduction 1999), in which the main arguments did not have to do with the status of the embryo but rather with the vulnerability of patients and the allegedly reckless and lawless practices of foreign clinics. Even medical commissions working on the subject of assisted human reproduction (ART)—such as the Assisted Reproduction Sub-Committee of the Institute of Obstetricians and Gynaecologists 1999—limited themselves in their reports to biological-technical observations. The term embryo does not appear anywhere, instead only the terms "pro-nuclear" and "zygote". Out of the guidelines of the Medical Council, the ban on freezing embryos was simply implicitly removed starting in 1998 without any ethical discussion taking place. Both sides, supporters and opponents of AHR, did not wish for a moral-ethical debate on the topic. The one side feared that a restrictive law could limit research and medical practice; the other side had concerns that a law could lead to a further liberalisation in the treatment of embryos.

As AHR was ethically controversial in Ireland and raised legal problems (among other things because the constitution gave special protected status to the embryo and the family) (Madden 2004, pp. 1–6), in 2000 the Department of Health created the *Committee for Assisted Reproduction* (CAHR). The CAHR was tasked with publishing a report on all legal, social and ethical factors in the handling of gametes and embryos, informing the public and submitting proposals for government regulations. However, no ethicists were initially among the over 20 members of the committee—it was only after protests that three were integrated into the sub-committees. The results of the commission were relatively liberal: semen, egg and embryo donations were allowed; IVF should be accessible to all, regardless of gender, sexual orientation or marital status. The principle of destroying embryos under certain circumstances was also accepted (Commission on Assisted Human Reproduction 2005). Less surprisingly, the CAHR report was strongly criticised by conservative and religious actors.[20]

[19]With respect to IVF, the handbook establishes that every fertilised egg should be used for a normal implantation and should not wilfully be destroyed. It is not made clear when the transfer should be made. The parents are also left free to decide to donate their egg cells if they have definitely decided they no longer wish to implant them.

[20]The critique is superficially based on the well-being of the child, which the CAHR recommendations allegedly took too little consideration of. In fact, Iona defends traditional (Catholic) family structures. Among other things, its position is that IVF should only be available to married heterosexual couples (Iona Institute 2008).

Until today, there remains no political ruling on various aspects of IVF, for example on surrogate motherhood, even though the CAHR report called for this. However, how little institutionalised bioethics is can also be seen in the fact that the term "bioethics" does not appear anywhere in the entire CAHR report. Furthermore, other than its widely read report, the commission hardly made a public appearance and the public hearings were primarily useful to the pro-life lobby in presenting its position.

When looking at the newest discussions of molecular biology and genetic technologies (embryonal human stem cells, clones, PID), the most noteworthy thing about it is how strikingly little of it there has been. Only a few universities hosted conferences and expert debates on the subject (Juncker-Kenny 2003). Some guidelines were given by the Irish Medical Council in 2009, when it banned medical practitioners from creating embryos specifically for research (Medical Council 2009). The Irish Council of Bioethics (ICB), set up by the government in 2002 to advise as well as consider opinions on scientific and technological issues that may affect human health and society, ceased to operate for financial reasons in 2010 (ICB 2005, 2008, 2010).[21] The ICB focused primarily on clinical research and the development of guidelines for ethics committees in research. ICB attempted to avoid an engagement with medical practices such as ART and remained strikingly quiet in public debate. In 2008, the ICB, like the Assisted Human Reproduction Report from 2005, recommended that a regulatory authority be set up to oversee the licensing and monitoring of research using embryos. Both reports recommended that embryos up to 14 days old could be used for research. Nothing has happened since. Another reason for the lack of discussion was the initial assumption that the production, use or destruction of (extracorporeal) embryos was forbidden according to the previously mentioned constitutional article of 1983 (Matthews 2007). However, already by the early 2000s it had become clear that the interpretation of the article with respect to embryonic research was controversial: the most important word in the constitutional amendment was the (unspecified) term "unborn". One side interpreted the statement of the constitution as a ban (as it viewed the embryo/fertilised egg as a person); the other side did not see surplus embryos (outside of the uterus) as included within the term. It remained questionable whether protection began at the moment of fertilisation or implantation.[22] Moreover, the present legal situation is uncertain because of the M.R. vs. T.R. case. In this "Frozen Embryos" case, the Irish Supreme Court in 2009 agreed with a 2006 Irish High Court judgment

[21]The Irish Council for Bioethics (ICB) is an independent national body, which discusses ethical questions in connection to newer developments in science and medicine. The council has two primary functions: (1) independently advising government and politics and (2) supporting the public debate of bioethical questions.

[22]The High Court determined in its ruling on abortion that "unborn" can be equated to an embryo that is implanted in the uterus, or with a foetus. Frozen embryos are therefore not "unborn" in the sense of Article 40.3.3. Ruling of the Chief Justice McGovern 15/11/2006; see The Court Service of Ireland. 2006. M.R. v T.R. [2006] IESC 359. http://www.courts.ie/Judgments.nsf/bce24a8184816f1580256ef30048ca50/e5617d292b7b6b26802572480032999 2?OpenDocument. Accessed 3 October 2015.

that embryos held in cryopreservation and created outside the womb are not protected under the Constitution.[23] There is furthermore no case law that explicitly deals with embryonic research.[24]

As on the subject of ART, the political actors have tried to avoid a broader discussion of embryonic research. A bill introduced in the Irish parliament in 2008 aimed to ban embryo research in Ireland and impose criminal penalties but was never passed into law. The bill proposed no regulations for stem cell research and assisted reproduction. Two other bills, including the Human Tissues Bill 2008, that sought to regulate the use, storage and removal of human tissue and materials similarly never became law, too.[25] A legal vacuum continues to exist today (2016), although the Irish government has stated its intention to regulate stem cell research in its recent government programme. Generally, biopolitical questions are handled with a contradictory ignorance: cloning is rejected by most politicians; however, it remains as little legally regulated as research with stem cells, human embryos and aborted foetuses. Political decision-makers' fear of a new "embryo struggle" could be seen not least in 2014 when the University of Galway opened the first Centre for Cell Manufacturing. All speakers—with the exception of the representative of the Irish Stem Cell Foundation—did their utmost to avoid the word "embryo" and only present research with adult stem cells. There is only an informal agreement that replacing damaged human cells by artificially grown cells has already proved its worth in a number of areas (e.g. in cancer treatment). It is also agreed that embryo cells, i.e. those "left over" after assisted fertility treatments, are effective in far more circumstances than those harvested from adult humans, and can be used in many more areas. Likewise, the public opinion concerning biotechnology is cautiously positive today.[26]

---

[23]The Court Service of Ireland. 2009. M.R. v T.R. [2009] IESC 82. http://www.courts.ie/Judgments.nsf/597645521f07ac9a80256ef30048ca52/0973cbd1fd5204028025768d003d60f7?OpenDocument. Accessed 3 October 2015.The Court Service of Ireland. 2006. M.R. v T.R. [2006] IESC 359. http://www.courts.ie/Judgments.nsf/bce24a8184816f1580256ef30048ca50/e5617d292b7b6b268025724800329992?OpenDocument. Accessed 3 October 2015.

[24]The Convention on Human Rights and Biomedicine (1997) was also not ratified by Ireland (Sheikh 2002, p. 33). On the international level, Ireland belongs to those countries who agreed with the 2005 UN declaration on human cloning. The declaration itself is vague in its position and largely "purged" of normative values; it also does not explicitly forbid cloning.

[25]Houses of the Oireachtas. 2008. Stem Cell Research (Protection of Human Embryos) Bill. http://www.oireachtas.ie/viewdoc.asp?fn=/documents/bills28/bills/2008/6008/document1.htm. Accessed 3 October 2015.Houses of the Oireachtas. 2008. Human Body Organs and Human Tissue Bill. http://www.oireachtas.ie/documents/bills28/bills/2008/4308/b3408s.pdf. Accessed 3 October 2015.

[26]Opinion polls on embryo research are uncommon and mostly refer generally to human tissue donated for research purposes, but their results show that most citizens have a positive attitude towards genetic research and its future possibilities for healing (70%). At the same time, considerable ethical reservations exist: 42% have the opinion that it is tampering with nature. Highest approval levels were reported for stem cell research using adult human tissue (49%) and for cloning human cells to combat disease (42%). Conversely, disapproval was highest for the development of genetically modified foods (52%) and stem cell research using human embryos (34%). A significant minority, from a quarter to a third, were undecided about each of these issues (Cousins et al. 2005, p. 38).

2.3 Religious Positions and Arguments in Biopolitical Debates in Ireland

Already in 1996 the Catholic Church had established the Bishops' Committee for Bioethics. The committee has published a number of relevant documents. Another widespread statement was the speech of the Archbishop of Dublin, Dr. Desmond Connell. On the thirtieth anniversary of the papal encyclical Humanae Vitae, he held a speech at St. Patrick's College (Life Society). He claimed that IVF, surrogate motherhood, embryonic research and family planning reduced the child to a (technological) product of its parents so that it was no longer perceived as a gift and a person.[27] In sum, the Catholic hierarchy rejected embryonic research with the same arguments it used in the case of abortion and IVF: humans are people from their very beginnings (the species, potentiality, identity and continuity arguments). Aquinas (and his argument of the soul), Pope John Paul (and his argument of inherent human rights and the Gospel of life), the psalms and the church tradition were used to explicate the relationship of God and humans. Science, by contrast, the Catholic Church stated, did not see humans holistically but rather only as objects/things serving the purposes of research (Murray 2006; Damschen and Schönecker 2003). However, the church's paternity and abuse scandals since the 1990s have forced them to present their public moral judgments somewhat more softly. The church's credibility as an institution and moral compass has weakened considerably since, although private belief in Catholic values/norms remains present in many parts of the population.

The most comprehensive document of the Irish bishops is a paper that defines guidelines on ethical judgments and on dealing with AHR techniques.[28] The document is primarily addressed to Church members and interested persons. As a first step, basic principles are laid out which should be used as a standard in making evaluations. These are the fundamental Catholic faith convictions with reference to the protection of unborn life, family, reproduction and sexuality. In the second step, various techniques and methods of AHR are depicted objectively. In the third part, the previously depicted AHR techniques are judged according to the basic principles established in the first chapter. Each AHR technique is judged according to whether it is in accordance with these principles and thereby whether it should be accepted or rejected. In the first and third part, the discussion focuses on lines of argument based in natural law and religion. Throughout the beginning part, religious arguments, Christian narratives or references to doctrinal documents (Donum Vitae) are used. These are then translated into general principles, which in turn are justified with reference to human nature or the state of nature (secularisation or translation of religious viewpoints into general normative principles of human community and

---

[27] Irish Catholic Bishops' Conference, Committee for Bioethics, at: http://www.catholicbishops.ie/bioethics. Accessed 24 June 2014. Speech: Irish Times, 8 March 1999, at: https://www.irishtimes.com/news/child-resents-a-parentage-based-on-power-1.160394. Accessed 24 June 2014.

[28] In February 2008, a revised version of the paper was re-published (Irish Catholic Bishops' Conference 2003b).

action). In part, there is an attempt to underpin the principles based in natural law with scientific insights. A considerable part of the arguments rests on the idea of the family as the natural place of reproduction and as a social institution.[29] These arguments are not per se religious but are often enriched with religious concepts such as the "mystery of love" or the "creative action of God" and at least their intention can be classified as religious. Numerous arguments about identity and continuity appear when the discussion turns to the protection of unborn life and the necessity of socialisation and reproduction taking place within an intact family.[30] These are based on religious concepts and can be classified as religious arguments. Further, the Christian narrative of the correct relationship with creation appears as an argument: that life is given from God and humans may not dispose of it. This is especially argued in connection with the mechanisation of AHR in order to show that humans are alienating themselves from the idea of a godly creation through the union of man and woman.[31] At some points, references are made to the "Guidelines for ethical conduct and behaviour" of the Medical Council, for example in order to prove the reprehensibility of storing and destroying superfluous IVF embryos. In research, embryos become a means to an end and are no longer individual living beings. The storage and destruction of embryos violates ethical principles.[32] According to our methodological scheme, there can be found here religious arguments of the types: conviction/intention, content/source and conventionality/validity.

The above-mentioned document was complemented by a position paper/press release of the Irish Catholic Bishops on the CAHR (Irish Catholic Bishops' Conference 2003a, b). The document lists 12 positions of the bishops without providing arguments or discussion. The reasons behind the positions can be ascertained from

[29]The Right to Life and bodily integrity (p. 3): "Although the right to life finds a particularly strong foundation in Christian faith, it is a right which is acknowledged by people of all faiths and none."... The Right to an identity of origin: (pp. 4–5): ... "It is in the family first and foremost that children discover their identity and their individuality, that they learn respect for themselves and for others. It is in the family that cultural and moral values are learnt. Any procedure which undermines the unity and integrity of the family also damages the fabric of society, because the institution of the family is the foundation on which society is built."

[30](pp. 3–4): "Once fertilisation is complete, the organism has become a human being. There is nothing else it can be. It continues to develop and grow, of course. But all development or change necessarily involves some continuity; something in which the change takes place. This "something" is the human individual. It has its own genetically unique body. It has its own substantial form, the human soul, which is its first principle of life. It is this principle of life which facilitates and directs the development of the person throughout the lifetime of the organism." [...].

[31]The essential meaning of human sexuality: (pp. 5–6): ... (p. 6): "The more dominant technology becomes, the more the personalistic dimension of human sexuality tends to be separated from the act of life-giving, and the more easily the creative act of God is obscured." [...].

[32](pp. 13–15): "... Every human individual is a value in himself or herself. As noted above, the practice of multiple embryo replacement involves accepting that the survival rate per embryo is in inverse proportion to the likelihood of achieving pregnancy. In effect, therefore, IVF contributes to an ethos which regards the embryo as a means to an end, rather than as a value in and for itself." [...].

the reference document "Assisted Human Reproduction: Facts and Ethical Issues". The language used is secular and the positions are concisely formulated. By looking at the reference documents, it is possible to prove, however, that the position is derived from religious arguments and convictions. In summary, the positions are: the embryo has a right to life and identity; egg or semen donations should be forbidden because the family is a social institution and the site of reproduction; the production of embryos for research and cloning should be forbidden; leftover embryos from IVF should be avoided; and the freezing of embryos should be forbidden. A further, similar statement was published by the Irish Bishops' Congress in 2006.[33] It is directed against the recommendations of CAHR, in other words against a procedural protection of life, the expansion of AHR treatments and the production of embryos for research purposes. In the subchapter "Towards a Creative Response to Infertility", the bishops identify a series of non-negotiable principles: "[...] respect for the right to life and bodily integrity, respect for the family; respect for the meaning and purpose of human sexuality". These principles are justified with arguments connected to the embryo's identity, potentiality, continuity and natural law. This is supplemented by social-psychological arguments about the well-being of the child. Children are not the property of their parents but rather persons with their own rights. Cloning should be rejected because human embryos are produced in the process. The Supreme Court's evaluation of the term "unborn" is considered by the bishops to be a modern interpretation, which is not in keeping with the intention of the law. Interestingly enough, the Catholic Church places special value on the claim that they do not wish to pressure the state to pass laws in keeping with its teachings, but rather that these principles have to be respected because they are "inalienable, fundamental rights of the person". The bishops represent the opinion that every person has the right to life from the moment of conception (in vivo and in vitro), as well as the right to grow up in a family with married spouses. This is depicted as the natural order ("[...] the essential natural link between the life giving role of parents, and their responsibility of care"). From the point of view of the church, they are not drawing on religious arguments, but on generally accepted facts. Another short press release from Archbishop Dermot Clifford informed the public of the official position of the church on stem cell research: "[...] human life is sacred from the moment of conception until natural death". Embryonal stem cell research is wrong and should be forbidden by law. This research is contrary to the ethical and moral teaching of the Catholic Church because it is contrary to the common good.[34]

---

[33]The document restates the bishops' response to the 2005 CAHR Report (Irish Catholic Bishops' Conference 2006).

[34]Archbishop Dermot Clifford statement concerning UCC Board decision regarding research on embryonic stem cells, 28 October 2008. At: https://www.catholicbishops.ie/2008/10/28/archbishop-dermot-clifford-statement-ucc-board-decision-research-embryonic-stem-cells/. Accessed 24 June 2014.

Some further statements from the Irish Bishops' Conference make reference to stem cell research at the EU level.[35] One statement—directed at the Irish Prime Minister—criticises the fact that Irish policy has not taken a direction which is vehemently and unambiguously against the suggestions of the European Commission supporting stem cell research and warns of the repercussions for unborn human life. The Irish delegation in the Council of the European Union should abandon its neutral position and vote against the recommendation of the commission: "[. . .] we believe that this is an issue so fundamental that neutrality is not an option [. . .] the Constitution of Ireland is known to afford protection to human life from its beginnings". The paper contains no unambiguously religious arguments or content. The choice of language is (consciously) secular. Arguments and reasoning appear infrequently in general or are not gone into at depth. It uses as its argumentation merely that stem cell research leads to the "deliberate destruction of human embryos" and cannot be justified; further, that embryonal stem cell research is "unethical and illegal" (paragraph 3, 5, 7). The document is not interesting with respect to religious arguments, but it does indicate that the Catholic Church attempts to exercise direct influence in politics. Addressed directly to the Prime Minister (Taoiseach), the document shows that connections to the religious camp exist at the highest political level (McDonnell and Allison 2006).

Alongside the Catholic Church, some smaller churches have also expressed themselves on embryonal research. In the opinion of the Protestant *Church of Ireland*, embryos should be treated with the same respect as a normal human being. Embryos may not be produced for research purposes, and there may not be any financial profit through stem cell research. At the same time, however, human beings have the right to use medical findings with the purpose of healing illness, although only within certain limits. The whole argumentation is within the context of Christian ethics. The main arguments are the potentiality argument, human dignity, the importance of safeguarding God's creation, the sanctity of the embryo and the holiness of humankind. The church additionally has a societal duty to act as the conscience of society (Church of Ireland, General Synod, Church in Society Committee, Medical Ethics, Science and Technology Sub-Committee 2006). Other documents in IVF and stem cell research use mainly medical and scientific arguments (Church of Ireland Medical Ethics Working Group 1998). The *Presbyterian Church* rejected cloning because it undermines human values, the understanding that man is made in God's own image, the Christian understanding of mankind, the uniqueness of the individual, the holiness of life and marriage as the natural place of reproduction and as a Christian institution. Cloning is not an appropriate way of treating God's creation, and its end is no justification for its means. Overall, the

[35]Catholic Bishops' letter to Taoiseach on embryonic stem cell research, Press release of 13th October 2003. A shorter statement was published in 2006. At: https://www.catholicbishops.ie/2003/11/13/catholic-bishops-letter-taoiseach-embryonic-stem-cell-research-public/. Accessed 24 June 2014. The government is called upon to support research with adult stem cells. Research with embryos, by contrast, is the destruction of human life. Bishops expressed concern in advance of an EU Parliament vote on funding research on human embryos, 14 June 2006.

entire argumentation relies upon fundamental religious/Christian principles and narratives (The Presbyterian Church in Ireland 2016). If we classify their patterns of argumentation according to our methodological scheme, the following arguments appear: (1) conviction/intent, (2) content, (3) source and (4) conventionality/validity. The *United Methodist Church* has likewise published a short paper, but it avoids religious arguments. Reproductive cloning and the production of embryos for research purposes are both rejected. Stem cell research with already existing embryos, however, is approved. This is justified with the argument that while the embryo can make moral demands upon us (as "potential life"), it cannot do so to the same degree as a developed human life. The future therapeutic uses of stem cell research justify it, so long as it takes place with caution and within certain limits (United Methodist Church 2008). The Pro-Life Movements, in particular the *Pro Life Campaign* (PLC), have also published an entire series of statements or leaflets on embryonic research. In the statements, the government is repeatedly directly called upon to issue laws to protect embryos and to become engaged at the EU level for the protection of life. Ministers who promote research-friendly laws are, in the opinion of the Pro Life Campaign, behaving in an undemocratic way against the will of the majority of the Irish. Researchers and supporters of research are cast as the "anti-life lobby" (Pro Life Campaign 2004, 2010a, 2011). During European and local elections, the PLC publishes lists with the names of the candidates and questions them on their position towards the protection of life. The voters are then directly told that they should only choose candidates who are pro-life (Pro Life Campaign 2014a, b, 2015a, b).

The main argument, which can be seen throughout all the statements of the Pro Life Campaign, is that the embryo is a human with full rights from the very beginning and that embryonic research entails the killing or destruction of humans or members of the human family. This position is depicted as an objective truth and scientific fact and is not justified or explained. The bioethical commissions CAHR and IBC are accused of following a false, undemocratic path and of taking part in the destruction of embryos (Pro Life Campaign 2008a, 2010b, c, 2014c; Pro Life Campaign. Stem Cell Research—Frequently asked questions). The latest judgments of the Supreme Court—that the protection of the embryo in vitro is not covered by the eighth constitutional amendment—are criticised by the PLC (Pro Life Campaign 2009a, 2010d, e). Also strongly attacked by the Pro Life Campaign has been the "Stem Cell Foundation", which has supported laws that would legalise embryonic research. The PLC's arguments are primarily medical: they state that all major scientific discoveries have been using stem cell research from adult cells. On the other hand, research with embryonal cells has up until now only led to meagre results. As a result, research efforts should be limited to adult stem cells. Adult cells include, moreover, an uncontroversial and ethical form of research. Further key arguments are the continuity argument and that life begins at the moment the eggs and sperm join together. Both these arguments are also used by the Catholic Church. A further actor attacked by the PLC are the universities, which take part in research with human embryos: even when universities import embryos and do not take part in killing them, they become complicity guilty. Here the medicinal arguments mentioned above are made use of. The argument is that the motivation to heal does not justify the destruction of life. The position is based on the potentiality argument and the argument that life and personhood begins at the moment of conception. At least in their intention, these arguments could be classified as religious (Pro Life Campaign 2008b, 2009b, 2014b).

3 Summary

Catholicism has been an important element of the Irish national identity since the nineteenth century. After the foundation of the Irish state, the influence of the church was so great that the political elite left questions of defining public morals to its judgment. Deference and clericalism in the two large parties FF and FG have decreased since the end of the 1960s. At the same time, a liberalisation of the law has taken place. In the last two decades, the direct influence of the church hierarchy on the political elite as well as its influence on public opinion and the behaviour of single individuals has changed. The church no longer possesses the sole power of defining Irish identity and moral values. When looking at topics that are sharply affected by religious beliefs (such as abortion), today a considerable divide can be seen in Irish society along the religious-conservative and secular-liberal lines. Nevertheless, Catholicism still has an important presence in Irish society. The support for a real separation of church and state continues to be rather limited in large parts of the population and the political elite. This can be explained above all by the general continued interest of the population for the Catholic religion and the legacy of religious nationalism.[36] Although Catholicism has been weakened in the last decades as a source of national identity, many Irish people retain a sense of pride in the century's long unifying band of Catholicism. However, the position of the church is seen today much more critically. Not least, the paternity and abuse scandals—which politicians largely ignored out of respect for the Catholic Church—have led to a loss of credibility and shown the necessity of an internal change in the Catholic Church. Ever more voters likewise disapprove of the attempts of the church to influence elections or political decisions.

Concerning biopolitical topics, it is obvious that the hostile attitude on the part of the Catholic Church and the pro-lifers has led to an avoidance of a public debate among politicians, journalists, doctors, legal scholars and even economic actors in Ireland. Legal regulations exist for none of the new biotechnologies, not least because the right to a legal, unpunished abortion remains highly contested. On the one hand, state and church attempt to hold on to the ideal of a (Catholic) pro-life nation, not least because they see it in terms of preserving the nation's collective identity. For these reasons, bioethical questions are tightly linked to questions of

[36]Although nationalism today is differentiated into several subgroups (e.g. traditional, liberal, revisionist, neo-traditional), only a small number of nationalists actively favour a secular ideology.

national identity, while the topic in the rest of Europe is more linked to questions of social diversity and cultural pluralism. On the other hand, economic arguments seek to modernise Ireland, making it attractive to foreign investors. Many are keen to avoid ethical debates reminiscent of the old materially impoverished and morally strict Ireland.

Looking at the discursive tradition, in the past ethical questions were not discussed by presenting pluralistic arguments but rather according to unquestionable conclusions based in natural law. With the help of natural law, religious value judgments are transferred to "neutral" state law. Prescription and proscription, rather than negotiation and dialogue, has been the modus operandi of the Catholic Church in Ireland. The position and arguments of the Irish Catholic Church represent the official papal position. The Catholic Church is opposed to artificial insemination and in vitro fertilisation because it is a subversion of the dignity and unity of marriage and of the integrity of natural and necessary parental relationships with the children. The church is against embryonic stem cell research because it involves the destruction of human embryos, and it is related to abortion, euthanasia and other attacks on innocent life. The arguments surrounding ART and embryonal research come from a position of moral absolutes: it is fundamentally and always wrong to destroy an embryo. Their arguments make use of the species argument, the continuity argument, the identity argument, the potentiality argument, natural law, the instruction "Donum Vitae" and the eighth constitutional amendment. In their position that "life starts at the beginning", they make no distinction between embryonal development stages. They do not even engage in debate with other philosophical or religious positions. Liberal theologians who support IVF, contraception and abortion are spurned. They stand for a wrong moral relativism; this relativism undermines the concept of an objective truth, in which there is an understanding of the intrinsic wrongness or evilness of such acts. The Catholic Church has directly demanded that the Irish government create regulations on ART, because the right to life of the unborn and the unique status of the family founded on marriage must be legally anchored. At the EU level, the Bishops' Committee calls on the government to maintain the Irish position on the protection of life. Additionally, the Catholic Church indirectly or directly influences the medical establishments under its control. Doctors employed there are obligated to observe Catholic norms. Through the ethics committees of hospitals, the Catholic Church is also able to maintain control over moral guidelines. Even medical training centres are largely Catholic-run. The entire debate on biopolitics takes place in a narrow institutional framework with limited participants and contents. The church operates like an institutional power player and would prefer to limit the discussion rather than to promote it. Neither the legal framework nor the social and ethical effects of the technologies are sufficiently opened up for discussion. The lack of public (and political) discussions and the very slow institutionalisation of bioethics are due to the weakness of the political elite, the historical position of the Catholic Church as moral authority and societal power and the divide within the Irish population along a conservative-rural (religious) and modern-urban (secular) fault line.

References

Anderson, John, ed. 2006. *Religion, democracy and democratisation*. London: Routledge.

Audi, Robert. 2000. *Religious commitment and secular reason*. Cambridge: Cambridge University Press.

Banchoff, Thomas, ed. 2008. *Religious pluralism, globalization, and world politics*. Oxford: Oxford University Press.

Bartlett, Thomas. 2010. *Ireland: a history*. Cambridge: Cambridge University Press.

Budge, Ian. 2002. Great Britain and Ireland. Variations of party government. In *Political institutions in Europe*, ed. Joseph M. Colomer, 17–55. London: Psychology Press.

Casey, James. 1995. Staat und Kirche in Irland. In *Staat und Kirche in der Europäischen Union*, ed. Gerhard Robbers, 205–228. Baden-Baden: Nomos.

Church of Ireland, General Synod, Church in Society Committee, Medical Ethics, Science and Technology Sub-Committee. 2006. Submission to the Irish council for bioethics: stem cell research.

Church of Ireland, Medical Ethics Working Group. 1998. In vitro fertilisation. Working Paper Number 2. https://www.ireland.anglican.org/resources/155/in-vitro-fertilisation. Accessed 24 June 2014.

Coakley, John. 1993. Society and political culture. In *Politics in the Republic of Ireland*, ed. John Coakley and Michael Gallagher, 22–48. Dublin: Folens/Psai Press.

Coakley, John, and Michael Gallagher, eds. 2004. *Politics in the Republic of Ireland*. Dublin: Folens/Psai Press.

Commission on Assisted Human Reproduction. 2005. Report of the commission on assisted human reproduction, Department of Health and Children. http://www.lenus.ie/hse/bitstream/10147/46684/1/1740.pdf. Accessed 24 June 2014.

Connolly, Sean J. 2007. Religion and nationality in Ireland. An unstable relationship. In *Religion und Nation. Katholizismen im Europa des 19. und 20. Jahrhunderts*, ed. Urs Altermatt and Franziska Metzger, 121–134. Stuttgart: Kohlhammer.

Conrad, Kathryn. 2001. Fetal Ireland: national bodies and political agency. Eire-Ireland. *Interdisciplinary Journal of Irish Studies* 36 (3): 153–173.

Cousins, Graine, et al. 2005. *Public perceptions of biomedical research: a survey of the general population in Ireland. Psychology reports paper*, 8. Dublin: Health Research Board.

Daly, Mary. 2000. Die normgerechte Familie. Katholizismus und Sozialpolitik im deutsch-irischen Vergleich. In *Religion und Politik. Zwischen Universalismus und Partikularismus*, ed. Heinz-Dieter Meyer, Michael Minkenberg, and Ilona Ostner, 253–285. Opladen: Leske + Budrich.

Damschen, Gregor, and Dieter Schönecker, eds. 2003. *Der moralische Status menschlicher Embryonen*. Berlin: De Gruyter.

Dillon, Michele. 2002. Catholicism, politics, and culture in the Republic of Ireland. In *Religion and politics in comparative perspective, the one, the few, the many*, ed. Ted Gerald Jelen and Clyde Wilcox, 47–67. Cambridge: Cambridge University Press.

Elvert, Jürgen. 1999. Das politische system Irlands. In *Die politischen Systeme Westeuropas*, ed. Wolfgang Ismayr, 255–287. Opladen: Leske + Budrich.

Fahey, Tony. 1998. The Catholic church and social policy. *The Furrow* 49 (4): 202–209.

Fox, Jonathan, ed. 2012. *Religion, politics, society & the state*. Oxford: Oxford University Press.

Gallagher, Michael, and Komito Lee. 1993. Dail deputies and their constituency work. In *Politics in the Republic of Ireland*, ed. John Coakley and Michael Gallagher, 150–166. Dublin: Folens/Psai Press.

Girvin, Brian. 1992. Moral politics and the Irish abortion referendums. *Parliamentary Affairs* 47.2 (1994): 203–221.

———. 1996. Church, state and the Irish constitution: the secularisation of Irish politics? *Parliamentary Affairs* 49 (4): 599–615.

Habermas, Jürgen. 2008. *Between naturalism and religion: philosophical essays*. Cambridge: Polity Press.

Hanafin, Patrick. 1996. Reproductive rights and the Irish constitution: from sanctity of life to the sanctity of autonomy? *European Journal of Health Law* 3: 179–188.
Heit, Alexander, and Georg Pfleiderer, eds. 2012. *Religions-Politik II. Zur pluralistischen Religionskultur in Europa*. Baden-Baden: Nomos.
Hemton, David. 2006. Religion and political culture in Britain and Ireland. In *Politik und Religion in der Europäischen Union. Zwischen nationalen Traditionen und Europäisierung*, ed. Hartmut Behr and Mathias Hildebrandt, 155–172. Wiesbaden: VS Verlag für Sozialwissenschaften.
Henry, Mary. 1999. Regulation of assisted human reproduction bill. Seanad Debates Official Report.
Iona Institute. 2008. Response to the report of the commission on assisted human reproduction. https://www.ionainstitute.ie/wp-content/uploads/2014/11/Making_Babies__Full-length_version.pdf. Accessed 24 June 2014.
Irish Catholic Bishops' Conference. 2003a. Publication of Catholic Bishops' submission to government commission on assisted human reproduction. https://www.catholicbishops.ie/2003/02/12/publication-catholic-bishops-submission-government-commission-assisted-human-reproduction. Accessed 24 June 2014.
———, Committee for bioethics. 2003b. Assisted Human Reproduction: Facts and Ethical Issues. https://www.catholicbishops.ie/2008/02/07/assisted-human-reproduction. Accessed 24 June 2014.
———. 2006. General meeting of the Irish Bishops' conference: press release, Diaconate. https://www.catholicbishops.ie/2006/12/07/december-general-meeting-irish-bishops-conference/. Accessed 24 June 2014.
———, Committee for bioethics. http://www.catholicbishops.ie/bioethics. Accessed 18 Dec 2014.
Irish Council for Bioethics. 2005. *Human biological materials, recommendations for use, collection and storage in research*. Dublin: ICB.
———. 2008. *Ethical, scientific and legal issues concerning stem cell research, opinion*. Dublin: ICB.
———. 2010. *Stem cell research: hope or hype? Exploration of the scientific and ethical questions*. Dublin: ICB.
Junker-Kenny, Maureen. 2003. Religion and bio-ethics in Ireland today. In *Religion and politics in Ireland at the turn of the millennium*, ed. James P. Mackey and Enda McDonagh, 166–184. Dublin: Columba Press.
Keogh, Dermot. 1986. *The Vatican, the bishops and Irish politics: Church and State in Ireland, 1919–1939*. New York: Cambridge University Press.
Laver, Michael, and Michael Marsh. 1993. Parties and voters. In *Politics in the Republic of Ireland*, ed. John Coakley and Michael Gallagher, 105–125. Dublin: Folens/Psai Press.
Madden, Deirdre. 2004. Assisted reproduction in the Republic of Ireland: a legal quagmire. Cardiff Centre for Ethics, Law & Society.
Mair, Peter. 1993. The party system and party competition. In *Politics in the Republic of Ireland*, ed. John Coakley and Michael Gallagher, 86–103. Dublin: Folens/Psai Press.
Matthews, Kristin. 2007. Overview of world human cloning policies, appendix, table I – national legislation.
McConnell, Michael W. 1999. Five reasons to reject the claim that religious arguments should be excluded from democratic deliberation. *Utah Law Review* 3: 639–657.
McDonnell, Orla, and Jill Allison. 2006. From biopolitics to bioethics: church, state, medicine and assisted reproductive technology in Ireland. *Sociology of Health & Illness* 28 (6): 817–837.
Medical Council. 2009. Guide to professional conduct and ethics for medical practitioners, 7th ed. http://www.medicalcouncil.ie/News-and-Publications/Publications/Guide-to-Professional-Conduct-and-Ethics/Guide-to-Professional-Conduct-and-Behaviour-for-Registered-Medical-Practitioners-pdf.pdf. Accessed 03 Oct 2015.
Millar, Michelle. 2008. Social inclusion and the welfare state: who cares? In *Contesting the state. Lessons from the Irish state*, ed. Maura Adshead, Peadar Kirby, and Michelle Millar, 100–119. Manchester: Manchester University Press.

Murphy-Lawless, Jo. 1993. Fertility, bodies and politics: the Irish case. *Reproductive Health Matters* 1 (2): 53–64.

Murray, Donal. 2006. *What is a person?* Dublin: Veritas Publication.

National Woman's Council of Ireland. 2010. Moving in from the margins – women's political representation in Ireland. https://www.nwci.ie/index.php/learn/article/moving_in_from_the_margins_womens_political_representation_in_ireland. Accessed 10 April 2013.

Nitsche, Susan. 2012. *Wertewandel und politische Parteien in der Republik Irland.* Chemnitz: Wißner-Verlag.

O'Halpin, Eunan. 1993. Policy making. In *Politics in the Republic of Ireland*, ed. John Coakley and Michael Gallagher, 190–206. Dublin: Folens/Psai Press.

O'Mahony, Eoin. 2010. Religious practice and values in Ireland. A summary of European values study 4th wave data. The Council for Research and Development, Commission of the Irish Catholic Bishops' Conference.

Pope John Paul II. 1993. Encyclical Veritatis splendor. Libreria Editrice Vaticana.

Porter, Elisabeth. 1996. Culture, community and responsibilities: abortion in Ireland. *Sociology* 30 (2): 272–298.

Pro Life Campaign. 2004. Harney should move against embryo research. Tuesday, January 06, 2004. https://prolifecampaign.ie/main/portfolio/detail/06-01-2004dr-ruth-cullen-harney-should-move-against-embryo-research/. Accessed 24 June 2014.

———. 2008a. Should the state legislate for embryonic stem cell research? The Irish Times 12th May 2008. https://prolifecampaign.ie/main/should-the-state-legislate-for-embryonic-stem-cell-research-the-irish-times-12th-may-2008/. Accessed 24 June 2014.

———. 2008b. PLC accuses UCC of "gross misrepresentation" over its embryo research stance, 28 Oct, 2008. https://prolifecampaign.ie/main/28-10-2008-plc-accuses-ucc-of-gross-misrepresentation-over-its-embryo-research-stance-2/. Accessed 24 June 2014.

———. 2009a. Supreme court decision in R- vs- R regrettable but creates an opportunity to unite ethics and science in favor of a win-win solution, 15th December 2009. https://prolifecampaign.ie/main/rvr/. Accessed 24 June 2014.

———. 2009b. PLC comments on FDA approval of embryo research clinical trials, 23. Jan, 2009. https://prolifecampaign.ie/main/23-01-2009-plc-comments-on-fda-approval-of-embryo-research-clinical-trials/. Accessed 24 June 2014.

———. 2010a. Embryonic stem cell research and the EU – revised EU directive on animal testing highlights Ireland's urgent need to ban human embryo-destructive research. 24th September 2010b. https://prolifecampaign.ie/main/24-09-2010-embryonic-stem-cell-research-and-the-eu-revised-eu-directive-on-animal-testing-highlights-irelands-urgent-need-to-ban-human-embryo-destructive-research/. Accessed 24 June 2014.

———. 2010b. Stem cell research – judge blocks US federal funding of embryo research – what Ireland must do, 27th August 2010. https://prolifecampaign.ie/main/27-08-2010-stem-cell-research-judge-blocks-us-federal-funding-of-embryo-research-what-ireland-must-do/. Accessed 24 June 2014.

———. 2010c. Pro life submissions far outnumber opposing side in public consultations, 26th March 2010. https://prolifecampaign.ie/main/26-03-2010-pro-life-submissions-far-outnumber-opposing-side-in-public-consultations/. Accessed 24 June 2014.

———. 2010d. Minister Harney should learn from California. Experience: money spent on embryonic stem cell research is money wasted. 11th February 2010. https://prolifecampaign.ie/main/11-02-2010-stem-cell-research-minister-harney-should-learn-from-californian-experience/. Accessed 24 June 2014.

———. 2010e. Pro Life Campaign welcomes reduction in abortion rate. https://prolifecampaign.ie/main/25-05-2010-abortion-rate-falls-pro-life-campaign-welcomes-reduction/. Accessed 24 June 2014.

———. 2011. PLC welcomes Geron Corp decision to cease embryonic stem cell research, 18th November 2011. https://prolifecampaign.ie/main/18-11-2011-plc-welcomes-decision-of-geron-corporation-to-cease-embryonic-stem-cell-research/. Accessed 24 June 2014.

———. 2014a. European Parliament Elections on 23rd May 2014. https://prolifecampaign.ie/main/european-elections-23rd-may-2014/. Accessed 03 April 2016.

———. 2014b. Promote results-based stem cell research says PLC. https://prolifecampaign.ie/main/portfolio/detail/30-10-2014-promote-results-based-stem-cell-research-says-plc/. Accessed 04 Dec 2015.

———. 2014c. Stem cell research – frequently asked questions. https://prolifecampaign.ie/main/stem-cell-research-frequently-asked-questions/. Accessed 24 June 2014.

———. 2015a. Carlow/Kilkenny by-election candidates stand on right to life issues. 15th May 2015. https://prolifecampaign.ie/main/portfolio/detail/14-05-2015-carlowkilkenny-by-election-use-your-vote-to-protect-human-life/. Accessed 03 April 2016.

———. 2015b. Use your vote to protect human life on 22nd may. Dublin West By-Election – 23rd May. https://prolifecampaign.ie/main/portfolio/detail/14-05-2015-carlowkilkenny-by-election-use-your-vote-to-protect-human-life/. Accessed 03 April 2016.

Quinn, Philip L. 1997. Political liberalisms and their exclusions of the religions. In *Religion and contemporary liberalism*, ed. Paul J. Weithman, 138–161. Notre Dame: Notre Dame Press.

Rawls, John. 1993. *Political liberalism*. New York: Columbia University Press.

———. 1999. The idea of public reason revisited. In *John Rawls: collected papers*, ed. Samuel Freeman, 573–615. Cambridge, MA: Harvard University Press.

Rowe, Paul S. 2012. *Religion and global politics*. Oxford: Oxford University Press.

Ruane, Joseph. 1995. Secularisation and ideology in the Republic of Ireland. In *La secularisation en Irlande*, ed. Paul Brennan, 239–254. Caen: University College Cork. http://books.openedition.org/puc/135. Accessed 03 October 2015.

Sheikh, Asim A. 2002. *Genetic research and human biological samples. The legal and ethical considerations*. Dublin: Health Research Board.

Sinnott, Richard. 1993. The electoral system. In *Politics in the Republic of Ireland*, ed. John Coakley and Michael Gallagher, 67–85. Dublin: Folens/Psai Press.

Smyth, Ailbhe. 1996. And nobody was any the wiser. Irish abortion rights and the European Union. In *Sexual politics and the European Union. The new feminist challenge*, ed. R. Amy Elman, 109–130. Oxford: Berghahn Books.

Taylor, Charles. 1998. Modes of secularism. In *Secularism and its critics*, ed. Rajeev Bhargava, 31–53. Delhi: Oxford University Press.

Taylor, Judith. 1999. Case X: Irish reproductive policy and European influence. *Social Politics* 6 (2): 203–229.

The Presbyterian Church in Ireland. 2016. Cloning (Social Issues & Resources Committee). https://www.presbyterianireland.org/getmedia/f687560a-edc2-460c-9bb4-1734c3aa56cb/Social-Issues-Cloning-2012.pdf.aspx?ext=.pdf. Accessed 15 August 2017.

United Methodist Church. 2008. Ethics of embryonic stem cell research. http://www.umc.org/what-we-believe/ethics-of-embryonic-stem-cell-research. Accessed 03 April 2016.

White, Timothy J. 2007. Catholicism and nationalism in Ireland: from fusion in the 19th century to separation in the 21st century. *Westminster Papers in Communication and Culture* 4 (1): 47–64.

Wolterstorff, Nicholas. 1997. Why we should reject what liberalism tells us about speaking and acting in public for religious reasons. In *Religion and contemporary liberalism*, ed. Paul J. Weithman, 162–181. Notre Dame: Notre Dame Press.

Part III
Protestant(-English) States with High or Moderate Level Religious Pluralism

Chapter 6
Biotechnology and the Nonreligious Uses of God Talk

John H. Evans

Abstract Evans begins by showing that there is a slight tendency for the American public to oppose the use of religious justifications for political arguments. Political theorists assume that the support or opposition to this "God talk" in the public sphere is due to its transcendent content. Evans claims that "God talk" is not just transcendent but is also a type of "thick" discourse that concerns the more fundamental and social aspects of an issue. Through an empirical examination of survey data, Evans demonstrates that, independent of the religious views of the respondent, those who are supportive of "God talk" are less supportive of genetic enhancement. Evans concludes that "God talk" is not only about God but is used to articulate "thick" concerns about biotechnology.

1 Introduction

In another paper, I show that there is a very small tendency for Americans to prefer that arguments in the public sphere be justified by science rather than religion (Evans 2018). I call these religious justifications "God talk." In that paper, I also found that religious Americans do not have a preference for God talk or science. However, the nonreligious public wants to avoid God talk. In this chapter, I build upon that other paper to test the presumption from the political theory literature that the only reason people like or dislike God talk is because of its transcendent or particularistic content. I ask whether people are averse to—or welcome—God talk because of nonreligious features of God talk.

In this contribution, I argue that people prefer God talk not *only* because of their own belief in religion but because God talk expresses a deeper debate about collective goals and a non-individualistic concern with the collective good—what I call a "thick" debate. To empirically test this claim, I start by showing that at least in the US public sphere, it is not possible to oppose human biotechnology without a thick argument. I then show that members of the public who are opposed to

J. H. Evans (✉)
Department of Sociology, University of California, San Diego, CA, USA
e-mail: jhevans@ucsd.edu

M. Weiberg-Salzmann, U. Willems (eds.), *Religion and Biopolitics*,
https://doi.org/10.1007/978-3-030-14580-4_6

enhancement of children are more likely to find God talk acceptable than are those who are opposed to enhancement. Critically, this is even true when analysis is limited to the nonreligious respondents, which suggests a nonreligious use of God talk.

2 Americans' Views of God Talk[1]

Giving reasons for arguments is the cornerstone of a liberal democratic public sphere. There has long been a debate about the acceptability of giving *religious* reasons for arguments in the public sphere, what I am calling "God talk." One position is that any public norm *against* using religious reasons is equivalent to discrimination against religious people, because they cannot express their actual beliefs. As Rogers Smith summarizes, using the term "public reason" for secular discourse: "Many critics . . . contend instead that these 'public reason' standards are normatively inappropriate because they discourage religious expression that should be accepted, even encouraged. Public reasons are often deemed to be undemocratic; to privilege secular views unfairly; to limit desirable alternatives in pluralistic societies; and to be politically counterproductive, breeding resentment, not social harmony" (Smith 2008:276).

A more classic reason to be normatively opposed to God talk is the assumption that people who hear it will ultimately be driven to violence because it violates their own deeply held religious beliefs. This concern predates the formation of the United States and was instrumental in the development of Enlightenment reason after the Thirty Years' War of the early seventeenth century. According to one theorist: "without citizens' allegiance to public reason and their honoring the duty of civility, divisions and hostilities are bound in time to assert themselves . . . harmony and concord depend on the vitality of the public political culture and on citizens' being devoted to and realizing the ideal of public reason" (Smith 2008:274). Another group of theorists write that "political liberals fear the diffidence and divisiveness that deep-seated and particularistic commitments might produce, and . . . believe that it is important to domesticate and to privatize such commitments in order to forestall such dangerous political consequences" (Isaac et al. 1999:230).

On the other hand, people who want to use religious justifications do not think that religious discourse leads to conflict and violence. For example, Nicholas Wolterstorff argues that "the slaughter, torture, and generalized brutality of our century has mainly been conducted in the name of one or another secular cause—nationalisms of many sorts, communism, fascism, patriotisms of various kinds, economic hegemony. In seventeenth-century Europe, human beings cared deeply about religion. In our century, most have cared more deeply about various secular causes" (Wolterstorff 1997:78, 80). Many theologians have concluded that a concern

[1]Much of the material in this section builds upon and is taken from Evans (2018).

about religious reasons is "a secularist ruse designed to reduce religion to insignificance" (Stout 2004:63).

Normative consensus among theorists that God talk is impermissible depends on which zone of the public sphere the God talk is occurring. In a bull's-eye metaphor, the center of the target is justifications by elected officials for policies. There is strong consensus among theorists that public officials should not justify policies with God talk. In the middle zone are campaigns for public office and interest group mobilization for the passage of laws and policies (Rawls 1993:215–216). This has not been focused upon by scholars. At the outermost zone are discussions among citizens that influence the center in a more indirect way (McCarthy 1994:50–54). Theorists of deliberation in the public sphere, such as Jürgen Habermas, have been concerned with this outer zone, and the norms of proper debate in this zone are much more contested. I therefore focus on this outer zone.

Political theorists make normative claims, but the public's views of God talk are unknown. American politicians infamously invoke God to justify their political claims, and to some extent this simply reflects a more common practice among the citizens. A noted social ethicist writes that "'God talk' as much as 'rights talk' is the way Americans speak" in politics (Elshtain 2001:39). On the other hand, at least an earlier generation of Americans were taught that avoiding God talk was a part of proper public etiquette. For example, etiquette promoters make declarations such as "never discuss politics or religion."[2] As a headline of a "Miss Manners" column that widely circulates through regional and local newspapers in the United States states: "politics and religion? Don't go there."[3]

I conducted a survey to determine whether Americans were actually averse to hearing God talk in a face-to-face conversation about politics with a friend, comparing God talk to justifications based in science (Evans 2018).[4] I chose an experimental instead of an observational design because it allows a stronger claim to

---

[2]Post, Anna. Avoid Political Pitfalls When Talking, Working and Posting. The Emily Post Institute. http://www.emilypost.com/on-the-job/workplace-relationships/1062-avoid-political-pitfalls.
Another etiquette advisor wrote: "Not everyone can pick a fish fork out of the cutlery line up, but most of America made it through childhood with a few ground rules of inviolable etiquette in place: say please and thank you; don't chew with your mouth open; cut the price tag off a gift; if you can't say anything nice, talk about the weather. Just don't, for goodness sake, ever talk about politics or religion at the dinner table!" Post, Anna. The Etiquette of Talking Politics. HuffPost. 21.02.2008. http://www.huffingtonpost.com/anna-post/the-etiquette-of-talking-_b_87893.html.

[3]Martin, Judith. Miss Manners: Politics and religion? Don't go there. The Washington Post. 14.04.2010. http://www.washingtonpost.com/wp-dyn/content/article/2010/04/13/AR2010041304109.html.

[4]These data come from a broader survey on religion and science fielded by GfK. GfK maintains an online panel that is representative of the US population, randomly recruited through probability-based sampling using addresses provided by the US Postal Service. GfK provides computer equipment or online access to prospective panel members, and once people have joined the panel, they are notified via email that they have been chosen to participate in a survey, which they can take at their convenience. Data from the GfK panel have been extensively used by social scientists. For example, one component of the 2012 American National Election Studies and the sixth wave of the American component of the World Values Survey were fielded by GfK.

causality and because the respondent does not see the comparisons being made so they cannot provide a more socially acceptable answer (Mutz 2011:9, 54, 56, 64). If I had simply asked respondents whether they prefer God talk or science talk, I might have had people recognizing that they should pick science due to American etiquette or their lack of religion or perhaps picking religion because of their own religious tradition.

I designed a vignette with three randomly assigned factors resulting in a total of eight conditions in a $2 \times 2 \times 2$ factorial design. Each respondent only saw one vignette. The vignette began with "Now we are going to describe an imaginary social situation, and then afterwards ask you some questions about it." The vignette read:

Robert is a friend of yours. He is trying to [F1: discuss his view with/convince] other friends that [F2: abortion/income inequality] is wrong because [F3: teachings in the New Testament of the Christian Bible tell/the latest published scientific research tells] us that [F2: human life begins at conception. Therefore, to be consistent, abortion should be illegal/unequal societies have conflict between the rich and the poor. Therefore, tax policy should be changed to lessen inequality.]

F1, F2 and F3 refer to the factors, split by the "/", that were randomly assigned to the respondents. There are then eight possible combinations of the factors. Given the focus of this contribution I ignore F1 and F2 and compare the 50% of the sample who saw the religious justification to the 50% who saw the scientific justification (F3), regardless of which of the F1 and F2 conditions they were randomly assigned to see.[5]

At the end of the vignette, the respondents were asked "If you were there with Robert, how interested would you be in continuing the discussion of his argument

---

Between December 27, 2013, and January 13, 2014, GfK surveyed 9900 respondents from the general public for the Religious Understandings of Science Survey, conducted by Elaine Howard Ecklund of Rice University (which was made possible by John Templeton Foundation Grant # 38817). The survey contained extensive measures of the religiosity of the respondents. Between August 27, 2014, and September 2, 2014, I fielded a survey with a sample drawn via simple random sampling from the respondents to the first study, resulting in 3587 responses. This latter survey contained questions about how the respondent defined a human being, as well as the vignette examined in this chapter. The questions about human beings are used in a different publication. The questions I use in this contribution about religious and political identity, as well as moral attitudes, were asked on the first survey, eliminating any concern about question-order or priming effects. I use the standard GfK post-stratification weights which include gender, race, education, income, and census region.

[5]To scholars, some of the arguments resulting from the combinations of factors may be more persuasive than others. For example, using the "is/ought" distinction would lead to the conclusion that science cannot tell us that life begins at conception, only what happens at conception, with the label "life" being a normative claim. However, the follow-up question to the vignette—and the topic of this study—concerns willingness to hear the argument, not agreeing with it, so the quality of the argument is less important. Moreover, in the public discourse, the "is/ought" distinction is regularly ignored. "Science" is consistently invoked for policy arguments without a corresponding normative claim. Moreover, claiming that science tells us "life" begins at conception is probably now the most common antiabortion argument. The antiabortion movement has tried to claim the mantle of science, and these activists adopted science precisely because it is a powerful justificatory language (Shields 2009).

with him? Would you be: Very interested, Interested, Somewhat interested, Not very interested, Not interested at all?" I consider this to be a measure of aversion to hearing a conversation about a social/political issue. These possible responses were coded into a five-point measure, with the "not interested at all" in further discussion receiving the highest number of five. When compared to the level of aversion among those who saw the science vignette, we have a measure of overall aversion to God talk compared to science talk.

I focused upon the specifically Christian God talk people are most likely to hear in the United States, but it was more of a challenge to select a type of discourse to compare God talk to. There is no neutral discourse in the public sphere, but we can imagine comparing God talk to justifications for arguments based on secular American values, common sense, personal experience, and much else. The political theory literature suggests that liberal theorists' ideal justificatory language should be, in principle, independent of particularistic belief and thus universal. This makes American values, common sense, and personal experience ineffective comparisons. Instead I selected justifications based on science, which John Rawls, the most influential of the political theorists, held up as a neutral justificatory language (Rawls 2001:91–92). Science portrays its findings as objectively true and thus independent of the beliefs of any group.

An even greater challenge was to select a political issue for Robert to be talking about. I wanted to select one that was conservative and one that was liberal so as to make a generalization across political position and focus on God talk itself. Ideally I would have selected the same issue but would have varied Robert's stance. However, I concluded that there was no one issue that has a truly visible liberal and conservative religious presence in the public sphere, and for which there are also credible religious and scientific arguments. Therefore, for the conservative stance, I selected Robert making antiabortion claims and, for the liberal, making claims about mitigating social inequality. The survey also contains demographic measures such as education level, gender and ethnicity, as well as very precise measures of the religious orientation of the respondent.

The earlier paper showed that if we consider the "averse" to be those who said they are "not very interested" or "not interested at all" in continuing the conversation, we see that 46.2% of respondents were averse to continuing the conversation if Robert was using God talk and 41.5% were averse if he was using science to justify his claims (Evans 2018). While this difference is statistically significant, it is also a difference of only 4.7% points. One way to think of the magnitude of this divide is to consider it in light of the most famous claims about God talk, which is that God talk is a "conversation stopper" (Rorty 2003:148). While I have no way of precisely establishing a threshold of noticeability, it seems unlikely that someone would notice if 46 of 100 people wanted to stop a conversation justified with religion and 42 of 100 wanted to stop a conversation justified by science. In sum, there is a statically significant aversion to God talk, but there is not a substantively significant aversion to God talk in the US public.

This generalization masks some differences by subpopulation. I examined the religiously active Christian Americans from all of the measurable groups: Catholics, biblical literalist conservative Protestants, non-literalist conservative Protestants, and mainline/liberal Protestants. Members of each of the first three of these groups were equally averse to a continued conversation with Robert if he was justifying his claims with God talk or with science. Mainline/liberals were more averse to a conversation based on *science*. Therefore, while one of the justifications for avoiding God talk is to avoid conflict between types of Christians—a lesson liberal theorists learned from the Thirty Years' War in Europe in the seventeenth century (Smith 2008:274; Isaac et al. 1999:230)—regular religious citizens do not seem to be concerned. Moreover, we might think that the religious have a preference for religious language, but they do not.

There is, however, one group that is clearly averse to God talk—the nonreligious. For the nonreligious, there is a 12.9% difference in not wanting to continue discussion when shown the religion vs. science vignettes. This certainly suggests that aversion to God talk is because of its religious content. At this point we should ask: what drives support for and aversion to God talk?

3 Support for and Opposition to God Talk

The research summarized above was not able to address *why* people would or would not be supportive of the use of God talk in the public sphere. However, theorists have offered many reasons why we should or should not use God talk. The first argument in favor of God talk is that this is what religious people actually believe, and to make them use arguments they do not believe in is not only tantamount to linguistic discrimination, it also results in a weaker democracy. As political scientist Hugh Heclo bluntly writes, insisting on secular public reasons "is telling religious people to shut up" (Heclo 2007:125).

A second argument among political and social theorists concerns people's ability to *understand* God talk. The origin of concerns about mutual understanding is that giving reasons that are accessible to all citizens is a cornerstone of liberal democracy and that these reasons can only be those "that others can reasonably be expected to endorse" which means avoiding "reasons based on religious doctrines" (Klemp 2010:5). As Robert Audi summarizes, "when there must be coercion, liberal democracies try to justify it in terms of considerations—such as public safety—that any rational adult citizen will find persuasive and can identify with" (Audi 1997:16). For theorists, a minimal precondition for persuasion is understanding the argument, and thus God talk is uncivil in the public sphere. Religion is thought to not be something you can understand or ascent to without believing in the transcendent claims.

In this literature, religious claims are problematic either because religion is transcendent, and thus is thought to engage more powerfully with people, or at least it is a comprehensive doctrine in the Rawlsian sense, and thus not universally held. But, critically, religious discourse and God talk are not just about God and not only about transcendent claims. God talk is part of a family of discourse, and it may well be as a representative of that family that leads to people supporting or not supporting its use.

3.1 Substantive and Formal Rational Discourse

The family of discourse that religion is in is substantively rational discourse. On an elite level, discourse about human genetic engineering and other biopolitical issues has been structured by a conflict between substantively rational discourse—which has largely been marginalized in public debates—and formally rational discourse. I take these terms from Weber via Habermas (1987, 1989). The basic insight in these two types of discourse is that a substantively rational argument about something like human genetic engineering concerns whether the means—the engineering technology—are *consistent with* the ultimate ends or values. That is, ends and means are debated as a piece. On the other hand, a formally rational argument asks whether the means maximize, compared to other possible means, predetermined ends. Elsewhere I have called these two forms of argumentation "thick" and "thin," respectively (Evans 2002).

What is most important about the characteristics of substantively and formally rational arguments in the elite debate about human genetic engineering is the calculability required in the link between means and ends. It is this calculability that drives the other features of this discourse. For the purposes of this chapter, there are six components of these arguments that are important: the link between means and end, the extent to which ends are debated, the number of ends considered legitimate in the debate, the commensurability of ends and the universality of ends, the ethical unit of analysis, and, related to all of these, the possibility of advocating more than a temporary moratorium on a technology.

First, the basic distinction raised by Weber and others regarding substantive and formal rationality is in the link between possible means and ends. Some thick forms of argumentation do not consider consequences, but rather evaluate a technology on whether it is consistent with the ends being pursued. That is, deontological arguments or arguments that a technology is inherently wrong are then substantive or "thick." Consequentialist reasoning is much more calculable and is typical of a formally rational link between means and ends. An example of this sort of argument was the late twentieth-century argument for germline genetic engineering, using the implicit and taken-for-granted end of maximizing human health, that "germ-line intervention is more efficient than repeating somatic cell gene therapy generation after generation" (Walters and Palmer 1997:81). Unlike substantive rationality, which does not even purport to calculate, by examining the predicted consequences of various means, the more efficient means can be calculated and selected.

Second, substantive and formally rational arguments are best distinguished by the extent to which the ends are considered to be a part of the debate. A substantively rational debate concerns not only the means—like biotechnology—but a debate about what goals we should be pursuing. In the Anglo-American elite debate about human genetic engineering, in the 1950s through the 1970s, people debated all sorts of goals that could be consistent with, or pursued with, genetic technology. Put simply, a substantively rational debate is about ends. In a formally rational debate, the ends are not debated, but are assumed. For example, in recent ethical declaration at the end of an international summit about the use of CRISPR-Cas9 in humans, essentially the only ends that were emphasized—but not even explicitly identified or defended—were the relief of suffering and the forwarding of reproductive autonomy.[6] Debating ends makes ethical debates less calculable because ends are incommensurable (Weber 1946:147).

Third, related to a debate about ends to distinguish the two types of arguments is the number of ends in the debate. As a debate becomes more "thin," the number of ends being considered decreases. For example, I earlier pointed out that what is now called bioethics, at least of the Anglophone variety, has been transformed into arguing whether technological means maximize four institutionalized, assumed ends (autonomy, beneficence, non-maleficence, and justice) (Evans 2012). An even more formally rational debate can be found among utilitarians, who only have one end. Similarly, a debate is more calculable and thus more formal or thin, if a range of ends can be commensurated into the limited ends in a debate. Again, utilitarianism has a commensurable end of the greatest good for the greatest number that a range of ends can be translated into "Thicker" ends resists this commensuration. Can "fidelity to God" as an end be commensurated to utility?

The fourth difference between thick and thin arguments is how universal the ends are considered to be. Formal thin debates assume universal ends—ones that are held by everybody—and try to exclude nonuniversal ends. Universal ends make calculation much easier because there is no need to account for disagreement. The most obvious example in contemporary biopolitical debates is the end or goal of the furtherance of human health and the relief of suffering. While this is probably close to being universally held, what is more pressing about these debates are the nonuniversal ends that are excluded, resulting in a very constricted debate, at least from a "thick" perspective.

The fifth difference is in the ethical unit of analysis: are the means thought to impact an individual or society? A typical substantively rational argument is that germline modification will change the way everyone in society views children by making them appear to be more like objects (Meilaender 2008:264). This argument is not very calculable because it is not clear how to represent society in these analyses. A typical formally rational argument is that germline modification will allow individual X to have a genetically related child that does not have a particular genetic disease. This is fairly calculable in that we could even talk with the individual in question, examine their interests and desires, and compare those with another individual.

---

[6]"On Human Gene Editing: International Summit Statement." Distributed at Summit, December 3, 2015.

The final difference is in the willingness to argue for a greater restriction on a technology than a temporary moratorium. Since in a substantive debate the means are often inherently wrong, independent of consequences, substantive arguments often end with a call for a permanent ban. However, in formally rational debate, no means is inherently wrong, only if it is not calculated to maximize the taken-for-granted ends. Since the ends and the nature of the means may change, technologies are only provisionally wrong and should be reassessed as more information comes in to help with the calculations.

3.2 *History of Debates About Biopolitics: Thick to Thin*

Let us then consider thick and thin (substantive and formal) to be two separate families of discourse in the public sphere. Social theorists have identified a general increase in formally rational (thin) arguments and a decline in substantively rational (thick) arguments in the public sphere. For example, Habermas states that institutions governed by formal rationality ("system") are colonizing the institutions governed by substantive rationality ("lifeworld") (Habermas 1987, 1989). The "lifeworld" is the location of free debate about ends, unencumbered by the biases introduced by formal rationality. Similarly, sociologist Robert Wuthnow concludes that public debates in advanced industrial societies are, due to social and cultural forces, increasingly limited to "questions of efficiency, practicality, cost-effectiveness, instrumental rationality, and expedience" (Wuthnow 1991:298).

Beyond this general trend, in earlier work, I found that the elite debate about human genetic engineering moved from thick to thin from the 1950s to the end of the twentieth century (Evans 2002). More to the point of this study, this also coincided with the decline of religious arguments in debates about human genetic engineering because the religious voices tended to be using substantively rational (thick) argumentation.

For example, viewing the ongoing thinning of the debate, theologian James Gustafson wrote in 1978 that he saw a typical debate in bioethics as "Should one cut the power source to a respirator for patient y whose circumstances are a, b and c? [which] is not utterly dissimilar to asking whether $8.20 an hour or $8.55 an hour ought to be paid to carpenter's helpers in Kansas City" (Gustafson 1978:387). That is, these debates had become calculative and thin, and thus not really subject to religious discourse. "It is not easy to give a clearly theological answer to a question that is formulated so that there are no theological aspects to it" he continued (Gustafson 1978:387).

Religious discourse, at least in the West, is more "prophetic," about moving people to action or setting goals or ends than about calculative arguments, so the theologians wanted to debate ends. Religious ends for at least contemporary religious speakers are not considered to be commensurable or universally held—the point is to convince others to adopt them. Religion is less about the individual and more about the collective. Religious discourse is more likely to be thick than thin.

3.3 Substantive Rationality Required to Oppose Biotechnology

In at least the American context, to oppose human biotechnology in the contemporary public sphere requires making substantively rational arguments. At present the public debate is extremely thin and is centered around the taken-for-granted and undebatable ends of autonomy, beneficence, non-maleficence, and justice (Evans 2012). How could someone oppose germline human genetic enhancement when those are the only acceptable ends? It is not really possible. If one end is autonomy, there is always going to be someone who wants to use the technology to enhance their children, so that end cannot be used to argue against germline enhancement. "Justice" is largely ignored in these debates and only is used to indicate that people without social power should not be experimented upon against their wishes. If raised at all, the argument would be that there needs to be equal access to enhancement. The real limits come from the ends of beneficence and non-maleficence, but this only stops germline modification if it is not considered to be safe and effective. Once it is safe and effective, these ends would not limit these means. The only arguments against germline enhancement that can be constructed using these ends are extremely weak, such as the claim that people who do not yet exist have not given their autonomous consent to be experimented upon.

So, to oppose germline enhancement, you need to introduce ends that are currently not allowed into the conversation. You need ends like species integrity or other substantive accounts of flourishing that are by definition not going to be shared. To introduce ends, you would need a substantive debate. Note that none of the arguments has to be religious—it is just that religious substantive arguments are some of the most prominent substantive arguments that exist in the public sphere.

Relatedly, to oppose germline enhancement, you would have to change your unit of analysis from the individual to society. When ethical debates are considered to be between individuals, they are clearly more calculable because we could even talk to the people with conflicting interests. Again, there are always going to be individuals who want to do something; the question is whether it is good for society, and an ethical debate with the individual as a unit of analysis cannot answer that question. To oppose germline enhancement, you would need an ethical analysis of the collective good. And, again, religious discourse is one of the few remaining public discourses about the collective good. There are of course "civic responsibility" and nationalism, but religion is not focused on what is good for the individual but rather that individual's obligations to something larger than themselves, be it God or society. Again, a social argument is necessary to oppose biotechnology, and religion is one of the sources of such arguments.

To return to the question that motivates this chapter: why would someone be supportive of God talk? In addition to reasons such as God talk being their natural language, I hypothesize that many people, including the nonreligious, like God talk because it expresses substantively rational arguments that they want expressed in the public sphere. Those people who want a substantively rational debate would not be averse to "God talk" because, even if they are not religious, it at least expresses substantively rational concerns. Those who do not want a substantively rational debate would not like God talk.

In sum, given that it is not possible to oppose human biotechnology with a thin, formally rational argument, I expect to find that those who are opposed to human enhancement will find God talk more acceptable than will those who are supportive of human enhancement, independent of the respondent's religious views.

4 Testing the Substantive Rationality Theory

Unfortunately, my survey lacks a question that measures "this respondent likes substantively rational discourse." I will instead use an imperfect measure, which is the extent to which someone supports or opposes "improving" the human species, again given that to make this argument someone would have to use a substantively rational argument. I will use additional analyses to eliminate alternative interpretations of this measure.

The measure of whether a respondent wants to improve the species through reproductive genetic technology combines two questions. The survey asked: "How morally acceptable do you personally believe each of the following is? (1) Creating a baby that is a clone of another person. (2) Using genetic engineering, that is, changing a person's DNA or genes, to create a baby that is smarter, stronger, or better looking." Respondents were asked to select a position on a five-point scale from "always morally acceptable" on one end to "always morally wrong" on the other. The answers to these two questions were highly related (Cronbach alpha = 0.809), suggesting they are indicators of a deeper idea that I will call "willingness to improve the species." Presumably, if you want to clone someone, you would clone someone with properties you desired. I placed each respondent to the survey on a scale of the extent to which they want to improve the human species through reproductive genetic technologies. This scale ranged from 2 to 10, with higher numbers indicating more opposition to improving the species. People who selected "I am not aware of this issue" or refused to answer either question were excluded from analyses, resulting in 3075 valid responses. Respondents were then split into "bioconservatives," the 55% of the respondents who scored 9–10 on the index, and the "bioliberals," the 45% who scored 2–8 on the index.

The first analysis is whether bioconservatives are opposed to God talk (compared to justifications based on science). Let us say that "averse" is defined as those who are "not very interested" or "not interested at all" in continuing the conversation with Robert. Analysis shows that 44.03% of bioconservatives are averse to God talk and 43.07% are averse to justifications based in science. This one-percentage-point difference is not statistically significant, leading to the conclusion that bioconservatives are not averse to God talk compared to other forms of talk. Analysis also shows that 49.76% of bioliberals are averse to God talk, but only 39.04% are averse to science talk. This 10% point gap is statistically significant, leading to the conclusion that bioliberals are averse to God talk.[7] Or, put differently, bioconservatives are more supportive of God talk than are bioliberals.

This is of course not surprising, and a simple correlation also shows that the bioliberals are disproportionately the nonreligious, and I have already shown that the nonreligious are averse to God talk. To support my hypothesis and argue against the idea that bioconservative lack of aversion is simply the result of their religiosity, I must move analysis into a more complicated statistical framework.

The first column of Table 6.1 shows an ordered logistic regression model predicting the degree of aversion of a respondent to a continued conservation. Higher values of the dependent variable indicate greater aversion. The "God talk" variable is positive and statistically significant, replicating the result I discussed above that this is a small aversion to continuing a conversation with Robert if he is using God talk instead of science to justify his claims. The enhancement index is not significant, which simply means that a respondent's stance on biopolitics does not determine their willingness to have political conversations in general.

I created an interaction variable with the dichotomous "Robert was using God talk" variable and the enhancement index. This variable is included in the model in the second numeric column. This is negative and significant, which means that respondents who were more bioconservative, and who saw God talk instead of science talk, were less averse to a conversation with Robert. This establishes the basic finding that bioconservatives like God talk in a debate more than do bioliberals. But is this simply due to the nonreligious being averse to God talk *and* being less likely to be bioconservatives?

To rule this out, I control for the religious affiliation of the respondent. To do this, I must also control for a number of variables that are associated with religiosity such as education, gender, and ethnicity. I created a series of dichotomous variables indicating affiliation and frequent participation in the dominant religious traditions in the United States, with the reference group being those who claimed their religious tradition to be "not religious/none," "agnostic," and "atheist" or who refused to answer the initial religion classification question. This is a total of 18.0% of the US population, a figure roughly consistent with other contemporary measures (Hout and Fischer 2014). For details of this coding, see the earlier paper (Evans 2018).

The third column shows the results of controlling for demographics and the religious affiliation and participation of the respondent. Most notably, the interaction effect is unchanged from the previous model. This means that bioconservatives, independent of their religious identity and practice, are less averse to God talk than are the bioliberals.

[7]For tests of statistical significance, I used all of the ordered categories of the variable, which provides for a more general test, and I calculated a measure of categorical association named Spearman's Rho. For the bioconservative analysis, the association between the five-point "continue discussion" measure and the dichotomous "saw the God talk version of the vignette" variable, with analysis limited to bioconservatives, was −0.018, which has a *p*-value of 0.460. For the bioliberal analysis, the Rho was 0.108, which was significant with a *p*-value of 0.000.

Table 6.1 Ordered logistic regression coefficients, predictors of aversion to God talk

Variables	(1) Aversion to God talk	(2) Aversion to God talk	(3) Aversion to God talk	(4) Aversion to God talk
Enhancement index	−0.0287	0.0311	0.0379	0.174**
	(0.0212)	(0.0289)	(0.0312)	(0.0652)
Robert is using God talk	0.160*	1.118**	1.113**	2.267***
	(0.0790)	(0.354)	(0.363)	(0.667)
Enhancement X using God talk		−0.117**	−0.115**	−0.229**
		(0.0419)	(0.0429)	(0.0878)
Bachelor's degree or greater			0.0106	−0.205
			(0.0925)	(0.211)
Has some college, but no degree			−0.001	−0.0620
			(0.104)	(0.246)
Age			0	−0.0056
			(0.0024)	(0.0056)
Woman			0.181*	0.358
			(0.0819)	(0.187)
African American			−0.252	−0.291
			(0.161)	(0.524)
Hispanic			−0.0030	0.171
			(0.158)	(0.365)
Other ethnicity			−0.275*	−0.406
			(0.128)	(0.236)
Biblical literalist conserv prot. attender			−0.641**	
			(0.200)	
Non-literalist conserv prot. attender			−0.760***	
			(0.155)	
Mainline protestant attenders			−0.374*	
			(0.153)	
Catholic attenders			−0.515***	
			(0.156)	
Other Christian attenders			−0.803***	
			(0.228)	
Low attending with Christian identity			−0.181	
			(0.118)	
Religious minorities			−0.591**	
			(0.197)	
Constant cut1	−2.251***	−1.766***	−2.049***	−1.091*
	(0.191)	(0.250)	(0.307)	(0.554)
Constant cut2	−1.126***	−0.638**	−0.906**	−0.0544
	(0.185)	(0.245)	(0.303)	(0.566)
Constant cut3	0.129	0.620*	0.378	1.339*
	(0.184)	(0.245)	(0.302)	(0.587)
Constant cut4	1.254***	1.749***	1.527***	2.287***
	(0.189)	(0.248)	(0.305)	(0.593)
Observations	3154	3154	3154	562

Robust standard errors in parentheses
$^{***}p < 0.001$, $^{**}p < 0.01$, $^{*}p < 0.05$

The fourth column provides a stronger test, where I only analyzed the nonreligious respondents. We see that among the nonreligious, those who are more bioconservative and who see the vignette where Robert is using religious reasons are more likely to want to continue the conversation than are those who saw the scientific justifications. Why would nonreligious bioconservatives like God talk? Evidence suggests that they like the thick discourse found in God talk, even if they are not religious themselves.

5 Conclusion

The American public has a statistically significant but substantively insignificant aversion to the use of God talk in the public sphere. The assumption of the political theory literature on God talk is that people are opposed to God talk because of its religious, transcendent, and particularistic qualities. I suggest another, additional explanation, which is that God talk is part of a broader family of substantively rational or "thick" discourse and certain people want substantively rational discourse in the public sphere—in whatever form it takes. I further make the case that at least in the US public sphere, it is not possible to oppose human biotechnology using a formally rational or "thin" argument. Therefore, if this is the case, those who are opposed to human biotechnology should be more in favor of substantively rational "thick" arguments.

Returning to the survey data, I show that those who are the more opposed to human genetic enhancement are more supportive of God talk. But, this could be because the religious respondents both like God talk for religious reasons and are also opposed to enhancement. To rule out this possibility, I control for the religion of the respondent and also conduct a separate model with only the nonreligious. Both show the same result that independent of religion, those opposed to human enhancement are more favorable toward God talk.

Of course, the data used in this contribution are not definitive because they do not directly measure the phenomena in question. I believe that it is suggestive enough that other researchers should consider the possibility that God talk is not just about God.

References

Audi, Robert. 1997. Liberal democracy and the place of religion in politics. In *Religion in the public square*, ed. R. Audi and N. Wolterstorff, 1–66. Lanham, MD: Rowman and Littlefield.

Elshtain, Jean Bethke. 2001. Faith of our fathers and mothers: religious belief and American democracy. In *Religion in American public life*, ed. A.Y. al-Hibri, J.B. Elshtain, and C.C. Haynes, 39–61. New York, NY: W.W. Norton.

Evans, John H. 2002. *Playing god? Human genetic engineering and the rationalization of public bioethical debate*. Chicago, IL: University of Chicago Press.

———. 2012. *The history and future of bioethics: a sociological view*. New York, NY: Oxford University Press.
———. 2018. Aversion to and understanding of god talk in the public sphere: a survey experiment. *Journal for the Scientific Study of Religion* 56 (3): 459–480.
Gustafson, James M. 1978. Theology confronts technology and the life sciences. *Commonweal* 105: 386–392.
Habermas, Jürgen. 1987. *The theory of communicative action: lifeworld and system: a critique of functionalist reason*. Vol. 2. Boston: Beacon Press.
———. 1989. *The structural transformation of the public sphere*. Cambridge, MA: MIT Press.
Heclo, Hugh. 2007. *Christianity and American democracy*. Cambridge, MA: Harvard University Press.
Hout, Michael, and Claude S. Fischer. 2014. Explaining why more Americans have no religious preference: political backlash and generational succession, 1987-2012. *Sociological Science* 1: 423–447.
Isaac, Jeffrey C., Matthew F. Filner, and Jason C. Bivins. 1999. American democracy and the new Christian right: a critique of apolitical liberalism. In *Democracy's edges*, ed. I. Shapiro and C. Hacker-Cordon, 222–264. New York: Cambridge University Press.
Klemp, Nathaniel. 2010. The Christian right: engaged citizens or theocratic crusaders? *Politics and Religion* 3: 1–27.
McCarthy, Thomas. 1994. Kantian constructivism and reconstructivism: Rawls and Habermas in dialogue. *Ethics* 105: 44–63.
Meilaender, Gilbert. 2008. Human dignity: exploring and explicating the council's vision. In *Human dignity and bioethics*, 253–277. Washington, DC: President's Council on Bioethics.
Mutz, Diana C. 2011. *Population-based survey experiments*. Princeton, NJ: Princeton University Press.
Rawls, John. 1993. *Political liberalism*. New York, NY: Columbia University Press.
———. 2001. *Justice as fairness: a restatement*. Cambridge, MA: Harvard University Press.
Rorty, Richard. 2003. Religion in the Public Square: a reconsideration. *Journal of Religious Ethics* 31 (1): 141–149.
Shields, Jon A. 2009. *The democratic virtues of the Christian right*. Princeton, NJ: Princeton University Press.
Smith, Rogers M. 2008. Religious rhetoric and the ethics of public discourse. *Political Theory* 36 (2): 272–300.
Stout, Jeffrey. 2004. *Democracy and tradition*. Princeton, NJ: Princeton University Press.
Walters, LeRoy, and Julie Gage Palmer. 1997. *The ethics of human gene therapy*. New York, NY: Oxford University Press.
Weber, Max. 1946. *From Max Weber: essays in sociology* (trans: Gerth, H. and Mills, C.W.). New York, NY: Oxford University Press.
Wolterstorff, Nicholas. 1997. The role of religion in decision and discussion of political issues. In *Religion in the Public Square*, ed. R. Audi and N. Wolterstorff, 67–120. Lanham, MD: Rowman and Littlefield.
Wuthnow, Robert. 1991. *Between states and markets: the voluntary sector in comparative perspective*. Princeton, NJ: Princeton University Press.

Chapter 7
The Political Debate on Embryo Research in New Zealand and the Role of Religious Actors and Arguments

David Gareth Jones

Abstract Embryo research is currently not allowed in New Zealand, although it is not prohibited by the Human Assisted Reproductive Technology (HART) Act. The unusualness of this legislative stance is accentuated by the ready availability of in vitro fertilization (IVF) and pre-implantation genetic diagnosis (PGD), both of which are deemed to be established procedures. Embryo research could, in principle, be approved by the Minister of Health acting on a recommendation by the Advisory Committee on Assisted Reproductive Technology (ACART). No Minister of Health has as yet accepted such a recommendation from ACART. Before making such a recommendation, ACART has to consult widely and has to take account of religious and cultural (Māori) considerations. Religious input has been diverse and has not pointed unequivocally either for or against embryo research, although a large number of submissions to ACART objected to any destruction of embryos, with some of this input opposing IVF and PGD as well as embryo research. Unfortunately, many submissions were little more than assertions that could not readily be integrated into ACART's decision-making processes. A dialogue model undertaken by Māori groups could provide a productive way forward by which religious and cultural voices could make themselves better heard in a pluralist society. However, this will only work if those with religious perspectives are prepared to accept that there is more than one religious viewpoint on embryo research.

1 Introduction

As a broad generalization, it can be stated that the standard of ethical debate within New Zealand society is not high. While some would undoubtedly contend that this is a debatable proposition, there are limited outlets for informed debate, especially within the biomedical area. This may in part be due to the politicization of public debate, leading to the adoption of polarized positions. Intense disquiet in the 1990s led to the establishment of a Royal Commission on Genetic Modification over 2000

D. G. Jones (✉)
Department of Anatomy, University of Otago, Dunedin, New Zealand
e-mail: gareth.jones@otago.ac.nz

M. Weiberg-Salzmann, U. Willems (eds.), *Religion and Biopolitics*,
https://doi.org/10.1007/978-3-030-14580-4_7

and 2001 (Report of the Royal Commission on Genetic Modification 2001). Among its recommendations was the establishment in 2002 of *Toi Te Taiao: The Bioethics Council*, 'To enhance New Zealand's understanding of the cultural, ethical and spiritual aspects of biotechnology and ensure that the use of biotechnology has regard for the values held by New Zealanders' (New Zealand Government 2002). In its 7-year existence, it undertook analyses of a range of topics, including prebirth testing with its report, *Who Gets Born?* (The Bioethics Council 2008). Housed within the Ministry of the Environment, and set up as a body to provide the government with independent advice, its guidelines and recommendations were not binding on the government, and in 2009 it was disbanded in order, it was claimed, to save money and because other government agencies were doing the same work.

It is notable that the Council's terms of reference were intended to take account of the 'cultural, ethical and spiritual aspects of biotechnology'. This was evident in the subtitle of the *Who Gets Born?* report, leading to the expectation that spiritual aspects of prebirth testing would be seriously incorporated into decision-making. While the recommendations of this report are not of direct relevance in the present context, it is relevant to enquire to what extent the spiritual aspects are manifested in the discussion of pre-implantation genetic diagnosis (PGD) and sex selection for social reasons. Although the Council adopted the methodology of deliberative consultation, and did consult widely, their four-view approach ('my choice, my right'; 'life is a gift'; 'tangata whenua'; and 'information, knowledge and public involvement') may have tended to minimize any potential religious input. Further, the relation between consultation and recommendations was obtuse (Jones et al. 2009), with the liberal presumption underlying the major recommendations suggesting that little note had been taken of either religious or disabilities groups. Consequently, one has to conclude that cultural and spiritual input had little sway in shaping the ultimate trajectory of the report.

This result is not surprising since the stipulation regarding 'cultural and spiritual aspects of biotechnology' is not specified, and as pointed out by one organization, these dimensions even increase the difficulties of ethical decision-making (NZORD 2004). However, in that instance, it was concluded that explicit acknowledgement of them highlights their potential role as important factors in ethical deliberations. Nevertheless, the lack of discernible criteria allows decision-makers to downplay or even ignore religious stances. These limitations and constraints apply with equal force to the legislation governing human reproduction, namely, the Human Assisted Reproductive Technology (HART) Act, with its oversight of embryo research.

2 Historical Relationship of Religion and Politics in New Zealand

Before proceeding with a description of the HART Act, it may well be useful to sketch briefly the relationship between religion and politics in New Zealand, a country with European origins. In the case of New Zealand, church and state have always been separate. As early as 1854, the first New Zealand Parliament enshrined what it described as 'a perfect political equity in all religious denominations' and sharply separated church and state. In line with this, the Education Act of 1877 described a nationwide system of free, compulsory and secular primary schools, and in a desire to secure harmony, religion was kept private. Notwithstanding these developments, religion has played an important role throughout society. For instance, until the 1960s, most people were religious, and the first Labour government (1935–1949) considered the introduction of the welfare state during its term of office as an outworking of applied Christianity (Stenhouse 2011).

A crucial factor in New Zealand society has been Māori Christianity, and until the mid-twentieth century, few Māori were secular. Some prophetic Māori religious movements developed with a strong focus on resisting the loss of Māori land, thereby moving them into the centre of political debate. Between the mid-1830s and early 1860s, there were serious conflicts over Māori rights, and it was at this time that a range of Christian missionaries (and Māori Christians) led the humanitarian campaign to uphold both these rights and the welfare of Māori. Other causes in which Christian groups were prominent emerged between the 1870s and 1930s, when a powerful evangelical coalition campaigned for Sabbath observance, votes for women and prohibition. More recently, religious influence has been felt more through the religious persuasion of individuals, such as leading politicians, and these have been able to exert a noteworthy influence on policies relating to abortion and homosexuality. Far more widely recognized and appreciated has been the influence of church-based social service agencies in filling gaps in the welfare state and also in advocating for many deprived groups in the community.

3 Human Assisted Reproductive Technology (HART) Act

In New Zealand, the context within which discussion of embryo research is framed is provided by the 2004 Human Assisted Reproductive Technology (HART) Act (HART Act 2004). This Act was intended to provide a 'robust and flexible framework for regulating and guiding the performance of artificial reproductive technologies (ARTs) and the conduct of human reproductive research'. The goal is to 'secure the benefits of these procedures for individuals and society in general' while protecting and promoting 'the health, safety, dignity, and rights of all individuals, but particularly those of women and children, in the use of these procedures and research' (s3(a)). It also seeks to 'prohibit unacceptable assisted reproductive procedures and unacceptable human reproductive research' (s3(b)).

To this end, the HART Act provides a legislative framework that prohibits some procedures, 'establishes' others and requires that others be regulated, that is, subject to ethical review. The prohibited actions include cloning, implanting a human or hybrid embryo, selecting an embryo's sex for non-medical reasons, genetic modification of gametes or embryos, obtaining gametes from a person under 16 years of age and selling or buying human embryos or gametes. These are proscribed by statute as criminal offences (s3(c), (e); schedule 1, cls 10–13; s8(4)).

The 'established' procedures (s5) are those considered to be routine procedures, including artificial insemination, in vitro fertilization (IVF), intracytoplasmic sperm injection (ICSI), PGD, collection of eggs or sperm for donation and egg, sperm and embryo cryopreservation (Part 1 of the Schedule to the HART Order 2005). Research on, and utilizing, human embryos fits into neither of these categories. Instead, the Advisory Committee on Assisted Reproductive Technology (ACART) must provide the Minister of Health with information, advice and, if it thinks fit, recommendations on the use of gametes and embryos in human reproductive research (s37(1)). While various pieces of research are referred to in the Act, including cloned embryos, donation of embryos and genetic modification of embryos, there is no mention of research per se in order to obtain additional information about, say, embryonic development. Until the Minister approves any of this research, embryo research cannot be classed as a regulated procedure. It remains in limbo.

Regulated procedures are actions that are restricted but are potentially permissible and are subject to ethical review. This is provided by the Ethics Committee on Assisted Reproductive Technology (ECART), the other committee established by the HART Act. However, ECART can only give approval if a guideline has previously been issued by ACART. ECART has to determine that the procedure and its specific application is consistent with the guidelines (s19(2)).

This tripartite arrangement of prohibition, establishment and regulation, forms a regulatory spectrum with the decision as to which category an ART falls into being a political one. Parliament determines whether or not an ART is a prohibited action or an established procedure by statute. Procedures falling into the regulated activities category are considered by Cleary (2013) to represent political indecision. By not classifying a procedure as either prohibited or established, 'Parliament and the Government are indirectly deciding that the performance of the ART should be controlled by ACART through the regulatory mechanism set out in the HART Act 2004' (Cleary 2013, p. 5).

The Minister of Health appoints members to both ACART and ECART, raising the possibility of political preference. ACART advises the Minister of Health on policy matters, as well as issuing guidelines and advice to ECART. In the absence of appropriate guidelines, ECART is unable to consider applications, and the procedures cannot proceed. While such an application has to be declined, ECART can refer it to ACART to determine whether it would be appropriate to formulate guidelines. In this situation, ACART has to operate according to the provisions in the

HART Act and make decisions 'against a background of existing legislation in related areas' (Jones and Whitaker 2012). These include abortion, IVF programmes with the production and eventual destruction of embryos and the protection of children.

The size and composition of the two committees central to the working of the HART Act are stipulated in the Act. Members' expertise has to cover assisted reproductive procedures, human reproductive research, ethics, Māori customary values and practice, consumer and disability perspectives, the interests of children and legal matters. There is no member stipulated to have special expertise in representing any religious voices or values within the community. ACART has no public educational role and is not resourced to carry out this role; its principal function is that of acting as a public policy-making body (Cleary 2013). Besides developing, maintaining and issuing guidelines to ECART, ACART can advise the Minister of Health if it believes that an existing established procedure should be modified or should cease to be an established procedure (s35(1)(b)(iii)). ACART can also advise the Minister if it believes that a procedure, treatment or an application of a treatment or procedure should be declared an established procedure (s6). In all instances, the final decision rests with the Minister.

4 Ethical Principles Stated in the HART Act

In addition to the explicit restrictions and prohibitions, the HART Act lists in section 4 the seven principles that are to guide all persons, including ACART members, who exercise powers or perform functions under the Act. The principles are:

(a) The health and well-being of children born as a result of the performance of an assisted reproductive procedure or an established procedure should be an important consideration in all decisions about that procedure.
(b) The human health, safety and dignity of present and future generations should be preserved and promoted.
(c) While all persons are affected by assisted reproductive procedures and established procedures, women, more than men, are directly and significantly affected by their application, and the health and well-being of women must be protected in the use of these procedures.
(d) No assisted reproductive procedure should be performed on an individual, and no human reproductive research should be conducted on an individual unless the individual has made an informed choice and given informed consent.
(e) Donor offspring should be made aware of their genetic origins and be able to access information about those origins.
(f) The needs, values and beliefs of Māori should be considered and treated with respect.
(g) The different ethical, spiritual and cultural perspectives in society should be considered and treated with respect.

These principles provide the most normative content for ethical decision-making found in the HART Act and are repeatedly referred to as the ethical underpinnings for its policy-making. However, they fail to provide substantive guidance on account of their vagueness. This observation is pertinent since ACART has to guide society through challenging ethical waters against a background of international best practice, New Zealand legislation, public concerns and attitudes and political perceptions. Since ACART is guided by what is described in the HART Act as 'a robust and flexible framework' for regulating ARTs and reproductive research, it has to take account of public opinion via public consultation. Unfortunately, there is no assurance that the submissions received are representative of the public at large nor that they contain overt ethical and/or religious reasoning. For instance, many of the submissions received on embryo research simply asserted one viewpoint or another in the absence of any explicit reasoning. ACART finds itself in an invidious position as it attempts to decide the weight to be placed on submissions of this ilk, and it is this that leads to criticisms by interested groups that their viewpoints have been overlooked.

The Act recognizes it may not be possible to reconcile the differing perspectives within society and so requires ACART to explore these perspectives in a conscientious and meaningful way. If moral pluralism is to be accommodated, viewpoints, including those representing religious voices based on non-shared beliefs, should not be allowed to dominate, although by the same token they should be given a fair hearing. Adopting an approach of cautious liberalism enables conflicting sides of a debate to pursue their differing ideas of what is and is not acceptable while accommodating their differences as much as feasible (Snelling et al. 2007). The end result is to secure the benefits of advances in assisted reproduction within a protective framework (s3(a)). In practice, however, it has to be queried whether these ideals are functioning in this manner, since the Minister of Health is not obliged to accept ACART's advice should it be contrary to the Minister's (and/or government's) preferred position (Jones 2014). The case of embryo research stands out as having proved intensely problematic from the perspective of recent governments.

4.1 Finding a Place for Embryos

While many important ethical values are encompassed by these principles, it is surprising that the HART Act contains no direct reference to the moral worth to be ascribed to human embryos (Jones and Whitaker 2012; Jones 2014). The centrality of embryos in so many of the ART-related debates highlights this omission, no more so than when the focus is on the potential use of embryos for research purposes. In spite of this omission, there are indirect allusions to embryos elsewhere in the Act, and these raise the possibility that developing humans, particularly early embryos, are not being given rights and protections equivalent to those given to children or adults (Jones and Whitaker 2012). Although embryo research is not allowed without future ministerial approval, the HART Act alludes to embryos in a number of places.

In vitro embryos cannot be allowed to develop beyond 14 days gestation. Theoretically, they may be used in human reproductive research (subject to prior approval by ECART, s16, s19(b)), imported and exported, and developed outside the womb, subject in all cases to guidelines and comprehensive ethical oversight and approval by the Minister of Health. The HART Act also restricts the maximum time of storage (ACART 2012). As it is extremely unlikely that all stored embryos will be implanted and thereby given a chance to become future individuals, this provision of the HART Act implicitly requires the destruction of human embryos (Akoorie 2014).

On the other hand, embryos appear to be given a special status warranting their protection. For instance, the Human Tissue Act (2008) notes that 'A human embryo or human gamete is not human tissue for the purposes of any provision of this Act', pointing to a protected status. Nevertheless, this does not amount to human rights as afforded to those who have been born nor the legal protections afforded to the foetus. It would appear then that embryos are being given an intermediate status that one might expect would allow limited embryo research to be approved. But this is not the case at present.

4.2 Place of Religious Input

The HART Act instructs ACART to take into consideration consultation with the public when formulating its guidelines (s41(1), s36(1)). However, since individuals and communities hold a range of ethical views, this instruction has to take account of the considerable diversity of opinion that exists in relation to assisted reproduction within society. While differing perspectives may not be reconcilable, the Act requires the exploration of these perspectives in a conscientious and meaningful way. If moral pluralism is to be accommodated, it can be argued that viewpoints based on non-shared beliefs should not be allowed to dominate. As pointed out by Snelling et al. (2007) cautious liberalism seeeks to accommodate the conflicting sides of a debate with their opposing perspectives, without unduly restricting their visions of the good life. However, in no way does it diminish the underlying purpose of the Act, namely, to secure the benefits of advances in assisted reproduction within a protective framework (3(a)) (Jones and Whitaker 2012).

The question is whether this approach is able to accommodate religious perspectives emanating from a religiously based framework that may be at odds with dominant secular ones that pay little, if any, attention to fundamental religious assumptions. The tenuous status of religious input was encountered in the report of the Bioethics Council on prenatal testing of embryos and foetuses. The HART Act once more recognizes a place for this input (s4(g)), along with considering the needs, values and beliefs of Māori (s4(f)), although there is once again lack of clarity surrounding their significance. In exploring this matter, Cleary (2013) contends that since the s4 principles are all worded very broadly and are prima facie equal, it is left to ACART to decide how to resolve conflict between the principles. This means it has to attach different weighting to the individual principles (pp. 8, 9). This includes the principles that are to take account of the needs, values and beliefs of Māori and the different ethical, spiritual and cultural perspectives in society. These positions and perspectives are to be treated with respect, leading to the view that ACART should attempt to understand them and elicit what influence they should have on its decision-making. This, in turn, places an onus on submitters to demonstrate the reasoning and therefore the assumptions, values and beliefs, underlying their position. It is not adequate to simply state agreement with, or opposition to, a proposal. An inability to make explicit why a particular spiritual/religious perspective should be taken into account places that stance at considerable disadvantage when ACART comes to assess its impact alongside that of competing views.

This is of considerable relevance when assessing ethical and religious issues surrounding such a highly contested area as embryo research. A failure to elaborate how religious insights throw light on embryo research militates against the rationale for having s4g as one of the principles of the HART Act. One way of partially alleviating this impasse would be if ACART were to be more active in obtaining the views of groups putting forward religious viewpoints (Cleary 2013). However, it would still be up to ACART to determine the weight to be given to these perspectives and the normative claims upon which they are based, for example, opposition to the destruction of embryos on the ground that embryos are human beings of equal moral value to adults and therefore should never be destroyed. While there is no simple way of resolving this ethical conflict, dual transparency is a sine qua non: accessible reasoning behind the religious basis for an advocated position and informed and unprejudiced decision-making by ACART. Both are required.

4.3 *Māori Voices*

The input of the Māori community into New Zealand responses to biotechnological issues is paramount, since Māori are the first people of the land. However, their tradition lacks a position on modern developments within the ARTs, including embryo research. These have to be formulated by the Māori community, and it is these attempts at such formulation that constitute their major contributions to the literature.

While Māori approaches to embryo research do not fit neatly under the banner of religious approaches, their cultural perspectives have religious overtones. They are also instructive in providing a model for how religious bodies with their traditional viewpoints might approach previously unexplored biotechnologies.

A ground-breaking study exploring Māori attitudes to the ARTs set out a series of recommendations regarding further research needed on infertility among Māori and implications for fertility services, health professionals and the New Zealand government (Glover et al. 2007; Glover 2008). In a related study, Glover and Rousseau (2007) spelled out what they viewed as the relevance of these technological developments for Māori concepts of *whakapapa* (ancestry) and for concepts of relatedness and reproduction within Māori communities.

A more directly focused dialogue on embryo research had the aim of investigating the cultural concepts relevant to this research (Hudson et al. 2010). In looking at the epistemological divide between Māori knowledge and science, this study consisted of a dialogue between ten Māori scientists and ten experts in *marauranga* Māori (Māori knowledge). These two groups emerged with similar themes, related to the identification of a purpose, relationship to existing knowledge, consequential spiritual and ethical implications and ability to participate in decision-making. Against this background, the cultural implications of embryo research were explored in relation to metaphysical, moral, social and technological dimensions.

Some of the major points to emerge in the metaphysical dimension were the interconnectedness of life, and the manifestation of spiritual imperatives through biological functions and physical forms, and protection of the unborn. In the moral area, differing values within Māori were acknowledged, but ethical concepts having broad agreement included do no harm, the potential for good, the *mana* (authority) of the embryo, *tapu* (sacred, sanctity), *mauri* (life essence) and the *tapu-utu-ea* (sanctity, balance, restoration) test. The social dimension stressed the importance of who decides (e.g. Māori or non-Māori, individuals or their *iwi* (tribe)) what gets ethical approval, since this was viewed as being as important as the decision itself. As part of the technological dimension, emphasis was placed on who defines the knowledge that is considered relevant for embryo research. In this context, the centrality of trust and listening to oral approaches emerged as vital.

The discussion on embryo research elicited neither blanket objections nor unqualified approval, accepting that this kind of decision-making will always be contentious, with Māori views varying as widely as those among the general public. But the significance of the spiritual world and its close integration with the material world emerged repeatedly as crucial to any Māori cultural construct, a link that requires further thought and elaboration. The place of *whanau* (family) in Māori decision-making points away from the frequently encountered stark individualism in much secular ethical decision-making.

Hence, while on the surface there are similarities with much general ethical thinking, this should not obscure the subtle differences. Ways of arriving at viewpoints on embryo research depend upon dialogue, and this is seen as a vehicle for the evolution of indigenous theory (Hudson et al. 2010). This is an attempt to take account of concerns and queries from a broad cross section of the community, paying attention to spiritual and ethical values but also asking how technological developments may benefit or harm both individuals and their *whanau*. Hudson et al. (2010) write: 'When dealing with new knowledge people have to figure out how it relates to and coheres with their existing knowledge' (p. 64). There has to be engagement with, and empowerment of, the indigenous culture, through the availability of 'negotiating spaces' to share ideas, concepts and values and to make room for innovative thinking. Primacy is not given to the technological possibilities, but neither is it dismissed.

4.4 Global Context

Regulations governing embryo research are generally considered in relation to the ability or otherwise to extract human embryonic stem cells (hESCs) from embryos, although this is not the sole reason for embryo research. These regulations have been divided into four dominant positions designated A to D by Towns and Jones (2006). Position A encompasses countries that prohibit all embryo research and therefore the extraction of hESCs. Position B confines the use of embryonic stem cells to those currently in existence, in that they were extracted prior to a specified date, thereby prohibiting the extraction of hESCs and utilization of hESCs derived in the future. Position C allows for the use and ongoing isolation of hESCs from surplus IVF embryos from IVF programmes. Position D allows the creation of human embryos specifically for research.

The Hinxton Group (2006) (An International Consortium on Stem Cells, Ethics and Law) again identified four groups: prohibitive (equivalent to A), restrictive compromise (B), permissive compromise (C) and permissive (D). The classification adopted by the European Science Foundation (European Science Foundation 2013) is similar but omits a position B equivalent. The groups are very restrictive (corresponding to A), permissive (C) and very permissive (D), with further categories of restrictions by default (where legislation is not explicit but national practices are quite restrictive in practice) and unlegislated (where there is no legislation on hESCs).

In assessing where New Zealand fits into these categories, Jones (2014) concluded that it falls into the 'restrictive by default' category on the basis of the 2012/2013 annual report of ACART (2013). In the section on advice to the Minister of Health on human reproductive research, it states:

Section 37 of the HART Act requires ACART to provide the Minister of Health with information, advice and, if it thinks fit, recommendations on certain matters in relation to the use of gametes and embryos in human reproductive research.

In June 2007 ACART provided the then Minister of Health with advice on human reproductive research following extensive public consultation in the 2006/2007 financial year. At the request of the Minister, ACART has not undertaken any work to develop guidelines or further advice. "The current *Guidelines for Research on Gametes and Non-Viable Embryos*, developed by the former National Ethics Committee on Assisted Human Reproduction, remain in force". (These allow research on sperm and eggs and also on embryos lacking the potential to develop into a foetus due to arrested growth, defects of their cells or other abnormalities.)

ACART is required to monitor developments in human reproductive research by Section 35(2) of the HART Act 2004 and so is fully aware of the human reproductive research currently approved in Australia and the UK. It also emerges from the above that in 2007 ACART recommended to the Minister that embryo research should be permitted in New Zealand. No approval was forthcoming.

The situation has continued in this vein, and in its more recent annual reports (2014/2015), ACART (2015a, b) stated that the present situation effectively limits any human reproductive research in New Zealand and that the current pre-HART guidelines are well overdue for revision. These matters had been discussed with the Associate Minister of Health in March 2015, but until a positive response was received from the Minister, it could only continue to monitor human reproductive research, including statutory requirements in other jurisdictions. Around this time ACART (2015a, b) provided a briefing report for the incoming Minister of Health outlining the current situation regarding the vacuum surrounding embryo research. In this it pointed to a lack of definition of the term 'viable' in the guidelines and that some embryos deemed 'non-viable' may in fact be viable. It also pointed to misleading wording of the guidelines with regard to viability and non-viability and that applications have to be turned down for work on embryos even when they will not be harmed in any way. Even studies comparing the media used in IVF have to be declined. In this briefing ACART (2015a, b) advocated the use of viable embryos since this would allow research to be conducted on three broad areas: improvements in embryo culture techniques, improvements in cryopreservation techniques and the development of ESCs.

In February 2016 ACART decided to draft a preliminary scoping document that would set out its views about the scope of human reproductive research, as a basis for deciding what would be in and out of scope in the potential guidelines (ACART 2016a). An update in June 2016 outlined how and why ACART's guidelines on human reproductive research should be amended (ACART 2016b). There is neither indication that religious perspectives were taken into account nor that they were considered to be an obstacle to an acceptance of embryo research.

At no point has ACART been given reasons for this lack of progress. In 2008, the then Minister of Health indicated that he had rejected ACART's advice to allow the use of embryos surplus to those required by people in IVF clinical programmes. It was further intimated verbally to the Committee that this rejection of ACART's advice was a political judgement. At that time ACART requested that it be allowed to publish its advice to the Minister, but that again was rejected. Consequently, ACART's reasoning in favour of allowing circumscribed use of embryos for research purposes has never been available for general discussion within the community. The overall tenor of correspondence with the Minister was one of prohibition. In response to my article (Jones 2014) critiquing the inconsistency of policy in the reproductive area in New Zealand, the Minister of Health's response was that he did not think there was public consensus on the matter and so would not be taking it any further.

It is unfortunate that the end result has been to bury discussion about embryo research, let alone public discussion of it. This is contrary to the intent of the HART Act, which is that the Governor General can place an 18-month moratorium on human reproductive research if the Minister is unhappy with ACART advice, with the possibility of a second period of similar length (s24). At the conclusion of these periods, ACART should be in a position to consult with the community again and provide the Minister with further advice. At the very least, this process keeps the issue alive in the minds of the community. In its absence, there can be no progress and no discussion of core issues within the debate.

Interfering with this process is of major significance since research on human embryos is not prohibited, but neither is it an established procedure nor is it one requiring ECART approval. It is in no man's land until a decision is taken to place it in one of these categories. Unwillingness to do so is a protective political stance, akin to abstaining from voting. The politicians with authority on this matter have sat on the fence, neither favouring nor objecting to embryo destruction. In practice, the end result is anti-research, since research on viable human embryos (as opposed to what are thought to be non-viable ones) is in limbo depending upon agreement of the Minister of Health to the positive recommendation from ACART to proceed, hence the designation 'restrictive by default'.

This satisfies no one (Jones 2014). Those favouring a very restrictive regime cannot be assured that this will remain the case, since the government is not ostensibly supporting this position. In the submissions made to the ACART discussion document, *Use of gametes and embryos in human reproductive research* (ACART 2006), individuals and groups with this viewpoint expressed 'opposition to the use of embryos in research on the grounds that they are human life and any manipulation, including in vitro fertilization (IVF) and pre-implantation genetic diagnosis (PGD) is akin to harming or killing a person' (ACART 2007). This goes well beyond the HART Act according to which IVF and PGD are established procedures.

Additionally, some responders to the ACART consultation went further and expressed 'opposition to the use of gametes in research on the grounds that they are human life'. These responses point to either a very extreme position in which research should never be carried out on sperm or eggs or signify confusion between gametes and embryos. A wall of silence on embryo research fails to answer the concerns of this constituency, with its opposition to IVF and destruction of embryos under all circumstances.

There is little doubt that much of this opposition to embryo destruction has a religious base, even if it has not always been made explicit. Of the 58 organizations that put in a submission in 2007, 32 were explicitly religious. While all these would not have been opposed to this research, it is probable that a majority was. However, there were few responses to the HART principle regarding the place of different ethical, spiritual and cultural perspectives, and what did emerge emphasized the diversity of such perspectives and the difficulty in developing policies aimed at respecting them (ACART 2007).

Those in favour of at least some uses of embryos in research did so 'on the grounds that they have a lesser moral status than persons who have been born, provided that such research has scientific merit and potential to benefit human health' (ACART 2007). Users of fertility services expressed the view that they wanted the choice to donate their surplus embryos for research purposes, in addition to current choices of donating them to another couple or discarding them ('allowing them to thaw and perish' ACART 2007).

A major problem for ACART in assessing the arguments put forward in the submissions was the lack of any argumentation in many of them. All too frequently, embryo research involving embryo destruction was condemned on grounds that it was ‘playing God’. This term came in some instances from religious groups, but this was not universal. While indicating opposition to a particular procedure, it failed to indicate why ‘playing God’ is to be decried. Faced with submissions along these lines (aggravated by a sizable number of submissions being very brief and almost identically worded), ACART had immense difficulty in being assured that they constituted a representative sample of the population. Additionally, there was little way of knowing to what extent the religious input was representative of the religious community nor the basic tenets upon which it was structured.

5 Opinion Surveys on Embryo Research

At the time of ACART’s work on the use of gametes and embryos in human reproductive research in 2007, it sought public submissions on the possibilities open to New Zealand society. It received 345 written submissions and 9 oral submissions, with 160 people attending 8 public consultation meetings (ACART 2007). In addition to the outcomes referred to in the previous section, the majority supported a view that life begins at conception, leading to the view that destroying embryos is akin to harming or killing a person. For others, some use of embryos in research is justified on the grounds that they have a lesser moral status than persons who have been born. A proviso was that such research has scientific merit and the potential to benefit human health. ACART also wished to have feedback on the sources of embryos for use in research. Unfortunately, few people answered this question, although some noted a moral difference between the use of surplus embryos and those created specifically for research purposes.

In 2003 focus groups were employed to investigate the use of embryonic stem cells as a postulated treatment for Alzheimer’s disease (Hunt et al. 2003). The groups were distributed throughout the country, and the acceptability of this hypothetical procedure varied across all degrees of acceptability from most to least. What emerged was that a dominant issue was the role of personal experience, those with relatives with Alzheimer’s disease being more open to the use of stem cells. For many others, their use was deemed abhorrent and should never be contemplated. What emerges from this small sample is that the potential of benefitting humanity is a crucial consideration and may override protection of the embryo. However, far more extensive studies need to be undertaken to provide a more adequate picture of the public responses to research using embryos, especially as their clinical potential becomes more evident.

6 Religious Responses to Embryo Research

Discussion about embryo research has a theoretical feel to it, and so it is commonly framed in terms of stem cells, specifically ESCs due to their derivation from embryos. Hence, the two may be used interchangeably, but it comes down to the same thing—the value placed on embryos. The following responses are to the ACART discussion paper on embryo research (ACART 2006).

Right to Life New Zealand objected on the basis that human life begins at conception, the point at which the embryo is endowed by its creator with human rights (Orr 2008). It is claimed that the human embryo is a unique and unrepeatable miracle of God's loving creation and that its destruction amounts to violation of the commandment against killing. Further, it is asserted that ESC research is part of a culture of death since it leads to the killing of innocent and defenceless unborn children by 'abortion'. The use of surplus embryos for research purposes is equated with the destruction of human persons, an act that can never be justified to promote medical research.

This stance has much in common with that of other religious groups for whom 'life begins at conception' leading to total protection of every embryo (at least in theory). Positions like this do not reveal how the moral significance of 'conception' (fertilization) has been arrived at, but it dominates much of the religiously inspired submissions in the public domain. While it would be foolhardy to equate this relatively straightforward position to that of far more theologically astute literature, there can be little doubt that its presence is keenly felt in public debates, especially in many church circles. Surprisingly perhaps, these objections were not extended to IVF and PGD. However, this is in line with an observation encountered elsewhere: the centrality of embryo/foetus destruction in abortion, and subsequently in the debate over ESCs, overshadows debate on the loss of embryos during IVF (Jones 2015).

The *Salvation Army* (2007) in its reflections on embryo research took a more nuanced protectionist position in that it started from its view of the biblical foundations. By referencing these it ended with the question: 'Would God's plan be that embryos should be used to alleviate human suffering? This raises spiritual and ethical questions that may never be answered satisfactorily'. Following an overview of scientific issues around embryonic development and the many potential contributions of gamete and embryo research in fundamental science, fertility and infertility and the prevention and treatment of disease, it turned to spiritual aspects.

> Christianity embraces a paradox in that Christians believe in a healing, redeeming God and are called upon to participate in God's work in relieving suffering, bringing healing and establishing justice. At the same time, Christians recognize that some suffering is unavoidable and that they are called to suffer with and for Christ and with and for others. Such paradoxes cannot be resolved; they must be held in tension.

Ultimately, the document adopted a moderate position, with its conclusion that 'the ethical justification of research projects using human embryonic stem cells will depend on the potential benefits of the research and the quality of the scientific questions being asked'. It sought to support those in need of healing, if the benefits outweighed the harms and if the use of stem cells or other approaches can be justified clinically. There was also awareness that embryo research is not to be seen in isolation of other procedures allowed within New Zealand. According to the document, 'these factors include the extent to which decisions on future gamete and embryo research should be influenced by precedents in current policy settings and legislation that take an implicit or explicit position on the moral status of the embryo'.

The *Interchurch Bioethics Council* is an ecumenical body supported by the Anglican, Presbyterian and Methodist Churches of New Zealand that makes submissions on bioethical issues within the community. Its submission on embryo research took a balanced perspective, not dissimilar to that of the Salvation Army. It was prepared to allow for embryo research under strict conditions and when it would make available information unavailable by any other means.

By far the most detailed account of a religious position on embryo research was that of the *Nathaniel Centre*, the New Zealand [Roman] Catholic Bioethics Centre (Nathaniel Centre 2007). Commencing with the fundamental premise of the inviolate dignity of the human embryo, any form of embryo destruction is regarded as morally illicit. To ignore this premise is to cross a fundamental moral line, since it amounts to the direct killing of human life in its earliest and most vulnerable stages. From the moment of fertilization, the human embryo has an absolute right to life. In other words, it has an intrinsic respect from which there can be no deviation.

This position closely mirrors that of traditional Roman Catholic doctrine. However, there are two points of interest. The first is that the document was written against a background of the possible derivation of ESCs, with the prospects being held out in 2007 for their utility in curing a range of human diseases. This led to what is described as a lack of congruence between the means employed (embryo destruction) and the outcome being sought (cure of diseases). Embryo research though is far broader than simply the derivation of stem cells, so that rejection of this one example does not exhaust the possible ends of the research, including embryonic development and the causes of congenital aberrations and infertility. This may not alter Roman Catholic teaching on embryonic inviolability, but it does call for a recasting of this stance to address the breadth of the New Zealand debate.

The second point of interest is the document's attempt to justify its opposition to embryo research but not to every facet of the application of the abortion law in New Zealand. According to the Nathaniel Centre (2007) document, the abortion law confirms that the norm is for human life, in that once conceived it is not to be destroyed. This is based on the premise that abortion is only to be carried out as a way of preventing harm to a specific person in a specific situation, namely, the mother. By contrast, it is averred that the destruction of prenatal life for a cause only potentially related to the generic good of society in the future (stem cells are in mind here) is completely different. In the case of surplus embryos, a distinction is being made between embryos that will eventually be destroyed via thawing (the 'intent' is not to destroy them) and destroyed via research conducted on them (the deliberate destruction of embryos). However, in both instances, embryos are being destroyed by human actions, and it could be argued that the second is more ethically defensible than the first. Alternatively, it can be argued that, in the case of embryo research, the intent is not to destroy them but utilize them for higher ends.

The arguments contained in the various Nathaniel Centre documents are serious contributions to society's debate on embryo research in New Zealand, and for this one has to be grateful. The sad reality is that the amount of public debate, let alone the quality of much of it, is impoverished. This applies to all aspects of it, including much of the religious input. One does not know whether this is because there are no concerns with embryo research as a procedure, too little is known about it, it is considered irrelevant for the churches and church bodies or it is esoteric and of limited, if any, spiritual significance.

7 Personal Perspective

My own perspective is peripheral to the thrust of this chapter, since it is neither derived from New Zealand nor does it pretend to mirror much religious thinking within New Zealand. However, as a member of ACART when it initially proposed the acceptance of embryo research using donated surplus embryos, it demonstrates that someone with a religious (Protestant Christian) stance is able in principle to support embryo research. Details of my arguments can be found in a number of places (Jones 2005, 2015). Some of their essential facets are as follows.

I approach the human embryo and reproductive technologies from dual starting points, that of a biomedical scientist and a Christian who looks to Scripture for theological guidance. For me these starting points are complementary and together provide a balanced appreciation of the potential scientific and clinical benefits together with any limitations, the centrality of informed ethical decision-making and a range of theological postulates and queries associated with technologies capable of intruding into the sanctum of prenatal life.

The following questions are crucial ones demanding answers in each instance. What is the scientific and clinical justification for any procedure that will destroy and/or manipulate embryos? What does the procedure tell us about our view of the value and status of these embryos? Are we exploiting them as if they are merely goods to be traded and disposed of, or are they being employed with serious intent likely to usher in beneficial prospects for mankind worthy of the destruction of these particular embryos? How far do the biblical data at our disposal take us, and are we convinced that these can be integrated with our scientific aspirations? Where does God as creator and redeemer fit into the endeavour?

Any attempt to address these questions is based upon a number of theological postulates (Jones 2005). The first of these is that prenatal human life is of considerable value and this should be assessed alongside the very considerable value of all human life. This prompts a search for equipoise between the respective values of human embryos and those likely to benefit in future from this research. Both should be of concern for all religious voices. Second, the realm of the human embryo and human procreation falls within the ambit of human creativity, as humans demonstrate elements of God's own creativity. They are also to recognize that there are limits to their abilities and that these can on occasion be misused as well as put to good ends. Third, theological vistas are to be grounded in the future as well as the past, in hope as well as fear, in God's good purposes and not overwhelmed by the possible consequences of human frailty. Therefore, it is fitting that attention is directed to potential benefits that may emerge from the research for the human community. It is at this point that my position deviates from that of many religious actors who look only to the past and to the destruction of embryos. However, the position I am adopting does not open the door to unlimited use of human embryos, since the fourth postulate emphasizes the ends to which it is directed. These are to ascertain the degree to which manipulating the early stages of human development will be beneficial or counter-productive, something that in turn will depend upon the goals to which it is directed. From a religious stance, this will be solely to benefit human welfare and never to serve unrealistic grandiose ends.

These postulates accept that the biblical and theological basis is tipped towards the legitimacy of humans exercising control over the early stages of human existence, therein reflecting their creation in God's own image. However, a balance is to be sought between the ever-present negative and positive repercussions, knowing that unethical exploitation can readily overtake legitimate human dimensions.

8 Critique of the New Zealand Legislation

While this chapter is directed towards the religious input into decision-making on embryo research in New Zealand, it has to be viewed within the context of the legislation that exists in New Zealand. As previously outlined, this is provided by the HART Act that accepts the permissibility of IVF and inevitably, if unintentionally, accepts the validity of embryo research in other countries where IVF was developed and continues to be refined (HFEA 2013). Consequently, IVF and embryo research are inextricably linked (Jones 2015).

Jones (2014) has proposed four models of embryo protection. Model 1 (consistent protection of embryos) rejects embryo research and therefore IVF and PGD. Model 2 (acceptance of existing policies) rejects embryo research but accepts IVF and PGD even though no embryo experimentation is allowed. Model 3 (acceptance of embryo research outside New Zealand) rejects embryo research in New Zealand but accepts IVF and PGD and explicitly accepts that research may be conducted in other countries. Model 4 (acceptance of embryo research) accepts IVF and PGD and accepts research on surplus embryos and possibly on embryos produced specifically for research purposes.

Of these models, 1 and 4 are consistent in their stances, prohibition in Model 1 and permission in Model 4. Models 2 and 3 are less consistent in that they accept to differing degrees the results of research on embryos, either in the past and/or on a continuing basis. While they aim to protect embryos in the future, they are prepared to benefit from data and procedures obtained from embryo destruction in the past and in the present in other countries.

The ethical inconsistency found in Models 2 and 3 raises questions about its legitimacy and has been termed 'performative inconsistency' (Holm 2007). On what grounds can it be acceptable to destroy (or allow to perish) embryos surplus to the requirements of a clinical fertility programme but refuse to allow the use of these about-to-be-destroyed embryos for research aimed at contributing to an understanding of the causes of infertility?

As with any clinical procedure, ongoing research on infertility is crucial since the welfare of children is at stake (Green 2001). This is of considerable relevance to the HART Act with its principle of protecting children. Very simply, a society that allows IVF also has a duty to be involved in ongoing research to increase the efficacy and safety of the procedures. Without this, one is entirely dependent upon research carried out in other societies. This precludes practitioners in New Zealand from improving the safety and protocols of their practice, an essential consideration when IVF is an accepted, let alone an established, procedure.

The relationship between embryos and future children is an intimate one whenever the existence of the latter depends upon a technological procedure like IVF. Without IVF they would not exist; with it not only do they exist but many aspects of their health and well-being depend upon the protocols employed in the clinic. The present situation in New Zealand means that it is impossible to conduct clinical trials on the efficacy of different IVF procedures, such as the cumulative pregnancy rates following transfer of embryos at day 3 (cleavage stage) compared with day 5 (blastocyst stage). Both are routinely used in clinical practice, without good evidence as to which is the better of the two (Jones 2014).

In view of this situation, it has to be asked whether religious actors can throw light on it or can they do no more than add to the confusion? Any viewpoint adopting a rigid pro- or anti-research stance will do the latter. However, I would hope that any religious voice would call for integrity, honesty and humility in the face of conflicting ends. If the HART Act is to prove a piece of legislation acceptable to those with religious stances, further work is needed on elaborating the criteria for concluding that IVF is an established and therefore generally accepted procedure. To this end, the consequences of this designation have to be spelled out, and as we have seen, these include research on embryos. If this is not accepted, IVF should not be designated an established procedure; more likely, it should be prohibited (or at the very least be designated a regulated procedure). Pointing out the stark consequences in this manner would be a positive, if unexpected, contribution of religious actors.

9 Concluding Comments

It is evident that there is a diversity of religious opinion within the New Zealand community, a not uncommon situation. However, themes emerge from these voices that wider society would do well to heed. While this will not result in any neat way forward, either in blanket acceptance or rejection of embryo research, they bring to the debate a cautionary perspective that may have implications for the source of embryos to be used in research and the extent of manipulations on embryos. This will satisfy neither end of the pro- and anti-research spectrum, but it provides fertile ground for dialogue and constructive assessment within the boundary of the HART Act. However, this will only be possible when a Minister of Health allows such debate to take place. This is where religious, as well as cultural, input can bring the HART Act to life.

Without openness to such input, progress on thinking about embryo research has reached an impasse in New Zealand, in spite of the continued efforts of ACART to bring the situation to the attention of the Minister of Health. The problem is that the final decision rests with the Minister of Health alone, and there have been four different Ministers since ACART first submitted its recommendations, and none has been willing to respond in an informed manner—regardless of the tenor of a response. Ten years have intervened since those recommendations were first submitted, and over these years, there has been no opportunity for ACART to consult again with the public on embryo research. The scientific milieu has continued to change, and those with expertise in science, ethics, culture and religion have continued to provide additional insights on embryo research. In addition, the membership of ACART has changed. Consequently, the original recommendations have a historic feel to them, and at the very least, they should be refreshed and reinvigorated, regardless of the conclusions a new report may reach.

The dialogue model adopted by one of the Māori studies (Hudson et al. 2010) could serve as a useful starting point for a project to elucidate religious input into the debate and how this can be integrated into the diverse concerns and interests of those in a pluralist society. If people wish to hear what religious voices have to say, those with these voices need to reflect together on the core thrusts of their varying positions and how these can best serve society at large and mankind beyond New Zealand society. But this will only occur when those with religious perspectives regard themselves as integral to society and capable of making a contribution that will stand alongside, and complement, a range of other perspectives.

References

ACART. 2006. *Use of gametes and embryos in human reproductive research: Determining policy for New Zealand*. A Discussion Paper, 2006. Wellington: Advisory Committee on Assisted Reproductive Technology.

———. 2007. *Use of gametes and embryos in human reproductive research: Summary of submissions*. Wellington: Advisory Committee on Assisted Reproductive Technology.

———. 2012. *Guidelines on extending the storage period of gametes and embryos*. Wellington: Advisory Committee on Assisted Reproductive Technology, 30 August 2012. http://acart.health.govt.nz/system/files/documents/publications/acart-guidelines-extending-storage-gametes-embryos-2012.pdf. Accessed 2 Nov 2016.

———. 2013. *Advisory Committee on Assisted Reproductive Technology annual report 2012/13*. Wellington: Advisory Committee on Assisted Reproductive Technology, November 2013.

———. 2015a. *Advisory Committee on Assisted Reproductive Technology annual report 2014/2015*. Wellington: Advisory Committee on Assisted Reproductive Technology, November 2015.

———. 2015b. Human reproductive research—Associated paper 6. Briefing to the incoming Minister of Health. https://acart.health.govt.nz/system/files/.../bim-background.docx. Accessed 19 Oct 2016.

———. 2016a. Minutes of fifty-ninth meeting of Advisory Committee on Assisted Reproductive Technology. http://acart.health.govt.nz/meetings/2016-agendas-and-minutes. Accessed 18 Oct 2016.

———. 2016b. Minutes of sixty-first meeting of Advisory Committee on Assisted Reproductive Technology. http://acart.health.govt.nz/meetings/2016-agendas-and-minutes. Accessed 18 Oct 2016.

Akoorie, N. 2014. Deadline forces tough IVF choice, *The New Zealand Herald*, January 20. http://www.nzherald.co.nz/news/print.cfm?objectid=11188978. Accessed 6 June 2014.

Bioethics Council. 2008. *Who gets born?* Wellington: Toi Te Taiao Bioethics Council.

Cleary, T. 2013. *(Un)Reasonable voices? The place of religious, spiritual and cultural voices in ACART's decision-making process*. MBHL dissertation, University of Otago.

European Science Foundation. 2013. *Human stem cell research and regenerative medicine: Focus on European policy and scientific contributions*. Strasbourg: European Science Foundation.

Glover, M. 2008. *Māori attitudes to assisted human reproduction: An exploratory study*. Auckland: School of Population Health, University of Auckland.

Glover, M., and B. Rousseau. 2007. Your child is your whakapapa; Māori considerations of assisted human reproduction and relatedness. *SITES New Series* 4 (2): 129–136.

Glover, M., A. McCree, and L. Dyall. 2007. *Māori attitudes to assisted human reproduction: An exploratory study. Summary report*. Auckland: School of Population Health, University of Auckland.

Green, R.M. 2001. *The human embryo research debates: Bioethics in the vortex of controversy*. Oxford: Oxford University Press.

HFEA. 2013. Research projects licensed by the HFEA between 1 April 2012 and 31 March 2013. http://www.hfea.gov.uk/166.html. Accessed 21 Mar 2014.

Hinxton Group. 2006. An international consortium on stem cells, ethics and law. World stem cell policies. http://www.hinxtongroup.org/wp_ao_map.html. Accessed 28 May 2014.

Holm, S. 2007. Policy-making in pluralistic societies. In *The Oxford handbook of bioethics*, ed. B. Steinbock, 153–174. Oxford: Oxford University Press.

Hudson, M., M. Roberts, L.T. Smith, M. Hemi, and S.-J. Tiakiwai. 2010. Dialogue as a method for evolving Matauranga Māori perspectives on the use of embryos in research. *AlterNative: An International Journal of Indigenous Peoples* 6 (1): 54–65.

Human Assisted Reproductive Technology Act (HART) Act 2004. Reprint 1 July 2013. Wellington: Parliamentary Counsel Office. http://www.legislation.govt.nz/act/public/2004/0092/latest/whole.html. Accessed 17 Oct 2016.

Human Assisted Reproductive Technology Order 2005. Reprint 25 March 2011. Wellington: Parliamentary Counsel Office. http://www.legislation.govt.nz/regulation/public/2005/0181/latest/DLM335192.html. Accessed 2 Nov 2016.

Human Tissue Act 2008. Reprint 1 March 2016. Wellington: Parliamentary Counsel Office. http://www.legislation.govt.nz/act/public/2008/0028/latest/DLM1152940.html. Accessed 2 Nov 2016.

Hunt, L.M., J.R. Fairweather, and F.J. Coyle. 2003. *Themes from the use of embryonic stem cells in the treatment of Alzheimers exemplar. Public Understandings of Biotechnology in New Zealand*. Christchurch: Lincoln University. Agribusiness and Economics Research Unit, Research Report No. 266, pp 69–84.

Jones, D.G. 2005. Responses to the human embryo and embryonic stem cells: Scientific and theological assessments. *Science and Christian Belief* 17: 199–222.

———. 2014. Where does New Zealand stand on permitting research on human embryos? *New Zealand Medical Journal* 127 (1399). https://www.nzma.org.nz/journal/read-the-journal/all-issues/2010-2019/2014/vol-127-no-1399/6236.

———. 2015. In vitro fertilization and the destruction of embryos. *Perspectives on Science and Christian Faith* 67 (3): 163–174.

Jones, D.G., and M.I. Whitaker. 2012. Ethical framework for ACART. http://acart.health.govt.nz/publications-and-resources/acart-publications/ethical-framework-acart. Accessed 17 Oct 2016.

Jones, D.G., M.R. King, and M.I. Whitaker. 2009. Who gets born? How did New Zealand's Bioethics Council arrive at its recommendations? *New Zealand Medical Journal* 122 (1294). http://www.nzma.org.nz/journal/122-1294/3585/.

Nathaniel Centre. 2007. The use of gametes and embryos in human reproductive research: Determining policy for New Zealand. http://www.nathaniel.org.nz/component/content/article/14-bioethical-issues/bioethics-at-the-beginning-of-life/152-the-use-of-gametes-and-embryos-in-human-reproductive-research-determining-policy-for-new-zealand. Accessed 6 Oct 2016.

New Zealand Government. 2002. Bioethics Council appointed. https://www.beehive.govt.nz/release/bioethics-coucil-appointed. Accessed 12 Oct 2016.

NZORD. 2004. NZORD contributes to Bioethics Council publication on transgenics. 12 January 2004. https://www.nzord.org.nz/news/newsletters?a=3868. Accessed 12 Oct 2016.

Orr, K. 2008. Embryonic stem cell research. Right to Life New Zealand. http://righttolife.org.nz/2008/12/20/embryonic-stem-cell-research/. Accessed 18 Oct 2016.

Royal Commission on Genetic Modification. 2001. *Report of the Royal Commission on Genetic Modification*. Wellington: Department of Internal Affairs.

Salvation Army. 2007. Gamete and embryo research. http://www.salvationarmy.org.nz/our-community/faith-in-life/christian-ethics/gamete-and-embryo-research. Accessed 18 Oct 2016.

Snelling, J., N. Peart, and M. Henaghan. 2007. Preimplantation genetic diagnosis; Testing the legal boundaries. In *Genes, society and the future*, vol. 1, 50–51. Dunedin: Human Genome Research Project.

Stenhouse, J. 2011. Religion and society. *Te Ara—the encyclopedia of New Zealand*. http://www.TeAra.govt.nz/religion-and-society. Accessed 26 Jul 2017.

Towns, C.R., and D.G. Jones. 2006. Stem cells: Public policy and ethics. In *The stem cell controversy: Debating the issues*, ed. M. Ruse and C.A. Pynes, 329–341. Amherst: Prometheus Books.

Chapter 8
The Political Debate on Embryo Research in Australia and the Role of Religious Actors and Arguments

Frank O'Keeffe and Kevin McGovern

Abstract Australian ethicists O'Keeffe and McGovern provide a comprehensive account of the move in Australia from a widespread view that destructive research on human embryos was ethically abhorrent, to the current, heavily regulated system whereby human embryos can be used or even created via somatic cell nuclear transfer (SCNT) for destructive research. The chapter focuses on events surrounding three landmark committees established by the Australian Government: the Andrews Committee in 2001, the Lockhart Review in 2005, and the Heerey Review in 2011. O'Keeffe and McGovern particularly detail the contribution of "religious actors" who almost without exception have continued to oppose destructive research on human embryos. The chapter contrasts the principled position of religious actors with the utilitarian ethic which now permeates Australian discourse on cloning and human embryo research.

1 Introduction

The political debate on embryo research in Australia and the role of religious actors and arguments therein have particularly been manifest in three landmark committees established by the Australian Government in 2001, 2005, and 2011. Named after their respective chairs, these are the Andrews Committee (2001), the Lockhart

F. O'Keeffe
School of Medicine, Sydney, NSW, Australia

The University of Notre Dame Australia, Darlinghurst, NSW, Australia

K. McGovern (✉)
Department of Moral Theology and Canon Law at Catholic Theological College, University of Divinity, East Melbourne, VIC, Australia

Australian Catholic University, Brisbane, QLD, Australia
e-mail: kmcg@powerup.com.au

M. Weiberg-Salzmann, U. Willems (eds.), *Religion and Biopolitics*,
https://doi.org/10.1007/978-3-030-14580-4_8

Review (2005), and the Heerey Review (2011).[1] These committees were appointed to report on the ethicality of research involving human embryos, embryonic stem cell (ES cell) research, therapeutic cloning, and reproductive cloning. Each of these committees invited public submissions, and in significant numbers religious actors—people of faith and their organisations and agencies—made submissions to each of these committees. Each of these committees also heard from expert witnesses, who included religious actors.

The Andrews Committee investigated the possibility of legalising embryonic stem cell research in light of the 1997 announcement that scientists in Edinburgh had successfully produced the world's first cloned animal ("Dolly" the sheep) and the 1998 revelation that scientists in Wisconsin had, for the first time, extracted stem cells from a live embryo. Moreover, use of assisted reproductive technology (ART) had developed in Australia beginning in the 1970s, and, by this time, there were thousands of frozen embryos in storage which were surplus to the requirements of those involved in IVF. Thus, the Andrews Committee was also asked to consider whether use of these embryos in ES and other research should be allowed. Two pieces of Australian legislation enacted in 2002 permitted—under licence and with strict conditions—the use of surplus embryos in research, including research seeking to extract ES cells from them, while not permitting the deliberate creation of an embryo through the process of fertilization, for the purposes of experimentation on it. This legislation prohibited reproductive cloning and placed a moratorium on so-called therapeutic cloning. The Lockhart Review was convened 3 years later. It recommended legislation that permitted—again under licence and with strict conditions—somatic cell nuclear transfer (SCNT) or cloning, not for reproduction but for the purposes of extracting ES cells and developing ES cell lines with the ultimate therapeutic goal of curing diseases. The Australian Government and all Australian states except Western Australia subsequently enacted legislation permitting this so-called therapeutic cloning. The Heerey Review was particularly shaped by the 2007 discovery of induced pluripotent stem (iPS) cells—adult cells that have been reprogrammed into cells with properties very similar to ES cells. The committee considered whether research into iPS cells could replace ES cell research and therefore whether SCNT should still be permitted. By majority, the committee recommended that the enacted permission for SCNT should not be removed.

This report will begin by detailing briefly the growth of assisted reproductive technology and the debates about embryo experimentation which had occurred in Australia prior to these three landmark committees. It will then elucidate what happened at and after these committees. It will particularly highlight the arguments that various religious actors put forward before these committees and in the developments which followed them.[2]

[1]Their chairs were Hon. Kevin Andrews MP, Hon. John S. Lockhart AO QC, and Hon. Peter Heerey QC. Kevin Andrews is well known as a staunch Catholic. After the Northern Territory Government legalised euthanasia in 1996, Andrews successfully led the campaign to overturn this legislation.

[2]All online documents cited in this report were accessed on 23 December 2016.

2 Background

In 1980—2 years after the birth of the world's first IVF baby Louise Brown in England—Candice Reed became the first Australian child to be born through IVF. Following this development, in 1982, Australia's National Health and Medical Research Council (NHMRC)[3] released its first guidelines on IVF and research on human embryos as Supplementary Note 4 to its *Statement on Human Experimentation*. While these guidelines were quite restrictive, many—probably most—Australians were disturbed that these guidelines did permit some research on human embryos.[4] In response to this, the state governments of all six Australian states established committees to enquire into, and report on, the various types of research which involved embryos.[5] This initial reaction in Australia to embryo experimentation indicated a widespread desire at this time, in accord with the concerns of many religious actors to recognise and protect the moral status of the embryo and to forbid any embryo experimentation that was not aimed towards assisting people who were experiencing infertility and reproductive problems.

In response to its own committee's recommendations, Victoria was the first Australian state to enact legislation which responded to the developments in reproductive technology. This was the *Infertility (Medical Procedures) Act 1984* (Vic). It stipulated a ban on cloning.[6] It generally forbade any "experimental procedure" that involved "carrying out research on an embryo of a kind that would cause damage to the embryo, would make the embryo unfit for implantation or would reduce the prospects of a pregnancy resulting from the implantation of the embryo".[7] However, it did permit that some of these experimental procedures could be authorised by a Standing Review and Advisory Committee.[8]

In 1985, the *Human Embryo Experimentation Bill 1985*[9] was introduced into the Australian Senate by Senator Brian Harradine, a Catholic politician who consistently opposed abortion, ES cell research, and pornography. This bill sought to prohibit experimentation on embryos, and its introduction caused controversy. Further debate followed in 1986 when the Australian Senate's Select Committee on Human Embryo Experimentation (the Tate Committee) produced its report, *Human Embryo*

[3]The NHMRC is Australia's leading expert body supporting health and medical research; developing health advice for the Australian community, health professionals, and governments; and providing advice on ethical behaviour in health care and in the conduct of health and medical research.

[4]"Brave New World – Down Under," The Interim, 15 May 1988, http://www.theinterim.com/issues/bioethics/brave-new-world-%E2%80%93-down-under/.

[5]Ibid.

[6]*Infertility (Medical Procedures) Act 1984* (Vic), S 6 (2) (a), Australasian Legal Information Institute, http://www.austlii.edu.au/au/legis/vic/hist_act/ipa1984311.pdf.

[7]Ibid., S 6 (3) and (4).

[8]Ibid., S 29 (6) (b).

[9]*Human Embryo Experimentation Bill 1985*, Australasian Legal Information Institute, http://www.austlii.edu.au/au/legis/cth/bill/heeb1985328/.

Experimentation in Australia, which also called for a ban on embryo experimentation.[10]

In 1992, the NHMRC Supplementary Note on IVF was not retained following the passing of the *NHMRC Act 1992* (Cth). This prompted a revision of Australia's ethical guidelines on embryo experimentation.[11] In 1996, the NHMRC issued its *Ethical guidelines on assisted reproductive technology*, which stated that "[e]mbryo experimentation should normally be limited to therapeutic procedures which leave the embryo, or embryos, with an expectation of implantation and development".[12] However, it did permit nontherapeutic research that involved the destruction of an embryo in "exceptional circumstances" and with the approval of an Institutional Ethics Committee. The "likelihood of significant advance in knowledge or improvement" in ART was a requirement for approval of such research.[13]

Up to this point, the primary concern of Australian legislation and guidelines was to regulate assisted reproductive technology. Research involving embryos—and even research which involved the destruction of embryos—had been considered, and there were some cautious mechanisms to authorise such research. However, Australian legislation and guidelines had not departed too far from the beliefs expressed by many religious bodies about the sacredness of human life. This was soon to change. Two developments proved to be watersheds in changing Australian conceptions on embryo experimentation. In 1997, scientists in Edinburgh successfully cloned a sheep named "Dolly". And shortly after, in 1998, scientists in Wisconsin succeeded in extracting stem cells from 14 embryos surplus to an IVF programme. In response to these developments, in 1998, the then Minister for Health and Aged Care, Hon. Dr. Michael Wooldridge MP, asked the Australian Health Ethics Committee (AHEC)[14] to report to him on the scientific, ethical, and regulatory considerations relevant to cloning of human beings. AHEC issued its report on 16 December 1998.[15] AHEC recommended that the Australian Government should reaffirm its support for the UNESCO *Declaration on the Human Genome and*

---

[10]The chair of this committee, Michael Tate AO, was then a member of the Senate. After leaving politics, he served as Australian Ambassador to the Netherlands and the Holy See. In 2000, he was ordained a Catholic priest.

[11]National Health and Medical Research Council, *Ethical guidelines on assisted reproductive technology* (Canberra: Commonwealth of Australia, 1996), iv, https://www.nhmrc.gov.au/_files_nhmrc/publications/attachments/e28.pdf.

[12]Ibid., 6.2.

[13]Ibid., 6.4.

[14]AHEC is one of the principal committees of the National Health and Medical Research Council. It is also Australia's health ethics peak body.

[15]Australian Health Ethics Committee, *Scientific, Ethical and Regulatory Considerations Relevant to the Cloning of Human Beings*, 16 December 1998, https://www.nhmrc.gov.au/_files_nhmrc/publications/attachments/e45.pdf. AHEC did not invite public submissions. However, it sought comment from 49 "learned individuals and organisations". Nine of these were Catholic. They included Cardinal Edward Clancy; Bishop Pat Power; Revd Dr. Gerald Gleeson; the Directors of the Catholic bioethics centres in Brisbane, Sydney, Melbourne, and Adelaide; and the Bioethics Working Party of the Catholic Women's League Australia.

Human Rights, in particular Article 11, which states that "[p]ractices which are contrary to human dignity, such as reproductive cloning of human beings, shall not be permitted".[16] AHEC further recommended that the remaining Australian states and territories enact legislation on human embryo experimentation "according to the principles set out" in the NHMRC *Ethical Guidelines on Assisted Reproductive Technology*.[17] Lastly, AHEC suggested that the Australian Minister for Healthcare should promote "informed community discussion on the potential therapeutic benefits and possible risks of the development of cloning techniques".[18]

In response to these recommendations, on 10 August 1999, Wooldridge asked the House of Representatives Standing Committee on Legal and Constitutional Affairs to review the report of the Australian Health Ethics Committee. This committee, known as the *Andrews Committee*, was headed by the Hon. Kevin Andrews MP. It received submissions from various parties to the debate, including religious actors, regarding the morality of embryo research, as well as therapeutic and reproductive cloning. The Committee's recommendations would then be submitted to the Australian Parliament, which was set to debate the issue of human embryo experimentation before passing legislation on the issue.

3 The Andrews Committee (2001)

The 2001 report of the Andrews Committee titled *Human Cloning: Scientific, Ethical and Regulatory Aspects of Human Cloning and Stem Cell Search*[19] was the result of 2 years of government enquiry investigating the complex issues surrounding human cloning and stem cell research. The religious actors and agents who made submissions to the Andrews Committee primarily included Catholics, Anglicans, other Christians, and Jews. Some of the prominent Catholic agents in the debate were the Australian Catholic Bishops' Conference (ACBC), the Catholic Archdiocese of Melbourne, the Caroline Chisholm Centre for Health Ethics, the Queensland Bioethics Centre, the Southern Cross Bioethics Institute (SCBI), and Australian Catholic bioethicist Nicholas Tonti-Filippini. Anglican actors and agents who made submissions to the Andrews Committee against embryonic stem cell research included the Anglican Church of Australia, the Social Responsibilities Committee of the Anglican Diocese of Melbourne, and Ridley College. The

[16] Ibid., v, 1, 5, 43.

[17] Ibid., v, 43.

[18] Ibid., vi, 43.

[19] House of Representatives Standing Committee on Legal and Constitutional Affairs, *Human cloning: scientific, ethical and regulatory aspects of human cloning and stem cell research* (Canberra: Commonwealth of Australia, 2001), http://www.aph.gov.au/parliamentary_business/committees/house_of_representatives_committees?url=/laca/humancloning/contents.htm. In our references to this report, we will refer to this document as AC—the report of the Andrews Committee.

Executive Council of Australian Jewry offered viewpoints on the ethicality of embryo research that differed from many Christian parties in the debate.

3.1 Reproductive Cloning

The Committee considered whether cloning a human being would be ethical. The Committee's final report stated that "[a]lmost universal opposition was expressed to the Committee about the use of cloning techniques for the purpose of creating, implanting and bringing to birth a human being—a human 'Dolly'" (AC x). However, religious actors opposed to ES research differed from pro-ES research activists even in their understanding of what is meant by the term "reproductive cloning". Supporters of ES research generally used this term to mean efforts to create a "live-born individual" or "another adult individual" genetically the same as someone else (AC 6.10). In contrast, religious actors argued that even the so-called therapeutic cloning creates a new human life and therefore that therapeutic cloning properly understood is also a form of reproductive cloning. Ridley College submitted that "the production of tissues or organs for transplantation... is really reproductive cloning" because "it involves the production of such a foetus (or embryo) but not with the aim of allowing this foetus to come to term and be born, but with the aim of using it for 'spare parts'" (AC 6.14).

As regards reasons for opposing reproductive cloning, the Committee noted that "[t]he most common reason given for regarding cloning for reproductive purposes as unethical was that it would be 'contrary to human dignity'" (AC 6.25). Religious actors expanded on this basic message in a variety of ways. Significant recourse was made by religious agents to human rights documents that enshrine something akin to a Christian notion of human dignity. Dr. Gregory Pike of the Southern Cross Bioethics Institute highlighted that "[i]t is this deep-seated inherent dignity which underscores the human rights documents and various codes of medical ethics which mark all human kind as worthy of the highest respect" (AC 6.26). Concern was raised about a utilitarian ethic that makes the human person "subservient to the needs of others" (AC 6.27). The Social Responsibilities Committee of the Anglican Diocese of Melbourne highlighted this by submitting that "human beings may never be treated merely as a means to an end, but only as an end". It cautioned that "[m]any of the suggested reasons for reproductive cloning... have a strongly instrumental character to them..." (AC 6.29).

Notably the Executive Council of Australian Jewry (ECAJ) diverged from the varying Christian perspectives, arguing that "cloning, which is based on pre-existing human genetic material, is not humans playing God but using God-given material, albeit not through normal methods of procreation..." (AC 6.33). While sometimes "medical intervention represents interference with Divine providence", the ECAJ stated that: "... Judaism does not therefore say that cloning is prohibited in itself ...[it] advises one to pause before one permits that which can lead down a variety of slippery slopes" (AC 6.33). From this perspective and based on these concerns, ECAJ did oppose reproductive cloning.

Concerns were raised by religious actors regarding whether reproductive cloning could "distort our understanding of human relationships" and therefore "what kind of identity a cloned person would have?" (AC 6.37). The Caroline Chisholm Centre for Health Ethics submitted that cloning "would deprive the child of the genetic basis of father, mother and other family relationships which are very significant and important for every human individual..." (AC 6.40). Catholic bioethicist Nicholas Tonti-Filippini underscored that "... cloning fragments the interconnectedness of human beings, because it allows a human being to be created without direct connections with a family" (AC 6.41). The Social Responsibilities Committee of the Anglican Diocese of Melbourne submitted that "[c]loning appears to undermine this structure of the family" (AC 6.42). It further contended that "[c]loning allows the separation of the sex act from the intimacy of the relationship, and brings a genetic difference from other humans who have genetic contribution from two parents" (AC 6.42).

The Australian Catholic Bishops' Conference also cautioned that the process of cloning "...is encouraged by the progressive depersonalisation of the generative act (introduced by the practice of extracorporeal fertilisation) which becomes a technological process making the human being an object to be used by anyone who can reproduce him in the laboratory" (AC 6.45). Concerns were also raised by religious actors regarding the safety of cloning procedures. In light of the multiple failed attempts to clone a sheep prior to the birth of "Dolly", the Social Responsibilities Committee of the Anglican Diocese noted that "[t]here are sufficient unknowns about physical problems in pregnancy with cloned sheep and cattle to suggest that human cloning experiments would violate normal medical ethics. There is no experiment that could be done to prove the safety of human cloning without causing serious risk to humans created in the process" (AC 6.57). Following the atrocities of World War II, Queensland Right to Life (QRTL) raised concerns regarding eugenics—which seeks to "improve" the human race by application of the principle of selective breeding. QRTL cautioned that "cloning can be used to select for various characteristics and potentialities" (AC 6.60). Ridley College submitted concerns regarding the role of women in the process of cloning and cautioned that "women's bodies would be required as sources of ova and of wombs for gestation of cloned individuals (whether they are allowed to develop to term, or sacrificed at some stage). A person cannot be isolated from her body, and therefore the 'use' of a woman's body is an exploitation of her whole person... There is a real danger of the commodification of women's bodies..." (AC 6.63).

3.2 *Therapeutic Cloning*

While reproductive cloning was almost universally condemned, a plurality of opinions was provided regarding the ethicality of seeking to harvest stem cells from cloned embryos for so-called therapeutic cloning. The Committee noted that "[t]he deliberate creation of embryos for research purposes is another possible source of material for research involving cloning technologies. Embryonic stem cells could then be extracted from such embryos. As noted above, the extraction would destroy the embryos" (AC 7.64). Pro-ES research activists supported cloning for this purpose. However, the practice of "therapeutic cloning" received almost unanimous condemnation from the religious actors who made submissions to the Andrews Committee.

The Social Responsibilities Committee of the Anglican Diocese of Melbourne submitted that "[t]he Church's position is that the moment of fertilisation should be considered as the unique human beginning" (AC 7.71). The Queensland Bioethics Centre saw an incongruity in allowing cloning for one purpose and not another: "If it is intended to allow the being to be nurtured and grow into an adult, then it is a human being and to be protected. If someone intends to use the organism for some other purpose then it is [seen as] either not human or not protected" (AC 7.82). Similar sentiments were uttered by the Catholic Archdiocese of Melbourne (AC 7.83), the Council for Marriage and the Family (AC 7.84), and NSW Right to Life (AC 7.85). NSW Right to Life submitted that "[t]his is treating a new human as a commodity..." (AC 7.85). Archbishop Barry Hickey of the Australian Catholic Bishops' Conference (ACBC) stated "[h]uman life is never disposable at any stage of its development. It should never be seen as a commodity ... nor is its worth and claim to protection dependent on age or utility to others..." (AC 7.10). The Catholic Archdiocese of Melbourne (CAM) chose to highlight the utility of scientific research that has focused solely on extracting adult stem cells, to avoid the ethically contentious issue of cloning human embryos. CAM submitted that "...adult stem cells or de-differentiated somatic cells would have all the therapeutic advantages of ES cells but not require the generation and dismembering of cloned human embryos ..." (AC 7.49).

Concerns were raised by religious actors that the practice of therapeutic cloning could result in the gradual acceptance of reproductive cloning. The Social Responsibilities Committee of the Anglican Diocese of Melbourne argued that "... if we agree that it is wrong to create cloned people, how can it be ethical to create a cloned embryo, knowing full well it must be destroyed to avoid ever growing to become a human being?" (AC 7.86).

3.3 Embryos Surplus to Assisted Reproductive Technology Programmes

A contentious issue was raised by the Andrews Committee: "In assisted reproductive technology programs (including IVF) more embryos are created than will be required to achieve children for those undergoing treatment..." (AC 7.54). The Committee noted that "[i]t has been suggested that these 'surplus' embryos are usually stored for a certain period of time and may then be discarded if unused" (AC 7.55).

Great disagreement was expressed between religious and nonreligious actors regarding the ethicality of extracting stem cells from embryos surplus to assisted reproductive technology. The Anglican Church of Australia opposed the use of surplus embryos. The Social Responsibilities Committee of the Anglican Diocese of Melbourne argued that the fact the "...tissue sources may come from 'spare' embryos or unwanted tissue does not alter the ethical status of that tissue. If a tissue exists or we have access to it we do not have a moral obligation to use it and there is no ethical imperative to ignore the source of tissue to achieve the ends desired" (AC 7.61). The Anglican institute Ridley College argued that "[u]sing the language of ES cell lines serves to mask the fact that the earliest form of human embryo, the blastocyst, must be destroyed in order to obtain these ES cells, which are extracted from the inner cell mass... One does not need to adopt the view that the early embryo has the same moral status as a developed human being, to nevertheless deny that it has no moral status and is not entitled to any protection or any respect" (AC 7.62).

3.4 Recommendations of the Andrews Committee

The Andrews Committee, in agreement with all parties in the debate on reproductive cloning, stated that "cloning for reproductive purposes is unacceptable" (AC 6.72). They therefore called for a "national legislative ban on cloning for reproductive purposes" (AC xi). All Committee members also called for "a ban on the deliberate creation of embryos for experimentation" (AC xii). The Committee stated that "given the number of surplus embryos resulting from assisted reproductive technology, the specific creation of new embryos for research purposes is unnecessary" (AC 7.108). Because of scientific uncertainty and ethical controversy, all Committee members also supported a 3-year moratorium on SCNT for the so-called therapeutic cloning. After 3 years, they suggested that the issue should be re-examined (AC xii, 7.119). A majority of the Andrews Committee (6–4) supported—in defined and limited circumstances—research involving surplus embryos. They also supported research on existing ES cells lines and any new ES cell lines created from surplus embryos (AC xii). The remainder of the Committee (including Kevin Andrews) held that research involving the destruction of human embryos should be prohibited. They permitted research involving the existing ES cell lines, unless it was discovered that these ES cells could develop into an embryo. If that was established, they then wanted research on all ES cell lines prohibited (AC xii, 7.123).

Finally, the Committee recommended that a national licencing body be established to regulate research involving embryos and ES cells in Australia (AC 12.43–12.46). Such research could only be undertaken in Australia once a licence had been granted. Licences could only be granted in limited and clearly defined circumstances, which included the likelihood of significant advance in knowledge or improvement in technology. Licences would also impose clear limits on what could be done and the number of embryos that could be used (AC 12.43).

3.5 *Aftermath to the Andrews Committee*

Despite the majority recommendation of the Andrews Committee that ES cells could be extracted from surplus embryos, initially it appeared that the Australian Government would desist from allowing experimentation involving embryos. On 25 February 2002, the federal cabinet (i.e. the council of senior ministers of the Australian Government) decided to ban the use of spare IVF embryos for research.[20] However, this decision was quickly reversed. On 5 April 2002, then Australian Prime Minister John Howard met with the Council of Australian Governments (COAG) to discuss a united approach to embryonic stem cell research. The Council agreed that the Australian Commonwealth, States and Territories would introduce nationally consistent legislation to ban human cloning and other unacceptable practices.[21] The Council also agreed that research would be permitted on existing excess ART embryos that would otherwise have been destroyed, under a regulatory regime which included requirements for the consent of donors, and that the embryos were in existence at 5 April 2002. Donors would be able to specify restrictions, if they wished, on the research uses of such embryos.[22] Notably, it was agreed that the proposed legislation would prohibit the creation of embryos via fertilization for the purposes of experimentation. Only embryos that were excess to ART could be used for research purposes.

The Australian Catholic Bishops' Conference (ACBC) criticised the COAG decision on embryo research citing that it "created, for the first time in Australian political and legal history, a class of human life which is statutorily expendable".[23] The ACBC further denounced the COAG finding as utilitarian in its ethic, and in conflict with previous ethical codes, such as the *Nuremberg Code* and the *Declaration of Helsinki*, that place "the interests of the research subject... over the interests of research of that of the wider community".[24]

In response to the April 5 COAG meeting, the *Prohibition of Human Cloning and Research Involving Embryos Bill 2002* (Cth) was drafted and introduced into the Australian House of Representatives on 27 June 2002. In July 2002, the Southern Cross Bioethics Institute (SCBI) responded to the proposed legislation, asserting that

---

[20]Louise Dodson and Darren Gray, "Embryo ban hits research," *The Age* (Melbourne), 26 February 2002, http://www.theage.com.au/articles/2002/02/25/1014471629519.html.

[21]Council of Australian Governments Meeting, 5 April 2002, *Communique*, 1.

[22]Ibid.

[23]Australian Catholic Bishops' Conference, "Submission to NHMRC Consultation into Human Cloning and Research Involving Embryos Bill 2002," 2.

[24]Ibid.

it would allow "destructive research on human embryos for a much wider range of purposes, the derivation of ES cells accounting for the smallest number of human embryos which have been, and will be, used".[25] SCBI highlighted that "it was made clear in the presentations to the Andrews Committee by those advocating ES cell research that for ES cell research to go ahead, few if any new human embryos would be required".[26] SCBI argued that following this concession it was later submitted by pro-ES cell advocates that "many more would be needed".[27] The SCBI added that "... the goalposts were shifted once more, with the claim that the rules were too restrictive by limiting access to embryos frozen prior to 5 April 2002." Several groups brought to the table concerns over the quality of frozen embryos, arguing that "high-quality" fresh embryos were needed, which begs the question: "What is the point of keeping IVF embryos in deep freeze for infertility treatment if they are 'low quality'?"[28]

On 29 August 2002, the *Prohibition of Human Cloning and Research Involving Embryos Bill 2002* was divided into two pieces of legislation: the relatively uncontroversial *Prohibition on Human Cloning Bill 2002* (Cth) and the much more controversial *Research Involving Human Embryos Bill 2002* (Cth). The *Prohibition on Human Cloning Bill 2002* (Cth) was easily passed by the House of Representatives and later passed by the Senate on 18 September 2002. On 25 September 2002, the *Research Involving Human Embryos Bill 2002* (Cth) was passed in the House of Representatives by a vote of 99 to 35. On 15 October, it was introduced into the Senate and referred to a committee which reported back on 14 November 2002.[29] On 5 December 2002, after making 14 amendments, the Senate passed this bill by a vote of 45 to 26.[30] On 19 December 2002, the two bills—the *Prohibition of Human Cloning Act 2002* (PHC) and the *Research Involving Human Embryos Act 2002* (RIHE)—having passed both Houses, received the Royal Assent and so became Acts of the Australian Parliament.[31]

On 16 April 2004, the NHMRC Licencing Committee, which had been established following the passing of the RIHE Act 2002, granted the first licences allowing research using excess ART embryos to Sydney IVF Ltd.[32] On 24 June 2004, Sydney IVF Ltd. became the first Australian institute to successfully extract

---

[25] John I. Fleming, Gregory K. Pike, and Selena Ewing, *Human Embryos: A Limitless Scientific Resource?* (Plympton: Southern Cross Bioethics Institute, 2002), 1.

[26] Ibid., 6.

[27] Ibid.

[28] Ibid., 7.

[29] Norman M. Ford and Michael Herbert, *Stem Cells: Science, Medicine, Law and Ethics* (Strathfield: St Pauls, 2003), 51.

[30] Ibid.

[31] Ibid.

[32] David Wroe and Tom Noble, "Embryonic stem cells on the way," *The Age* (Melbourne), 17 April 2004, http://www.theage.com.au/articles/2004/04/16/1082055648663.html.

stem cells from a human embryo.[33] Following this, in September 2004, the NHMRC issued a revision of its 1996 Ethical Guidelines to accommodate the legislative changes, allowing for access to excess ART embryos.

4 The Lockhart Review (2005)

Pursuant to two sections of the *Prohibition of Human Cloning Act 2002* (Cth) and the *Research Involving Human Embryos Act 2002* (Cth), a review of Australia's legislation on cloning and embryonic stem cell research was to be submitted to the Australian Parliament by 19 December 2005, the third anniversary of the day on which the two Acts received Royal Assent. In June 2005, the Minister for Ageing, the Hon. Julie Bishop MP, appointed a six-member Legislation Review Committee headed by John Lockhart AO QC. This became known as the "Lockhart Review" (LR). The Lockhart Review deliberated on whether the existing Australian regulations regarding human embryo research should continue and whether they should be amended in any way. Above all, the Committee considered the ethicality of creating embryos through SCNT with the intention of then destroying these embryos to harvest their ES cells. The central ethical question was whether it was morally permissible to destroy "specially created" embryos in order to facilitate research on stem cells.

4.1 Reproductive Cloning

The Lockhart Review briefly touched upon the issue of reproductive cloning. However, because most respondents supported the continued prohibition of reproductive cloning, little attention was devoted to the issue. The Committee noted that the reproductive cloning of human beings is "considered unacceptable throughout the world because of ethical concerns about the social and psychological implications of creating a copy of a living or dead person, and safety issues associated with the technology" (LR 6.1).

4.2 Therapeutic Cloning

A considerable hurdle that religious agents to the debate regarding therapeutic cloning had to overcome during the Lockhart Review was to demonstrate how

[33]Susan Williamson, "Sydney IVF produces stem cells," 24 June 2004, http://www.labonline.com.au/content/life-scientist/news/sydney-ivf-produces-stem-cells-258421422.

therapeutic cloning was morally impermissible in a culture that permitted destructive research on embryos surplus to ART. An essential part of their argument against therapeutic cloning was therefore to re-present their arguments that destructive research on surplus embryos was also ethically unacceptable.

For example, the submission of the Australian Catholic Bishops' Conference (ACBC) focused strongly on human dignity. The bishops insisted that respect for human dignity is "of paramount importance to our integrity as a nation" and "a fundamental community standard".[34] They noted that some people do not accord "the dignity of humanity" to human embryos. "But the onus is upon such people to prove such a claim".[35] In the absence of "irrefutable proof that a human embryo is not human, a human embryo must be accorded human dignity and thus protection".[36] Respect for human dignity would "not allow the cloning of a human" either for reproductive or therapeutic purposes, "not permit destructive research on human embryos", and "definitely not permit further research than currently permitted by law".[37] The Australian Christian Lobby (ACL) did not prescind from the moral unacceptability of any research which involved the destruction of human embryos. For example, they noted that "therapeutic or non-reproductive cloning still involves creating a human clone and then destroying the embryo", adding that "[t]his embryo is still potentially a viable human being".[38] They stated clearly that they "would prefer that no harmful or destructive research involving human embryos occurred". They therefore called for the "current restrictions on research involving embryos" to be "maintain[ed] and/or extend[ed]"[39] to exclude the existing permission for the use in research of excess embryos. However, the ACL submission focused primarily on the pragmatic argument that ES cell research was not bringing its promised benefits. They argued that community acceptance of embryo research was "predicated on the belief that this form of research may be acceptable if it yielded appropriate results; a sort of ethical *quid pro quo*". They concluded that "[g]iven the failure of embryonic stem cells to yield socially beneficial results, research of this type should be curtailed".[40] In making this argument, they pointed to the successes of adult stem cell research,[41] the risk that ES cells might cause cancers, and the fact that ES cell treatment would be "prohibitively expensive".[42] They therefore concluded that the

[34]Australian Catholic Bishops' Conference, "A Submission from the Australian Catholic Bishops' Conference To The Lockhart Review of The Prohibition of Human Cloning Act 2002 and The Research Involving Human Embryos Act 2002," 2.

[35]Ibid., 5.

[36]Ibid., 7.

[37]Ibid., 2.

[38]Australian Christian Lobby, "Public Submission: Legislation Review of Australia's *Prohibition of Human Cloning Act 2002* and *Research Involving Human Embryos Act 2002*," 5.

[39]Ibid., 3.

[40]Ibid., 8.

[41]Ibid., 7.

[42]Ibid., 8.

research focus should be not on ES cell research with its "ethical complexities and minimal social benefit"[43] but on "already demonstrated and proven routes other than embryonic research that can be pursued to cure diseases and illnesses".[44]

In its submission, the Southern Cross Bioethics Institute (SCBI) also began with the 2002 decision to permit research on excess embryos. This permission was given, the Institute argued, because the "community believed they were agreeing to the use of human embryos for possible medical cures via ES cells".[45] Against this, the Institute noted that "there have been no treatments, clinical trials or even solid research indicating the application of ES cells in the foreseeable future, but also [that] the majority of embryos licensed for use have been for purposes unrelated to ES cells or cures". SCBI concluded that the "public might justifiably feel misled".[46] The reason why experimentation was restricted to embryos already created on 5 April 2002, SCBI noted, was so that "no embryos should be deliberately created for research". It recognised this as one of the "key principles" of the legislation and notes that "there is no evidence that this community standard has changed".[47] "[I]f therapeutic cloning were to be allowed, the deliberate creation of cloned human embryos for the express purpose of their destruction would immediately undermine the community standard that embryos only be created for infertility treatment".[48] SCBI recognised that the "central ethical question" involved "the destruction of embryonic human life".[49] From this perspective, they recommended "that the ban on human cloning be maintained and that no research that is detrimental to human embryos be permitted".[50] Queensland Right to Life (QRTL) argued that already inherent within ART was the commodification of human life. Continuing to follow this path through therapeutic cloning, QRTL argued, could lead eventually to reproductive cloning (LR 6.3).

The calls by religious and other actors not to allow therapeutic cloning received valuable support in March 2005, when it was announced that Griffith University in the Australian state of Queensland had achieved a world first by growing adult stem cell lines harvested from the human nose.[51] Research conducted on these stem cells demonstrated that they possessed the potency to develop into new brain cells, liver cells, heart cells, kidney cells, and muscle cells.[52] The research conducted by Griffith

---

[43]Ibid., 9.

[44]Ibid., 3.

[45]Southern Cross Bioethics Institute, "Submission to the Lockhart Review Committee," 3–4, http://www.bioethics.org.au/Resources/Submissions/Lockhart%20Review%20Submission.pdf.

[46]Ibid., 2.

[47]Ibid., 4.

[48]Ibid.

[49]Ibid., 3.

[50]Ibid., 2.

[51]Tess Livingstone and Cameron Atfield, "Adult stem cells grown in world first research," *The Courier Mail* (Brisbane), 22 March 2005, 4.

[52]Ibid.

University had been partly funded by a $A50,000 grant from the Catholic Archdiocese of Sydney. The Archdiocese had made this money available to contribute in a practical way to the debate about stem cells by highlighting the promise of adult stem cells to provide the hoped-for cures of degenerative diseases.[53] These developments gave the then Catholic Archbishop of Sydney Cardinal George Pell a most significant place in the debate about therapeutic cloning. Cardinal Pell highlighted the Church's support for "ethically acceptable forms of biotechnology that would protect and promote the health and well-being of every member of the human family, and give proper expression to the dignity of human life".[54] With this development of yet another alternative to therapeutic cloning, Cardinal Pell also argued that there was "no strong scientific case for expanding access to ART embryos by allowing the deliberate manufacture of human embryos for research purposes".[55]

4.3 Recommendations of the Lockhart Review

The Lockhart Review issued its report on 19 December 2005.[56] It did support therapeutic cloning. It stated that "[h]uman somatic cell nuclear transfer should be permitted, under licence, to create and use human embryo clones for research, training and clinical application, including the production of human embryonic stem cells, as long as the activity satisfies all the criteria outlined in the amended Act and these embryos are not implanted into the body of a woman or allowed to develop for more than 14 days" (LR Recommendation 23). The Committee acknowledged the argument from pro-life interlocutors that the acceptance of therapeutic cloning could result in the eventual acceptance of reproductive cloning (LR xvii). However, the Committee argued for a safeguard to prevent this contending that "development of human embryos created by any method not involving fertilisation of an egg with sperm beyond 14 days, or implantation of such an embryo into a woman's reproductive tract, should continue to be prohibited to ensure that such embryos are not used for reproductive purposes" (LR xvii). The Committee further asserted that it was irrelevant that an embryo created by SCNT which was implanted into a woman's uterus could "develop into a human being". They argued that such an embryo is "not intended to be implanted, but is created as a cellular extension of the original subject" (LR xvii). Following on from the majority of the Andrews Committee, the Lockhart Review concluded that the compelling

---

[53]Ibid.

[54]David Govorcin, "Cardinal tells Govt of embryo research fears," *The Catholic Weekly* (Sydney), 2 October 2005, 2.

[55]Ibid.

[56]Lockhart Review Committee, *Legislation Review:* Prohibition of Human Cloning Act 2002 *and* Research Involving Human Embryos Act 2002, 19 December 2005, https://www.nhmrc.gov.au/about/nhmrc-committees/embryo-research-licensing-committee/human-embryos-and-cloning/review-human-cl. We refer to this document as LR.

factor regarding whether research should be conducted on human embryos was that "the production and destruction of such an embryo is not dissimilar to the production and destruction of excess ART embryos" (LR xvii). In this, the Lockhart Review went further than the Andrews Committee by asserting that there is no distinction between the destruction of excess ART embryos and the creation and subsequent destruction of embryos created by SCNT. Thus, the Committee stated that "...to permit one (production and destruction of ART embryos) but not the other (production and destruction of nuclear transfer and other bioengineered embryos) would be inconsistent and appear to attach more importance to the treatment of infertility than to the treatment of other diseases and conditions that could be helped as a result of this activity" (LR xvii). However, the Committee did not recommend that embryos could be created through the process of fertilization for the purposes of experimentation.

In a particularly controversial decision, the Lockhart Review also recommended reducing the need for human oocytes or eggs by permitting SCNT involving animal oocytes and human somatic cell nuclei. Recommendation 24 stated that "[i]n order to reduce the need for human oocytes, transfer of human somatic cell nuclei into animal oocytes should be allowed, under licence, for the creation and use of human embryo clones for research, training and clinical application, including the production of human embryonic stem cells, as long as the activity satisfies all the criteria outlined in the amended Act and these embryos are not implanted into the body of a woman or allowed to develop for more than 14 days" (LR Recommendation 24).

5 Aftermath to the Lockhart Review

The Catholic Archdiocese of Sydney's grant for research into adult stem cells—$A100,000 since 2005—has been awarded every second year beginning in 2003, with the seventh such grant having been awarded in 2015.[57] By coincidence, the winner of the second of these grants was announced on 16 December 2005, less than a week before the Lockhart Review released its report.[58] This provided the backdrop to a statement by Cardinal Pell on 22 December 2005 strongly criticising the Lockhart Report. The Cardinal objected to its recommendation of "cloning using

[57]Catholic Communications, Sydney Archdiocese, "Applications Invited for Archdiocese's $100,000 Grant for Adult Stem Cell Research," 18 September 2015, https://www.sydneycatholic.org/news/latest_news/2015/2015918_428.shtml; Catholic Communications, Sydney Archdiocese, "Archbishop Fisher announces winner of $100,000 Adult Stem Cell Research Grant," 9 December 2015, https://www.sydneycatholic.org/news/latest_news/2015/2015129_630.shtml.

[58]Catholic Communications, Archdiocese of Sydney, "Cardinal Pell announces winner of $100,000 Adult stem cell research grant," 16 December 2005, https://www.sydneycatholic.org/news/media_releases/2005/20051216_1443.shtml.

human genetic material and animal eggs".[59] He criticised its distinction between reproductive cloning which would continue to be prohibited and the so-called therapeutic cloning which it would allow. On this, he stated, "All human cloning is reproductive because all human cloning creates new human life. It is never 'therapeutic' to destroy human life..."[60] He added, "There is a strange logic at work in the report which prohibits cloning when it is reproductive but allows it when it is destructive".[61] Above all, Cardinal Pell criticised the Committee for "manufacturing a specific sub-class of living human beings solely for use as research material".[62] He also pointed instead to human stem cells "taken from umbilical cord blood and the body tissues of children and adults. This harms no one and has repeatedly yielded good results for treatment".[63] Pell concluded his statement by calling upon all Australians and the Australian Parliament to reject the recommendations of the Lockhart Review.[64]

Initially it appeared that the then Australian Government, led by then Prime Minister John Howard, was set to disregard the recommendations of the Lockhart Review regarding therapeutic cloning.[65] Prime Minister Howard dismissed the suggestion that such an issue should be put to a conscience vote in Parliament.[66] However, following party room pressure and public outcry calling for a conscience vote, Mr. Howard allowed the ruling Liberal and National Parties a conscience vote on this issue.[67] Kevin Rudd, leader of the opposition Australian Labor Party, also allowed his party a conscience vote on this issue, despite being personally opposed to the practice of therapeutic cloning.[68] On 14 September 2006, the Australian Democrats Senator Natasha Stott Despoja and Labor Senator Ruth Webber tabled the *Somatic Cell Nuclear Transfer (SCNT) and Related Research Amendment Bill 2006* in Parliament.[69] On the same day, the Senate's Standing Committee on Community Affairs was asked to conduct an inquiry into the legislative responses

---

[59]Cardinal George Pell, "Statement on the Lockhart Report," 22 December 2005, https://www.sydneycatholic.org/news/media_releases/2005/20051222_306.shtml.

[60]Ibid.

[61]Ibid.

[62]Ibid.

[63]Ibid.

[64]Ibid.

[65]Kristy Horsey, "Australian states may go their own way on stem cells," BioNews, 17 July 2006, http://www.bionews.org.uk/page_12793.asp.

[66]Jess Buxton, "No 'conscience vote' on embryo cloning in Australia," 7 August 2006, http://www.bionews.org.uk/page_12816.asp.

[67]"PM to allow stem cell conscience vote," Australian Broadcasting Corporation, 15 August 2006, http://www.abc.net.au/news/2006-08-15/pm-to-allow-stem-cells-conscience-vote/1239518.

[68]Fernando Herrera and Justin Corfield, "Australia," in *Encyclopedia of Stem Cell Research*, ed. Clive N. Svendsen and Allison D. Ebert (Thousand Oaks: Sage, 2008), 18.

[69]Kerry Ross, Susan Dodds, and Rachel A. Ankeny, "Conscience Votes in Australia: Deliberation and Representation," in *Big Picture Bioethics: Developing Democratic Policy in Contested Domains* (Switzerland: Springer, 2016), 51.

to the recommendations of the Lockhart Review. While the inquiry was proceeding, and after much cross-party dialogue between proponents of SCNT, a more conservative bill was introduced by Liberal Senator Kay Patterson on 19 October 2006 titled *The Prohibition of Human Cloning for Reproduction and the Regulation of Human Research Amendment Bill* (the "Patterson Bill").[70] The Inquiry's report was tabled on 30 October 2016.[71] The Patterson Bill sought to amend both the *Prohibition of Human Cloning Act 2002* and the *Research Involving Human Embryos Act 2002* by implementing some of the Lockhart recommendations, which included:

- A continued ban on reproductive cloning, but a provision allowing human cloning for therapeutic purposes
- Allowing the production of hybrid animal-human embryos for research
- Prohibition of the production of human embryo clones except under NHMRC licence[72]

On 20 October 2006—the day following the introduction of the Patterson Bill—Catholic bishop Anthony Fisher was among a number of witnesses who appeared before the Australian Senate's Standing Committee on Community Affairs. Bishop Fisher insisted that the "cloning of human beings is ethically abhorrent:"[73] "…To clone them is a failure of respect for the human being who is manufactured and the denial of universal human dignity. To create human beings by this method in order to destroy them only aggravates the offence…"[74] Thus, the bishop concluded, "…So-called therapeutic cloning is much more unethical than so-called reproductive cloning because not only does it create human life in an immoral way but it does so with the object of killing that person for parts."[75] Bishop Fisher also noted that therapeutic cloning would violate the established principles of medical research, "international instruments regarding research ethics", and "the general tenor of medical research in [Australia]".[76] It would violate "the foundational principle of

---

[70]Ibid.

[71]Senate Standing Committee on Community Affairs, *Legislative Reponses to the Recommendations of the Lockhart Review*, 30 October 2006, http://www.aph.gov.au/Parliamentary_Business/Committees/Senate/Community_Affairs/Completed_inquiries/2004-07/leg_response_lockhart_review/report/index.

[72]Geoffrey Maxwell Bullock, "Submission 76 to the Inquiry into the Legislative Responses to Recommendations of the Lockhart Review," 3 October 2006, http://www.aph.gov.au/Parliamentary_Business/Committees/Senate/Community_Affairs/Completed_inquiries/2004-07/leg_response_lockhart_review/submissions/sublist.

[73]Anthony Fisher, *Hearing before the Australian Senate's Standing Committee on Community Affairs*, CA 52, 20 October 2006, Commonwealth of Australia, Official Committee Hansard, http://www.aph.gov.au/Parliamentary_Business/Committees/Senate/Community_Affairs/Completed_inquiries/2004-07/leg_response_lockhart_review/hearings/index.

[74]Ibid.

[75]Ibid.

[76]Ibid., CA 62.

medical research ethics, *primum non nocere*, 'first do no harm'..."[77] It would conflict with the Nuremberg Code, the Declaration of Helsinki, and the 2005 UN Declaration on Human Cloning which Australia had supported.[78] Bishop Fisher warned that if therapeutic cloning was permitted, a "moral Rubicon"[79] would be crossed.

Many people of faith and religious agencies also made submissions to the Senate inquiry. These included the Australian Catholic Bishops' Conference, Catholic Health Australia, the Southern Cross Bioethics Institute, the Caroline Chisholm Centre for Health Ethics, the Queensland Bioethics Centre, the Catholic Women's League Australia, the Social Issues Executive of the Anglican Diocese of Sydney, the Lutheran Church of Australia Commission on Social and Bioethical Questions, Evangelicals for Life, the Australian Christian Lobby, and so on.[80] In its submission, the Australian Christian Lobby (ACL) noted that cloning "requires large numbers of eggs per clone". Arguing that "altruism alone is unlikely to result in enough donated eggs", the ACL foresaw perverse incentives such as "discounted IVF" for egg donation—incentives which many couples desperate for IV would find "hard to resist". They also noted that this large-scale production of eggs would involve "hyper-stimulating women's ovaries"—a procedure which involves "risks to their health".[81]

Following much discussion, the Patterson Bill was subsequently amended to remove the permissibility of producing of hybrid animal-human embryos or chimaeras for research, before being passed by the Senate on 7 November 2006 in a close conscience vote (34–32).[82] On 6 December 2006, the Patterson Bill was debated in the House of Representatives, where the Bill was passed in another conscience vote (82–62).[83] On 12 June 2007, the *Research Involving Human Embryos Act 2002* and the *Prohibition of Human Cloning Act 2002* were successfully amended by the passing of the *Prohibition of Human Cloning for Reproduction and the Regulation of Human Embryo Research Amendment Act 2006*, to allow SCNT where a licence is granted, but to continue the ban on reproductive cloning, and to continue not to permit the deliberate creation of a human embryo through fertilization of a human egg by a human sperm for the purposes of experimentation. One amendment included renaming the *Prohibition of Human Cloning Act 2002* the *Prohibition of Human Cloning for Reproduction Act 2002*. Most Australian states

[77]Ibid., CA 52.

[78]Ibid., CA 61–62, CA 53.

[79]Ibid., CA 52, CA 62.

[80]All the submissions to the Senate Inquiry are online at http://www.aph.gov.au/Parliamentary_Business/Committees/Senate/Community_Affairs/Completed_inquiries/2004-07/leg_response_lockhart_review/submissions/sublist.

[81]Australian Christian Lobby, "Submission 90 to the Senate Inquiry into Legislative Responses to the Recommendations of the Lockhart Review," 1.

[82]Ross, Dodds, and Ankeny, 51–52.

[83]Ibid., 52.

passed legislation that mirrored the federal legislation in allowing SCNT for the so-called therapeutic cloning. Western Australia was the only Australian state not to pass legislation that would make SCNT legally permissible.[84]

The Parliament of Western Australia (WA) has a cross party, loose coalition of politicians who are interested in human life issues. Some but not all of them are people of faith (mostly Christian). From time to time—as they did on this issue—they work together on bills which cause them concern. In this case, they provided education on the issue by eminent scientist Dr. Jack Martin and Dr. David van Gend from Australians for Ethical Stem Cell Research.[85] They debated with those who supported the legislation and discussed the issues with their colleagues who were undecided. They also recorded their concerns in speeches to Parliament. In this case, two of the most important of these speeches were given in the WA Legislative Council on 10 April 2008 by the Hon Ed Dermer and the Hon Kate Doust. Both recorded their ethical objection to the legislation. For example, Kate Doust stated "I cannot deal with the concept that a human life can be created purely for the purpose of destruction. . .".[86] Both also spoke about the recent discovery of induced pluripotent stem cells as an ethical alternative to embryonic stem cell research. For example, Ed Dermer stated that "it is highly probable that the new method, which does not involve embryos but allows skin cells to be transformed into pluripotent stem cells, will achieve everything that can be claimed for a clone-produced stem cell".[87] The relevant Western Australian legislation was the *Human Reproductive Technology Amendment Bill 2007* (WA). It was defeated in the WA Legislative Council on 6 May 2008, by a vote of 18–15.[88] In 2007, the NHMRC issued a revision of their 2004 ART guidelines to incorporate the changes brought about by the 2006 legislation.

On 16 September 2008, the first licences permitting therapeutic cloning were granted by the Embryo Research Licensing Committee of the National Health and

[84]Australian Stem Cell Centre, "Fact Sheet 6: Ethics & Law of Stem Cell Research," 1, http://www.stemcellfoundation.net.au/docs/fact-sheets/fact-sheet-6%2D%2D-law-and-ethics-of-stem-cell-research.pdf?sfvrsn=13.

[85]Hansard: Western Australia, Legislative Council, 2008, *Parliamentary Debates*, 10 April, 2151b–2161a, Hon Kate Doust, http://www.parliament.wa.gov.au/Hansard/hansard.nsf/0/20426b570f5ae5c6c8257570000fae2a/$FILE/C37+S1+20080410+p2151b-2161a.pdf.

[86]Ibid.

[87]Ibid., Hon Ed Dermer.

[88]Hansard: Western Australia, Legislative Council, 2008, *Parliamentary Debates*, 6 May, 2337c–2341a, http://www.parliament.wa.gov.au/Hansard/hansard.nsf/0/f42953e737d3bbfec8257574000b29e0/$FILE/C37+S1+20080506+p2337c-2341a.pdf. We are grateful to Catholic priest Revd Dr. Joseph Parkinson from the Perth-based L J Goody Bioethics Centre for information about what took place in the Western Australian Parliament.

Medical Research Council.[89] Sydney IVF Limited was granted three licences to derive human embryonic stem cells from embryos utilising SCNT techniques.[90]

6 The Heerey Review (2011)

Pursuant to two legislative provisions in both the amended Commonwealth Acts, a further review of Australia's legislation on embryo research was to be conducted in 2010, 3 years after the amendments to the two acts had come into effect. In December 2010, therefore, a Legislation Review Committee was again appointed to undertake an independent review of the two acts. This review was headed by the Hon Peter Heerey QC (the "Heerey Review"). The Heerey Review (HR) comprised of five persons, who included Catholic priest Reverend Kevin McGovern, co-author of this chapter and then Director of the Melbourne-based Caroline Chisholm Centre for Health Ethics. The Heerey Review received 264 written submissions. These included many short submissions from private individuals, along with longer submissions from other individuals and various organisations. Of the submissions, 112 stated that they did not support human cloning, while 188 said that they did not support the use of human embryos for research (HR 3.4). Among the religious agents who made submissions were the Australian Catholic Bishops' Conference; the Australian Christian Lobby; the Southern Cross Bioethics Institute; the Life, Marriage and Family Centre of the Catholic Archdiocese of Sydney; St Vincent's Institute of Medical Research; St Vincent's and Mater Health in Sydney; Dr. Nicholas Tonti-Filippini; the Ad Hoc Interfaith Committee; Saltshakers; Cherish Life Queensland; NSW Right to Life; the Institute for Judaism and Civilisation; and Australians for Ethical Stem Cell Research.[91] After receiving written submission and hearing from expert witnesses, the Heerey Review presented its report on 27 May 2011.[92]

[89]National Health and Medical Research Council, "First licences granted for the derivation of human embryonic stem cells from cloned embryos," 16 September 2008, https://www.nhmrc.gov.au/media/nhmrc_updates/2009/first-licences-granted-derivation-human-embryonic-stem-cells-cloned-embryos.

[90]Ibid.

[91]All the submissions are listed in Schedule 1 of HR on pages 85–95.

[92]See Heerey Review Committee, *Legislation Review:* Prohibition of Human Cloning Act 2002 *and* Research Involving Human Embryos Act 2002, 27 May 2011, https://www.nhmrc.gov.au/guidelines-publications/hc38. We refer to this document as HR. Our report here focuses on what was arguably the most significant debate of the Heerey Review. Of course, other matters were also debated. For a more detailed report on the Heerey Review, see Kevin McGovern, "Australia's Cloning and Embryo Research Laws," *Chisholm Health Ethics Bulletin* 16, no. 4 (Winter 2011): 1–4, http://chisholmhealthethics.org.au/system/files/bulletin_16_4.pdf.

6.1 Reproductive Cloning

The Heerey Review devoted scant attention to the question of reproductive cloning because no submissions made to it recommended that the Australian ban on reproductive cloning should be lifted (HR 5.1). Thus, the Heerey Review recommended that, "Cloning of a human being for reproduction should remain a criminal offence" (HR 5.1).

6.2 Therapeutic Cloning

The debate surrounding embryonic stem cell research during the Heerey Review was particularly shaped by a landmark scientific breakthrough. In 2007, a team led by Japanese scientist Shinya Yamanaka discovered how to "reprogramme" more specialised human somatic cells into induced pluripotent stem cells (iPS cells), returning them to a primordial state in which they regained most if not all of the capacities of embryonic stem cells.[93]

In its submission to the Heerey Review, the Australian Christian Lobby (ACL) restated that it opposed both the "destruction of excess ART embryos" and "all forms of cloning" either for reproduction or research.[94] It insisted that "cloning creates a living human being" and particularly opposed the "creation, and deliberate destruction, of human life".[95] It also reiterated its concern that eggs might be sought from women through "emotional and financial manipulation", while "harvesting eggs is dangerous for women".[96] In particular, it argued that developments since 2006—including the "groundbreaking work" by Yamanaka—had removed the need for therapeutic cloning. If there had once seemed to be pragmatic reasons for permitting SCNT, "subsequent developments in adult stem cell research have removed any justification which may have existed in 2006".[97] The ACL pointed out the very "limited success of 'therapeutic' cloning as a means of sourcing embryonic stem cells".[98] It noted that "adult stem cells are much easier to generate than embryonic stem cells", and, because they can be matched to the recipient, "there is significantly less risk of rejection".[99] The ACL also quoted authoritative sources who said that the

[93]Junying Yu et al., "Induced Pluripotent Stem Cell Lines Derived from Human Somatic Cells," *Science* 318, no. 5858 (21 December 2007): 1917–1920.

[94]Australian Christian Lobby, "Submission to the 2010 Legislation Review of the *Prohibition of Human Cloning for Reproduction Act 2002* and the *Research Involving Human Embryos Act 2002*," 1.

[95]Ibid., 3.

[96]Ibid., 9.

[97]Ibid., 2.

[98]Ibid., 11.

[99]Ibid., 7.

future lay with not with SCNT but with adult and induced pluripotent stem cells. These included Professor Ian Wilmut who created Dolly the Sheep, Dr. James Thompson who first discovered human embryonic stem cells, Sir Martin Evans of the Cardiff School of Biosciences, and Australian scientists Dr. Martin Pera and Dr. Jack Martin.[100] For all these reasons, the Australian Christian Lobby therefore argued for the "repeal of those provisions in the Acts that allow the cloning of human beings" and for the restriction of all "future scientific and medical research, and associated public and private funding" to the "advancement of medical technologies related to adult stem cell research".[101] Most other submissions from religious actors made very similar points. The submission from the Australian Catholic Bishops' Conference (ACBC) also pointed out that "all cloning and destructive research on human embryos"[102] contravenes "the international standards on human experimentation and the secular Hippocratic medical tradition",[103] the *United Nations Universal Declaration of Human Rights*,[104] and the UN General Assembly Declaration on Human Cloning 2005.[105] Indeed, as the ACBC noted, the latter called on all member states "to prohibit all forms of human cloning inasmuch as they are incompatible with human dignity and the protection of human life".[106]

In contrast with other religious submissions, the Executive Council of Australian Jewry (Submission 112) supported therapeutic cloning, recommending that while the current prohibition on reproductive cloning should be maintained, "a defining barrier is established around implantation and incubation".[107] On the other hand, the Institute for Judaism and Civilisation (Submission 283) opposed all forms of cloning, both reproductive and therapeutic.

6.3 Recommendations of the Heerey Review

The Heerey Review did not recommend against SCNT for the so-called therapeutic cloning. Instead, by majority, it recommended that "the provisions in the current legislation regarding SCNT should not be amended" (HR 53). It did note the "lack of progress in SCNT research". It added that this "must impact on the Licensing

[100]Ibid., 7–8.

[101]Ibid., 1.

[102]Australian Catholic Bishops' Conference, "A Submission from The Australian Catholic Bishops' Conference To The Heerey Review of The Prohibition of Human Cloning and Research Involving Human Embryos Acts 2002 as amended," 2.

[103]Ibid., 7.

[104]Ibid., 3.

[105]Ibid., 2.

[106]Ibid., 10.

[107]Executive Council of Australian Jewry, "Review of the Prohibition of Human Cloning for Reproduction Act 2002 and the Research Involving Human Embryos Act 2002," 3.

Committee's interpretation of its statutory obligation" whereby it can only approve research which has a "likelihood of significant advance in knowledge or improvement in technologies for treatment... which could not reasonably be achieved by other means" (HR 53). By law, SCNT research can only be licenced if there is the likelihood of significant advances. However, the lack of progress to date made this less likely. The Heerey Review therefore called the Licensing Committee to be rigorous in its application of its statutory criteria for the licencing of SCNT research. In a minority report to this recommendation, Dr. Faye Thompson, one of the five members of the Heerey Review Committee, stressed that she had only supported this recommendation "on the basis that the Licensing Committee place emphasis" on these standards "when it is considering any future application for a licence to undertake research involving SCNT" (HR 54).

The Heerey Review gave many reasons for this majority recommendation. Based on "the most recent surveys", it suggested that "the majority of the Australian community supports stem cell research even when it involves donated human embryos and therapeutic cloning (SCNT)" (HR 49). It quoted a number of scientific reports which suggested that there might be subtle but significant differences between ART-derived ES cells, iPS cells, and SCNT-derived ES cells (HR 44–48). It quoted US geneticist Dr. Francis Collins—"well known for his conservative Christian ethical views"—who stated that "...not enough [is] yet known about [iPS] cells to guess whether they have the same therapeutic potential as embryonic stem cells... So it's not 'either/or' that we should be pursuing. It's 'both/and'" (HR 43). The majority of the Heerey Review also recognised that "it is not possible for anyone to predict what may occur in this area of research in the future" (HR 44). "Given the uncertainty about future outcomes", it argued that all types of ES research should be allowed to continue (HR 48). Along with the minority report by Dr. Faye Thompson, there was another minority report by Revd Kevin McGovern. He noted that the Lockhart Review had stated that "where benefits are not yet established, or where there is widespread and deeply held community objection, then total prohibition through the legal system may be justified". He argued that this standard had now been reached for SCNT: "...The proposed benefits of SCNT research... seem not entirely convincing, sometimes rather small, and largely theoretical. On the other hand, SCNT involves the most profound of ethical concerns. It is the creation of human life which will be used in research and then destroyed" (HR 54). Revd McGovern noted that in the majority recommendation of the Heerey Review, the "most serious of ethical concerns has been judged less significant that the mostly theoretical benefits which might come if research into SCNT is allowed to continue". He therefore wondered "whether the ethical concerns about SCNT research are ultimately being given anything more than lip-service" (HR 54).

7 Conclusion

Since the conclusion of the Heerey Review in 2011, no further legislative reviews have been convened in Australia to assess the Australian laws relating to research involving human embryos, reproductive and therapeutic cloning, and embryonic stem cell research. Currently, the Australian laws on these matters remain the *Prohibition of Human Cloning for Reproduction Act 2002* (Cth) and the *Research Involving Human Embryos Act 2002* (Cth), as amended by the *Prohibition of Human Cloning for Reproduction and the Regulation of Human Embryo Research Amendment Act 2006* (Cth), along with matching legislation in most Australian states (except Western Australia). The issue of reproductive cloning remains the only area where a broad and enduring religious and secular consensus has been reached in Australia. The arguments set forth by religious actors in the debate on reproductive cloning during the Andrews Committee (2001) were affirmed by pro-ES research advocates. Religious parties were able to appeal to international human rights documents which state that cloning would be contrary to human dignity. Moreover, emphasis was also placed on the effect that reproductive cloning may have on a person's sense of individuality and identity and on their relationships with family and significant others. The strength of this consensus is reflected in the scant attention that was paid to reproductive cloning in the two successive legislative reviews, the Lockhart Review and the Heerey Review. However, the ethical sentiments uttered by religious interlocutors in the debate about destructive research on excess embryos and SCNT for the so-called therapeutic cloning have not coincided with the opinions put forward by advocates of ES cell research.

Prior to the widespread development of assisted reproductive technology (ART) in Australia, the views of most Australians on these matters would have accorded with the views still held by most religious actors. Destructive research on human embryos would have seemed ethically abhorrent to them. Deliberately creating human embryos by whatever means for destructive research would have seemed doubly abhorrent. The widespread development of ART in Australia created large numbers of stored embryos, which were in excess of the needs of those infertile couples who were involved in ART.

The cloning of animals and the discovery of human embryonic stem cells in the late 1990s created an environment in which research on these excess embryos seemed to have at least some pragmatic justification. Many Australians have had difficulty in understanding why destructive research on excess embryos is morally wrong. What is more, those of us who are convinced of the wrongness of this—be we religious or secular—have in the main not been successful in helping our fellow Australians to understand this. The reasons for this merit careful reflection. Perhaps we have not had the resources for extensive, ongoing public education programmes. Or perhaps we have not yet learnt how to convey these ethical insights in ways which are engaging and convincing. With much acquiescence if not approval of destructive embryo research, it then became even more difficult for those of us who oppose the so-called therapeutic cloning once again to assist other Australians to see the moral wrongness of this. In all this, what has developed over time has been a utilitarian ethic which has permeated the Australian discussion on cloning and human embryo research—an ethic which the opponents of destructive embryo research have not been able to counter.

It must be said that religious actors have been very involved in these debates in Australia. There has been ecumenical cooperation among the Christian Churches. There has been some interfaith cooperation with other faiths. Many religious leaders have been involved, along with many rank and file believers. Many submissions have been put forward, ranging from very significant statements from key organisations, agencies, and individuals to simpler but no less heartfelt contributions from other groups and individuals. We have been well-informed. We have been diligent. We have been assiduous in the debates over many years. However, in the main, we have not yet been successful.

References

Act 2002 and Research Involving Human Embryos Act 2002. 27 May 2011. https://www.nhmrc.gov.au/guidelines-publications/hc38.

Australasian Legal Information Institute. *Infertility (medical procedures) act 1984* (Vic). http://www.austlii.edu.au/au/legis/vic/hist_act/ipa198431 1.pdf.

———. *Human embryo experimentation bill 1985*. http://www.austlii.edu.au/au/legis/cth/bill/heeb1985328/.

Australian Broadcasting Corporation. 2006. PM to allow stem cell conscience vote. August 15. http://www.abc.net.au/news/2006-08-15/pm-to-allow-stem-cells-conscience-vote/1239518.

Australian Catholic Bishops Conference. 2002a. *A submission from the Australian Catholic bishops Conference to the Lockhart review of the prohibition of human cloning act 2002 and the research involving human embryos act 2002*.

———. 2002b. *Submission to NHMRC consultation into human cloning and research involving embryos bill 2002*.

———. 2002c. A submission from the Australian Catholic bishops conference to the Heerey review of *the prohibition of human cloning and research involving human embryos acts 2002* as amended.

Australian Christian Lobby. Public submission: legislation review of Australia's *prohibition of human cloning act 2002 and research involving human embryos act 2002*.

———. 2010. Submission to the 2010 legislation review of the *prohibition of human cloning for reproduction act 2002 and* the *research involving human embryos act 2002*.

———. 2006. *Submission 90 to the senate inquiry into legislative responses to the recommendations of the Lockhart review*.

Australian Health Ethics Committee. 1998. *Scientific, ethical and regulatory considerations relevant to the cloning of human beings*, December 16. https://www.nhmrc.gov.au/_files_nhmrc/publications/attachments/e45.pdf.

Australian Stem Cell Centre. Fact sheet 6: ethics & law of stem cell research. 2010. http://www.stemcellfoundation.net.au/docs/fact-sheets/fact-sheet-6%2D%2D-law-and-ethics-of-stem-cell-research.pdf?sfvrsn=13.

Bullock, Geoffrey Maxwell. 2006. *Submission 76 to the inquiry into the legislative responses to recommendations of the Lockhart review*. October 30. http://www.aph.gov.au/Parliamentary_Business/Committees/Senate/Community_Affairs/Completed_inquiries/2004-07/leg_response_lockhart_review/submissions/sublist.

Buxton, Jess. 2006. No 'conscience vote' on embryo cloning in Australia. *BioNews*, August 7. http://www.bionews.org.uk/page_12816.asp.

Catholic Communications, Sydney Archdiocese. 2005. Cardinal Pell announces winner of $100,000 adult stem cell research grant. December 16. https://www.sydneycatholic.org/news/media_releases/2005/20051216_1443.shtml.

———. 2015a. Applications invited for Archdiocese's $100,000 grant for adult stem cell research. September 18. https://www.sydneycatholic.org/news/latest_news/2015/2015918_428.shtml.

———. 2015b. Archbishop Fisher announces winner of $100,000 adult stem cell research grant. December 9. https://www.sydneycatholic.org/news/latest_news/2015/2015129_630.shtml.

Council of Australian Governments Meeting. 5 April 2002. *Communique*.

Dodson, Louise and Darren Gray. 2002. Embryo ban hits research. *The age* (Melbourne), 26 February. http://www.theage.com.au/articles/2002/02/25/1014471629519.html.

Executive Council of Australian Jewry. *Review of the prohibition of human cloning for reproduction act 2002 and the research involving human embryos act 2002*.

Fisher, Anthony. 2006. *Hearing before the Australian senate's standing committee on community affairs*, CA 52, October 20. Commonwealth of Australia, Official Committee Hansard. http://www.aph.gov.au/Parliamentary_Business/Committees/Senate/Community_Affairs/Completed_inquiries/2004-07/leg_response_lockhart_review/hearings/index.

Fleming, John I., Gregory K. Pike, and Selena Ewing. 2002. *Human embryos: a limitless scientific resource?* Plympton: Southern Cross Bioethics Institute.

Ford, Norman M., and Michael Herbert. 2003. *Stem cells: science, medicine, law and ethics*. Strathfield: St Pauls.

Govorcin, David. 2005. Cardinal tells Govt of embryo research fears. *The Catholic Weekly* (Sydney), October 2.

Hansard: Western Australia, Legislative Council. 2008a. *Parliamentary debates*, 10 April, 2151b–2161a, Hon Kate Doust. http://www.parliament.wa.gov.au/Hansard/hansard.nsf/0/20426b570f5ae5c6c8257570000fae2a/$FILE/C37+S1+20080410+p2151b-2161a.pdf.

———. 2008b. *Parliamentary debates*, 6 May, 2337c–2341a, http://www.parliament.wa.gov.au/Hansard/hansard.nsf/0/f42953e737d3bbfec8257574000b29e0/$FILE/C37+S1+20080506+p2337c-2341a.pdf.

Heerey Review Committee. 2011. *Legislation review: prohibition of human cloning*.

Herrera, Fernando, and Justin Corfield. 2008. Australia. In *Encyclopedia of stem cell research*, ed. Clive N. Svendsen and Allison D. Ebert, 18. Thousand Oaks, CA: Sage.

Horsey, Kristy. 2006. Australian states may go their own way on stem cells. *BioNews*, July 17. http://www.bionews.org.uk/page_12793.asp.

House of Representatives Standing Committee on Legal and Constitutional Affairs. 2001. *Human cloning, scientific, ethical and regulatory aspects of human cloning and stem cell research*. Canberra: Commonwealth of Australia. http://www.aph.gov.au/parliamentary_business/committees/house_of_representatives_committees?url=/laca/humancloning/contents.htm.

Livingstone, Tess, and Cameron Atfield. 2005. Adult stem cells grown in world first research. *The courier mail* (Brisbane), March 22.

Lockhart Review Committee. 2005. *Legislation review: prohibition of human cloning act 2002 and research involving human embryos act 2002*. December 19. https://www.nhmrc.gov.au/about/nhmrc-committees/embryo-research-licensing-committee/human-embryos-and-cloning/review-human-cl.

McGovern, Kevin. 2001. Australia's cloning and embryo research laws. In *Chisholm health ethics bulletin* 16(4) (winter 2011): 1–4. http://chisholmhealthethics.org.au/system/files/bulletin_16_4.pdf.

National Health and Medical Research Council. 1996. *Ethical guidelines on assisted reproductive technology*. Canberra: Commonwealth of Australia. https://www.nhmrc.gov.au/_files_nhmrc/publications/attachments/e28.pdf.

———. 2008. First licences granted for the derivation of human embryonic stem cells from cloned embryos. September 16. https://www.nhmrc.gov.au/media/nhmrc_updates/2009/first-licences-granted-derivation-human-embryonic-stem-cells-cloned-embryos.
Pell, Cardinal George. 2005. Statement on the Lockhart report. December 22. https://www.sydneycatholic.org/news/media_releases/2005/20051222_306.shtml.
Ross, Kerry, Susan Dodds, and Rachel A. Ankeny. 2016. Conscience votes in Australia: deliberation and representation. In *Big picture bioethics: developing democratic policy in contested domains*, ed. Susan Dodds and Rachel A. Ankeny, 51. Heidelberg: Springer.
Senate Inquiry. 2004. Submissions to: http://www.aph.gov.au/Parliamentary_Business/Committees/Senate/Community_Affairs/Completed_inquiries/2004-07/leg_response_lockhart_review/submissions/sublist.
Senate Standing Committee on Community Affairs. 2006. *Legislative Responses to the recommendations of the Lockhart review*. October 30. http://www.aph.gov.au/Parliamentary_Business/Committees/Senate/Community_Affairs/Completed_inquiries/2004-07/leg_response_lockhart_review/report/index.
Southern Cross Bioethics Institute. Submission to the Lockhart review committee. http://www.bioethics.org.au/Resources/Submissions/Lockhart%20Review%20Submission.pdf.
The Interim. Brave new world – down under. 15 May 1988. http://www.theinterim.com/issues/bioethics/brave-new-world-%E2%80%93-down-under/.
Williamson, Susan. 2004. Sydney IVF produces stem cells. 24 June. http://www.labonline.com.au/content/life-scientist/news/sydney-ivf-produces-stem-cells-258421422.
Wroe, David, and Tom Noble. 2004. Embryonic stem cells on the way. *The Age* (Melbourne), April 17. http://www.theage.com.au/articles/2004/04/16/1082055648663.html.
Yu, Junying et al. 2007. Induced pluripotent stem cell lines derived from human somatic cells. In *Science* 318(5858) (21 December 2007): 1917–1920.

Part IV
Protestant(-Scandinavian) States with Low Level Religious Pluralism

Chapter 9
The Status of the Human Embryo: A Case Study of Embryo Experiments and Embryo Research in Denmark

Jacob Dahl Rendtorff

Abstract The problem of the status of the embryo (fertilized eggs in the test tube) is central when considering genetic and reproductive technology. It is mainly about what is ethically justifiable in experiments with embryos at the earliest stages. However, ownership of the embryo, decisions about donation, and freezing of fertilized eggs are also important problems to discuss. Can embryos and fetal tissue be used for transplantation or in the pharmaceutical industry? The answer to these questions depends on the embryonal ethical and legal status of the embryo. This problem is one of the most controversial bioethical issues in Europe, where for religious and cultural reasons, there is a big difference between the perceptions of the various countries. Rendtorff argues that the embryo must be considered as a "third thing" between people and things. The legal policy discussion on embryo status shows that in the specific legal development, despite major differences between the perceptions of the status of the embryo, a pragmatic consensus on embryonal legal status between persons and things in the European countries can be traced. After having presented the ethical and philosophical issues relevant to embryo experiments and embryo research, he goes on to present a case study of the Danish debate about embryo experiments and embryo research. With this case example, he gives an overview of relevant issues concerning embryo experiments and embryo research in bioethics and biolaw, which concern the legal regulation of bioethical issues.

J. D. Rendtorff (✉)
Department of Social Sciences and Business, Roskilde University, Roskilde, Denmark
e-mail: jacrendt@ruc.dk

M. Weiberg-Salzmann, U. Willems (eds.), *Religion and Biopolitics*,
https://doi.org/10.1007/978-3-030-14580-4_9

1 Introduction

The problem of the status of the embryo (fertilized eggs in the test tube) is central when considering the genetic and reproductive technology.[1] It is mainly about what is ethically justifiable in experiments with embryos at the earliest stages. However, ownership of the embryo, decisions about donation, and freezing of fertilized eggs are also important problems to discuss. Can embryos and fetal tissue be used for transplantation or in the pharmaceutical industry? The answer to these questions depends on the embryonal ethical and legal status of the embryo. This problem is one of the most controversial bioethical issues in Europe, where for religious and cultural reasons, there is a big difference between the perceptions of the various countries. I would argue that the embryo must be considered as a "third thing" between people and things (Rendtorff 1997, 1998, 1999, 2001, 2002). Embryonal biological development suggests a gradualist understanding of the required protection embryo. At the same time, the embryo should be protected as unborn human life, a potential person who has symbolic significance for society's moral order. I begin by outlining some biological, theological, and philosophical aspects of the discussion on the status of the embryo. This forms the basis for the preparation of arguments for and against experiments with fertilized eggs. The legal policy discussion on embryo status shows that in the specific legal development, despite major differences between the perceptions of the status of the embryo, a pragmatic consensus on embryonal legal status between persons and things in the European countries can be traced. After having presented the ethical and philosophical issues relevant to embryo experiments and embryo research, I go on to present a case study of the Danish debate about embryo experiments and embryo research (Rendtorff 1999, 2003). With this case example, I hope to give an overview of relevant issues concerning embryo experiments and embryo research in bioethics and biolaw, which concern the legal regulation of bioethical issues (Rendtorff 2004, 2008, 2014).

2 The Biology of the Embryo

Biological arguments play a central role in the justifications of the embryo's ethical and legal status (van der Burg 1995; Arnoux 1995). Legislation and case law about the embryo must be based on a biological understanding of embryo development as the basis for defining the legal status of the embryo (Hare 1992). With regard to such

[1] It has become commonplace to describe the fertilized eggs in the test tube as "pre-embryo," and the discussion in this chapter will be about the legal status of the pre-embryo. In order not to establish an absolute distinction between pre-embryo and embryo, I do not go into this division, but continue to use the term embryo to describe the early human life both before and after the 14-day limit. This concept of pre-embryo has especially gained ground in the Dutch and the English discussion of the status of the embryo. However, one should be cautious in the way that would define itself out of the controversial ethical issues.

"natural legal" justification, legal positivists (Alf Ross) will, as with any other use of the natural law, argue that it serves as "a harlot, which is available to everyone." It is true that biology cannot in itself be the basis for a normative standpoint, but on the other hand, one cannot escape the fact that a biological detection of the stages of fetal progressive development is essential to the definition of the embryo's ethical and judicial status.

There is a biological and genetic continuity between a person's conception and their growth to adulthood, but this biological "identity" cannot be considered identical to the perceived bodily identity, only as a prerequisite for it. The unity in question is of a purely biological nature and is about the body's organic genesis during a progressive fetal development. With artificial insemination, the embryo may be up to 14 days in the test tube before being implanted into the mother's uterus. The test-tube embryo does not consist of more than 2–8 cells. After 16 days, the beginning of the backbone can be traced. After 40 days, the first brain and nervous traits are beginning to emerge. After 5 months, the mother can feel the baby's first movements. At this stage the organism appears increasingly like a biological prerequisite for future living human being. After fertilization a progressive cell differentiation and the progressive development of the organism occur as the bases for the later formation of the person (Ford 1988).

Although the egg and sperm already constitute life, we can still maintain that the new life began only at the moment of conception. In addition, if the human being presupposes the lived body, as well as a sense of belonging to the world and a self-awareness, it is biologically extremely difficult to argue that the embryo at any time before birth is a human being (Thévoz 1990: 188). If you want to argue with the ability to feel pleasure and pain, we may refer to the nervous system and brain development as the basis for the presence of touch or the capacity to be a "rational soul," i.e., a sentient being with moral status. In that case, the human being's genesis can, according to some points of view, be traced back to 21 days after conception, when the first parts of the nervous system appear.

A biological definition as a descriptive requirement for legal protection of the embryo can therefore be summarized in the following steps: (1) The first 14 days when the embryo is in the test tube outside the uterus and where the cell differentiation and genetic individualization are taking place; (2) the embryo in the uterus during development of the nervous system as the basis of experience of desire, pain, and self-awareness; (3) the embryo at a later stage of development with the possibility of a life of relationship that may involve basic experiences, movements, and external factors; and (4) the moment of birth when the fetus or a born baby achieves autonomy from the mother's body and thus has its own independent existence. The essence of the biological prerequisites for graduation is the forward-moving continuous development of a human life from fertilization to become the independent individual (Dunstan and Seller 1988). If the fetus cannot be understood as a human being at all stages of development, the degrees of development become crucial to its legal status. The definition of the fetus as an individual human life in each particular phase can function as a normative biological prerequisite for the required respect for the dignity of a human life, which must be sanctioned in the legal system.

3 The Position of the Catholic Church

A key reason for the controversy in the European debate on the status of the human embryo is that it provokes strong religious feelings. Not least, the Catholic Church is eager to defend the respect for unborn life. Already in Paul VI's *Humanae Vitae* in 1968, which was continued in *Donum Vitae* of 1988, the church considered conception of the embryo as a living human being, a human being from the moment of conception to be the critical basis for its negative assessment of genetic and reproductive technology (The Catholic Church 1997). The church believes that the embryo is a human life created by God at the moment of conception and that in all its development phases it must be protected ethically and legally as a person with full moral status. This is an unavoidable consequence of the conception of the embryo because of a divine creator's will. "Anti-probalistic argument," i.e., that we cannot prove otherwise and therefore must assume that the embryo is a human being, plays a major role in the Catholic Church's argument (Coughlan 1988: 7). And when the embryo is considered as a human person from the moment of conception, it is also a full member of the human species and has the full rights and obligations of human beings.

Michael Coughlan believes we can reconstruct the Church's argument in *Donum Vitae* in the following way: When the life of the fertilized egg is neither father nor mother, we must define it as a new life. This new life is considered a human life, for it would not be human if it were not already human. The new human life is consequently an individual whose identity is created as a human being at the moment of fertilization (Thévoz 1990: 198). The biological organism cannot therefore be considered as a potential human, but rather as a human being with a potential. These arguments follow Thomas Aquinas' philosophy that claims that the embryo possesses a sensitive soul and then little by little develops a rational nature. According to the Catholic Church's view, the fetus' soul develops in parallel with the body's gradual shaping and development.

The Catholic Church claims that all life begins with conception and that it is a continuum of life. This point of view is based on the theory of the genome and in consideration of the genetic individuality of the earliest cell formation as a basis for the definitions of pre-embryo, embryo, and fetus. The main thing is the requirement that society must respect the human being as a human being from conception. The Catholic Church can, on this basis, fight to protect the human organism's integrity and its functions and determine the personality's bodily identity as unique and singular. According to *Donum Vitae*, the embryo with cells and with a specific genome is a person who is already present at conception and at the same time contains all the coming person's psychophysical properties and potentials.

4 Is the Embryo a Human Being?

The living body's potential being in the biological life after fertilization is the essence of the church's definition of the person. This brings it into conflict with another definition of the human being, which emphasizes the ability to relate to the outside world. The question of the status of the embryo also involves the question of what it means to be a person. Here, I would not argue that the embryo is a person in the same sense as an adult human being, but I would instead put forward a theory that takes its point of departure from the embryo's relational and symbolic being between people and things as the basis for its ethical and legal status.

Such a theory would have to decide whether (1) the person becomes a person at fertilization or (2) at birth, which means that the fetus cannot be a person; (3) whether the embryo is a potential person who in a continuous movement from fertilization to birth becomes a real person; and (4) if the embryo is a possible person. That means that it is only a future possible person with no certainty that it will become a real person. In contrast to the Catholic position, it is true that the genetic individuality is present from the moment of conception and that gene structure is the same in the adult human (Dustan and Seller 1988: 97). This is not a static condition but a dynamic development process in which the human becomes him or herself through a progressive continuity as the basis for individuality. The Catholic Church is right to say that the living body cannot be separated from self-consciousness, but simplicity is its material basis, so that fetal development from its beginning has the potential to become the later human being.

Yet it is impossible to understand the embryo as a person at the level of an adult human being. Here the arguments of more utilitarian- and rationalistic-minded theorists are very difficult to ignore (Singer 1993; Tooley 1983; Harris 1985). They define a person as "a rational self-conscious creature," a subject with no immediate interests as an individual able to evaluate their own life. In addition, utilitarianism claims that it is a condition of having a moral status that implies a being who can feel pleasure and pain and having interests and preferences. On those grounds, we cannot claim that the early embryo that has no body or nervous system and cannot feel or anticipate the future is a person and can be assigned moral status. Contrary to the Catholic Church, the utilitarian person makes the claim that the embryo in the earliest stages is not a person. It is only a piece of vegetative life, neither is a person nor as a whole a being that can be subjected to moral harassment. In *Practical Ethics*, Peter Singer is very close to conceiving the early embryo as a thing (Singer 1993). He refers to life's progressive development, arguing that the early embryo has neither self-consciousness nor the ability to feel pleasure and pain, and therefore it cannot be a human being, according to rational thinking.

It is also extremely difficult to attribute the early embryo rights as an ordinary person. It has no self-awareness and cannot be attributed with interests and rights in the same way as a child or an adult. The fact that the embryo is a human life and the biological basis for genetic individuality is not enough to maintain that it is a person. Bonnie Steinbock similarly believes that the embryo from the moment of fertilization cannot be a person—it has no interest (Steinbock 1992). Only individuals who have interests and can anticipate the future can be defined as human beings, but the embryo does not even have any consciousness. In fact, there is a radical discontinuity between the early embryo's vegetative life and the adult person's ability to relate to the world.

Similarly, there may also be some hard-hitting arguments made against the notion that the fertilized egg has the potential to be an actual person. John Harris has demonstrated some problems in the potentiality argument (1985, 1993). The problem is that you can also understand eggs and sperm understood as potential persons, and contraception becomes morally problematic, since the aim is to prevent potential human beings from coming into existence and in advance wipe out their ability to exist and the life of future people. Potential persons may in fact not be separated from possible people, but do we all have the obligations to the possible people who never become actual persons? We can do evil against future actual people, but not against future potential people who are entirely fictional. Nonexistence cannot get any person to suffer. One cannot possibly do evil to any human being that will never come to exist.

One option to avoid this deadly criticism is simply not to refer directly to the embryo's status as a person or leave the idea of personhood as a criterion for immoral status of the embryo. You could instead take the point of departure in the embryo's status between persons and things and argue for a gradualist understanding of its development to the human being by distinguishing between real people and their embryological origin. In *Soi-même comme un autre*, Paul Ricœur presents a dialectic between two types of identity that Thévoz continues to use in his understanding of the human embryo, namely, between identity as idem and identity as ipse (1990: 29). The gap between them underlines the contrast between the early embryos as genetic material of individual character (idem) facing the living body's identity developed under human existential being-in-the-world (ipse) (Thévoz 1990). The embryo can be understood as a "pre-person," any living human tissue, which is a potential person and participant in the community. Thus, the embryo has moral status as human tissue and body parts between people and things in the human community. The embryo is not and cannot be perceived as a human being, but this does not exclude it from the society's moral order and does not make it legally without the right to protection. It is given a symbolic value, as something that can gradually become a human being. This theory presents a third way between positions which argue that the embryo is a person who should always be treated as an end in itself and those who argue that the embryo is a vegetative life, which cannot claim moral status (Thévoz 1990).

When focusing on embryonal symbolic and relational dimensions, we should not entirely understand potentiality genetically, but the idea of potentiality must be seen in the light of the existential reality. It is the body's overall organization of interaction with the environment, which constitutes and creates the actual person. The genetic potentiality is mediated through the human cultural world, and precisely in this context, the embryo receives a symbolic meaning as an expression of human beings' natural conditions in the culture: "l'homme est un être culturel parce qu'il est un être naturel par culture" ["Man is a cultural being by nature because he is a natural being by culture"] (Thévoz 1990). Society's relationship with the embryo is communicated through language and culture, and here it turns out that we treat embryos as potential and actual people, though in strict logical and genetic sense they are not. The cultural symbolic status of the embryo as a person requires moral respect, and therefore the embryo must be legally and ethically protected.

The cultural and societal understanding of potentiality must be reconciled with the gradualist view of embryonic development. Here we see the embryo potential to be a person from its symbolic status as part of the human community. The unborn life is like a vegetative human life, an expression of future generations as part of the evolution of humanity. The perception of life as a gradual process in which the embryo is developed into a coherent "chain of being" and can become an adult makes it possible to be critical of a utilitarian theory that argues that only sentient people can have moral status (Steinbock 1992). Life may well have a moral status, though it cannot be attributed with interests because it lacks self-awareness. This applies to both the unborn life and future generations. The bioethical understanding of life's own value and the embryo's symbolic meaning makes it, in other words, possible to set ethical limits of society's treatment of human embryos. It is society's cultural understanding of embryo's dignity as human life and not the granting of interests or rights to embryos that is the driving force of the formulation of the embryo's moral status as a potential person (Steinbock 1992: 41). This symbolic value is not identical to the value we grant born and already existing human beings, but can still provide some basis for formulating guidelines for social responsibility for unborn human life.

5 Arguments for and Against Embryo Research

The fierce debate on embryo research must be seen in this perspective. The purpose here is to balance society's utilitarian interest in new knowledge through embryo research with respect and protection of the dignity of unborn human life. The main reason for the interest in embryo research is represented by the big opportunities for new knowledge for use in the treatment of disease, genetic and reproductive technology, etc. (Dunstan and Seller 1988: 26). This is because these tests can provide such a basis for new techniques for gene therapy and the improvement of fertility treatment. Furthermore, the embryos can be used for further development of fertilization outside the womb, for cloning by embryo separation, and for various forms of gene therapy. The question is the degree to which society can allow the use of embryos left over from fertility treatments for research, if one must create embryos solely for the purpose of using them for experiments and research and if so whether such practices and embryo research should be allowed and encouraged.

The English physician Edwards, one of fertility treatment pioneers, advocated the use of embryos in research (Dyson and Harris 1990: 41). He believed that embryos are necessary and invaluable for progress in research. When the embryo is just a piece of vegetative life, there is no question of violation of personal dignity. It is therefore perfectly all right to use tissue from aborted fetuses or embryos for research when they are left over after IVF treatment. Likewise, John Harris argues in a utilitarian perspective for the use of embryos for research (Dyson and Harris 1990: 73). It is, according to him, a waste of resources not to use embryos for research. When we otherwise throw away the leftover residual embryos, we might as well use them for research. It is untenable speciesism to regard the embryo as human life because society refuses to assign higher animals the same protection by experiments. Harris does not believe that the embryo's gradual biological development can be used as an argument against the use of an embryo that actually is not a person for research. One cannot claim the embryo's right to self-determination, as it requires a person with the ability to decide their own life to be autonomous. And the embryo is not independent (Harris 1983, 1993). In the early stages, it cannot feel pleasure or pain, and thus no pain can be caused to the embryo during embryo research. We allow dissection of corpses and using organs and other body parts from adult humans for experiments, so why not let the state decide over the use of embryos for research and experiments. The embryos may be used in science and research for humanistic purposes considered high and valuable. The argument is that it is part of the responsibility of society to promote research, and therefore the embryos must not go to waste (Dyson and Harris 1990: 55). Why should the use of embryos and fetal tissue lead to a development that we cannot control? Because society allows some experiments that have high human value, it does not allow other experiments, including cloning, which will completely change the human condition.

Critical arguments against embryo research can in contrast to this be based on the defense of the embryo's symbolic status, which means that human life cannot be reduced to a collection of cells. Although the embryo cannot be considered as a human being from the moment of fertilization, embryo research and experiments will be an intervention in the genetic and biological basis for the person's individualization and dignified life. For this reason, it is important to protect unborn human life. Another type of skepticism focuses on abuse. The claim that it becomes difficult to avoid a slippery slope, as a reification of the embryo, will be indispensable to embryo research. Maybe science will soon begin its own production of embryos for experiments, because there are not enough embryos left over after IVF treatment. This can lead to the commercialization of embryos and international traffic in embryos. The argument is that it will be much easier to make scientifically dishonest research experiments that go beyond 14 days in the test tube and aims at cloning, the creation of hybrids and chimeras. Such embryo research is the first step toward total mastery of the living and a step toward the new world. Criticism of the implicit reification of embryos in embryo research also aims at using the embryo solely as a means for other purposes and not as an end in itself when it is usually destroyed immediately after the trial. Direct production of embryos for experiments and research is the ultimate consequence of this scientific objectification of the human life. In combination with genetic and reproductive technology, embryo research enhances dominion over the living world and fluid boundary between experiments and eugenics, and private companies and public institutions will have increased opportunities to manipulate the properties of future generations.

When comparing these arguments for and against embryo research, it must be emphasized that the case in principle is to respect the embryo dignity as human life. It can be argued that embryo research is necessary to promote human health and well-being, and experiments of therapeutic character that respect the embryo as an end in itself are not particularly problematic. If in addition the research does not destroy the embryo, it can even be argued that it respects the potential human being. In general there are indeed very few objections to research with leftover embryos if they serve important medical and therapeutic purposes and it does not harm the development of the embryo as human life.

6 How to Treat the Embryo?

Against this background, it is the task of legislation to find the right balance between the interests of society in embryo research, the parents' rights to their embryos, and the respect for the embryo itself as human life. The legislative task is to steer research fairly in relation to future generations and the symbolic status of the embryo as a potential person between people and things. This means that legal policy must be highly critical of the "homo faber" axiom that everything that is technically possible is progress and must consequently be realized. Instead of dealing with the metaphysical problem, whether the embryo is a human being, the legal policy should be about finding the right way to treat the embryo, so that the unborn human life's integrity and dignity are not violated.

This can be seen in Mary Warnock's answer to John Harris in connection with a discussion following the famous Warnock report's release in the United Kingdom (Warnock 1983: 241). Questions about the status of the embryo changed here from biological and ontological questions to a question of "How it is right to treat the embryo?" (Warnock 1983). Warnock agrees with John Harris that it is difficult to say with certainty when human life begins and that the embryo cannot simply be regarded as a person or a potential person. She does not believe that there can be a definite answer to this question, but that is not at the core of the debate. The person question confuses more than it benefits. Instead of discussing the human being and its rights, it is important to find some lasting principles for the protection of unborn life. A society's humanity is reflected in the moral limits, which it sets for its own actions. This also applies to manipulation of unborn human life, regardless of how one understands it. Mary Warnock appeals, in other words, to society's civilizational boundaries. If ethics is to have meaning, there must be some basic actions that individuals or society will not do, regardless of whether the political situation sets the stage for them or not and they can have potential positive impact (Warnock 1995: 247). There are some fundamental ethical norms that cannot be violated without a sense of civilizational collapse pressing on the individual. The European populations' general rejection of free embryo research could be an indication that the experiments provoke civilization borders. Legal policy, which takes account of the embryo's culturally determined symbolic status, does not have to argue for the status of a person of the embryo but can be built on the respect of the law of the unborn human life and on the ethical discomfort in otherwise coming to transcend civilization's boundaries. The instrumentalization of the human embryo simply cannot be allowed, as long as we understand human life as an end in itself.

Contrary to the Catholic Church, which protects the embryo as a person from the moment of conception and facing the utilitarianism that will not attribute moral value to the embryo, this analysis presents a completely different basis for defining the legal status of the embryo. The starting point here is embryonal intrinsic value as human life without granting the embryo's rights as general legal persons. From here, from the perspective of the gradualist, the embryo's growing demands for legal protection during fetal development must be defined and an ethically reasoned legal framework for the embryo's legal status be set in order to regulate genetic and reproductive technology and in particular embryo research. As human life, embryos must in other words be considered "legal subjects" with an increasing degree of dignity according to their symbolic being as potential persons.

Consequently, it must be acknowledged that there may be moral reasons for not doing something that threatens human life, that is, a person. Steinbock argues that the distinction between the moral status and moral values can be used to give the embryo legal protection, although it could not be contributed rights or supposed to have interests as human persons (Steinbock 1992: 161). Moral status requires an agent who both suffers and acts, and here the embryo falls outside this definition, but that does not mean that it cannot be understood as a symbol of human life and thus have a value in itself, which we must deal with ethically or morally. Embryo Research Regulation and the handling of embryos can then be justified in the public interest as legal protection of human life in itself without having to take into account the terms of the embryo's fictional rights and interests and not by balancing different fetal interests. The debate on the status of the embryo is more about values than about interests. The question is what cultural and symbolic value society must give the embryo if it is to remain within its own civilizational boundaries. Steinbock says: "A human fetus, even a preconscious one, is a potential person and a powerful symbol of humanity and as such should be treated with respect" (Steinbock 1992: 190). Or as the US Ethics Advisory Board advocate (1979), it is important to respect the embryos as human life. In this way, the legal protection of the embryo is based on the basic values of respect for human life itself. Thus, there will be a realization of the ideas about the body's integrity and dignity in relation to unborn life. The embryos are "legal subjects," although they cannot be considered as persons. The legal policy discussion on embryo protection becomes a debate about specific legal rules for handling embryos during each stage of development from conception to birth.

7 Strategies for Legislation About Embryo Research: Research Optimism Versus Fear of Eugenics

Against this background, we can now formulate some legal-political strategies, which I think has played a role in the legal regulation of embryo research in Europe and in determining the status of the embryo (Eser 1988; Rendtorff 1999). These strategies range between the extremes: research optimism and fear of eugenics.

The medical-therapeutic strategy for understanding perceives embryo research as an extension of the legal regulation of other medical experiments (Trappenburg 1993). On the one hand, there is a recognition of the needs of society for scientific research, and on the other hand, a democracy cannot afford to sacrifice innocent life on society's altar. To control the embryo research, the society mobilizes the distinction between therapeutic and non-therapeutic experiments and formulates the opposition between social utility and human dignity. The embryo is considered an incompetent and weak entity needing special legal protection, and the biomedical committee system is used to regulate research to ensure scientific freedom and avoid unwanted damage to the embryo. This approach is limited by the fact that the regulation of medical experiments on human beings based on informed consent and respect for patient autonomy cannot be applied in the case of the embryo since the embryo has no self-determination. Therefore, this kind of strategy needs to look for other ethical norms and standards for determining the boundary between respect for and violation of the embryo's integrity.

Another strategy emphasizes human sanctity of life and assumes that there is human life and a human person from the moment of fertilization. This applies, for example, in relation to the nature of the judicial personalistic position formulated by the Catholic Church, according to which the necessary judicial protection of the embryo must be at the same level as that of an adult. Any kind of experiments on embryos, which are not directly to the embryo's benefit, must be banned. The problem of this approach is that the embryo can hardly be said to be a legal entity in line with the general concept of legal persons.

The strategy that focuses on embryo dignity as human life without perceiving it as a human being is based on the body's position between persons and things and respects human life as human life from conception but proposes a gradualistic legal protection of unborn life. The embryo is awarded an intrinsic value from the moment of fertilization without proposing a definition of its specific metaphysical status. The differences in the understanding of the embryo as a person a priori are avoided in this strategy.

The symbolizing strategy focuses on legislation's expressive communication function in order to protect the embryo's value as a special form of life according to the basic democratic notions of the human being. One can, according to Engelhardt, say that the embryo appears here as a "social person," a possible participant in today's society that symbolically anticipates a real participation in future society (Engelhardt 1986). The embryo is here of greater value than an animal at the same level of development, because as any human being it is included as a member of the social order. The embryo's increasing degrees of dignity must play a crucial role in the actual legislation and case law according to the symbolic value that they are attributed by society. On the other hand, the symbolic argumentation for legal protection is far more a matter of legal judgment and much weaker than the position that wants to give the embryo the status of a human being.

The civil legal strategy is based on civil law regulation of reproductive technology as the basis for determining the status of the embryo in relation to embryo experiments and embryo research. The judicial control follows from the right of parents to decide over their embryos in fertility treatment. They must have autonomy and be able to give their consent to the donation of embryos for experiments. In addition, the question of the status of the embryo can be seen in addition to the extension of the existing legal regulation of abortion, but the strategy falls short facing the contemporary biotechnology development, which leads to the fact that the embryo will have an independent status outside the woman's body and thus it will limit the parents' self-determination when ethical and legal decisions must be made to take care of the embryo's own value.

Finally, I should mention a strategy that includes embryo experiments in its criticism of genetic engineering and eugenics and a general criticism of the reification of the embryo following the horror scenarios of genetic manipulation with human beings, which is perceived here as the only possible result of experiments with embryos. Consequently, it is argued that society should ban any embryo research and fight against eugenics and discrimination. This strategy shows a legitimate skepticism of science's increased mastery of the living but at the same time contributes to rejecting embryo research in cases where it could benefit particular people and future generations.

In summary, this discussion of the status of the embryo proposes that the law should emphasize respect for unborn human life's intrinsic value and emphasizes the symbolic importance of the ethical boundaries of democratic cultural tradition as a significant horizon of understanding for the determination of legal regulation of embryo experiments. The question is whether these views are in fact not also actually dominant in the specific legal development.

8 The Case Example of the Danish Debate About Embryo Research: From the Welfare State to Biopolitical Neoliberalism

Now we can consider the Danish political debate on stem cell research, cloning, and preimplantation genetic diagnosis (PGD) in Denmark from the perspective of the abovementioned elements of the bioethical debate on the status of the human embryo (Etiske Råd 2001, 2002, 2003a, b, 2005a, b, 2010, 2013a, b, 2014, 2015; Etiske Råd and Dyreetisk Råd 2007). We can see this debate as a case of a confrontation of strategies, focusing on arguments for human dignity and respect for human beings on the one hand. On the other hand, we see the emergence of strong arguments of consequentialist and pragmatic character. This happens in the process of moving from the welfare state to the competition state in the shadow of biopolitical neoliberalism.

In Denmark the civil liberal strategy concerning the use of stem cells has been dominant (Etiske Råd 2001, 2002, 2003a, b, 2005a, b, 2010, 2013a, b, 2014, 2015; Etiske Råd and Dyreetisk Råd 2007). The debate about the status of the human embryo started with abortion. The leading argument has been that stem cells from aborted fetuses must be allowed to be used for stem cell research. The law says that stem cells can be taken from fetuses that are 4–6 days old, i.e., 4–6 days from the time of fertilization. Stem cells can also be taken from fertilized eggs in vitro that will not be used. Moreover, stem cells can be used to replace sick cells with healthy cells, and stem cells can be used to grow tissue for transplantation. This is the case, for example, when researchers want to look into the possibilities of the treatment of Alzheimer's and Parkinson's diseases.

The arguments in the debate about stem cell research, which are critical and against it, are the following. They are based on the dilemma between the concern for treatment and the struggle with diseases versus the required respect for the life of the fetus as an unborn human life. In this context the argument is that our highest aim is not research, but respect for human dignity. Critical voices argue that human dignity is not respected in stem cell research since human life is used only as a means, but not as an end in itself. Following this, stem cell research can sometimes also be compared to cloning. In cloning, there is no respect for the other human being, since a human being is only used as a means for production of an identical copy. Moreover, since humanity begins at conception, manipulation in cloning and stem cell research represent a violation of the humanity of the human being. The argument is that a human life cannot be used only as a means of treatment of another human being. We have personal responsibility for our health, and we cannot use stem cells to solve the problems of a bad lifestyle. Thus, we can say that the arguments against stem cell research focus on the protection of the dignity of human life.

The people who use such kinds of arguments in Denmark are very different actors. Conservative Christian groups would argue against abortion and say that human beings should be fully respected as human life from the beginning of conception. These groups are minority groups of right-wing Lutheran Christians who also mobilize strong political activism against abortion which they consider as murder. Less conservative and more moderate groups are people and academics who are inspired by a Kantian philosophy and who favor human dignity as an important concern in human life. This approach has been important for the National Council on Ethics, and many expert members of this council from theology, philosophy, and humanities have been favoring this position. From this point of view even though the fetus and stem cells cannot be considered as a full human being, our general concern for the respect for human life and human rights and dignity requires very careful use of stem cells for research. They would allow use of stem cells in specific well-defined research situations, for example, the use of stem cells from aborted fetuses, but consider it a violation of human dignity to produce stem cells directly with the only aim of using them for research.

This moderate position of concern for human dignity in stem cell research has been important in the public debate, and a large portion of the population favor this position. However, there is also a strong pragmatic and utilitarian position that is present in Denmark. From this perspective, stem cell research is important for the future of humanity, and therefore it should be allowed. Even though they would agree that it may be wrong to produce embryos directly for research, many people, including professional scientists who favor this approach, argue for the importance of the use of stem cells from aborted fetuses for research and experimentation.

The most important arguments in the Danish debate favoring stem cell research focus on the importance of free donation of stem cells to research and treatment (Etiske Råd 2001, 2002, 2003a, b, 2005a, b, 2010, 2013a, b, 2014, 2015; Etiske Råd and Dyreetisk Råd 2007). The objective is to heal other children and adults and thus improve their lives through experiments and research. In this sense, the human life of stem cells serves a dignified purpose of helping other human beings. Since the aborted fetuses, fertilized eggs, and stem cells are leftovers from medical treatment and would otherwise not be used for anything, it would be better if they were not lost. Since they are not full human beings or a person in any sense, they can be used for research. So the argument is that lives are saved by donating fetuses and eggs to stem cell research and that a good deed is performed for humanity by using the stem cells for research and experimentation. Therefore, the people in favor of stem cell research argue that by allowing the use of leftover embryos' stem cells for research and treatment, you contribute to the good of humanity.

The ethical concepts that are important in the stem cell debate include the confrontation between different concerns, which are considered important for determining the justification of concern in relation to research on embryos and the use of embryos for stem cell research. Indeed the problem of implicit reification and instrumentalization of the fetus and the early embryo is an important argument. There is the danger that the fetus and embryo will be commodified by the process of making them into objects for use. So embryo experiments and embryo research do not respect the dignity and intrinsic value of human beings as ends in themselves.

The concern for human autonomy states that there is the danger that autonomy will not be respected in embryo research and experiments. There is no autonomy when the embryo is not allowed to decide himself or herself whether his/her cells should be used for research. The respect of autonomy is nonexistent when human beings cannot decide themselves about the use of their stem cells and their bodies.

In contrast to focus on reification and violation of self-determination, beneficence promotes the good for humanity. From this perspective, it is possible to argue that donating fetuses and embryos for stem cell research and treatment is a good and beneficent action for the good of the community. Therefore, embryo research is valuable since it promotes the good of humanity.

A further topic that has been important in the Danish debate is the concern for naturalness. People who are critical toward embryo research argue that such interventions are not natural. Therefore, they are not good for our society, and embryo research is against the truth of human nature.

These ethical concepts of concern for reification, self-determination, and naturalness in contrast to the argument for beneficence of embryo research have also been present in the Danish debate about the cloning of human beings.

The debate on human cloning (somatic cell transfer or separation to create twin cells) was restarted in Denmark by the cloning of the sheep, Dolly, in 1997 by a Scottish research team. The sheep, Dolly, was genetically identical with the individual from where the cell came and not created from sexual reproduction. This opened the possibility of beginning research on cloning of human beings. In Denmark, the cloning debate was about whether another human being should be created by using the same technology. The background was that cloning could be used as a technology for artificial reproduction and artificial insemination to create genetically identical individuals.

Strong arguments against human cloning have been present in the Danish debate. A major argument is that we live in a society that has become so narcissistic that it eliminates the sexual difference. Reproduction by cloning is the ultima ratio of reproductive technologies—it is the dissolution of sexual difference into the same. The fantasy of techno-cosmos is the idea of genetic incest where reproduction no longer occurs at the human level but is identical reproduction where the individual is reproduced without "the other." Moreover, an argument related to the genetic identity and integrity of the individual puts forward that cloning would imply that the individual would lose its own sense of uniqueness and autonomy since it is only a genetic reproduction.

Other arguments against human cloning in the Danish debate include the argument that human cloning would violate individual human dignity and integrity and the idea of the primacy of the human person. Moreover, human cloning would also violate family integrity since there would be no relation to grandparents and no family history but only reproduction of the same. In addition, cloning would violate the integrity of children since they would not be themselves but only reproductions of their parents. Indeed, we should also mention the practical problems of experimentation regarding the number of experiments that are needed before cloning is a success.

Accordingly, with regard to cloning there have been few good arguments in favor of permitting human cloning in Denmark, even though animal cloning seems to be generally accepted. Denmark has therefore signed the international conventions and legislation to prohibit cloning.

The positions of the actors in the debate about human cloning in Denmark are similar to those in the debate on stem cell research. On the one hand, we have conservative Christians who are against all modification of human fetuses. In the middle, we have the moderate principalists who want to respect human life, and on the other hand, we have the pragmatic utilitarian believers in science and technology. However, in opposition to the stem cell debate where many people favor pragmatic use of aborted fetuses for experiments and research, very few people favor cloning for pragmatic reasons. Here, arguments of human dignity in general are considered to be stronger than the pragmatic arguments for the future of sciences.

With regard to the debate concerning prenatal diagnosis, for a long time in Denmark, there has been use of prenatal diagnosis in order to screen embryos for genetic deficiencies and disabilities. Mainly women over 35 and families with genetic disorders are screened. The debate about prenatal diagnosis has centered on the question whether it is possible to present a list of permissible abortions on the basis of prenatal diagnosis and prenatal screening that is acceptable from the point of view of human dignity and human integrity. As early as 1984, a list of possible reasons for justification of abortion after prenatal diagnosis was proposed and included the following:

1. The defect means that there cannot be life after birth.
2. The child is born ill and dies in the first years of life. The child is healthy at birth but will become sick and die in adulthood.
3. During its whole life, the child will have a disease and a disability that demands increased care and help.
4. The child will have a disability that cannot be healed.
5. The child will be born with a disposition for a disease that will break out later in adult life and will eventually shorten the span of life. The child is healthy but has dispositions for a genetic disease.
6. The child is healthy but has certain dispositions that the parents do not like.

The last two possibilities are considered particularly difficult to justify in a democratic society.

In the Danish context in the public debate, focus has been on a number of problems with extended use of prenatal diagnosis. Indeed, screening in order to perform eugenics is contrary to the spirit of a democratic society. Here we can mention that prenatal diagnosis must not lead to cost-benefit calculations of the life of disabled people and the lack of respect of disability in society. Moreover, the autonomy of the woman to decide over her own body and eventually have the disabled child should be respected. Indeed, an ethics built on respect for human dignity would be critical toward prenatal diagnosis. However, prenatal diagnosis creates a tension between universal respect for human dignity and individual ideas of the good life. We can also ask the question about how prenatal diagnosis relates to the self-determination of the future child. It is also a question whether it is morally irresponsible to allow a child to live that would suffer very much. What about the risk to the fetuses, also the healthy fetuses, involved in the use of prenatal diagnosis? Is it fair to take that risk with extended use of this diagnosis? Here prenatal diagnosis seems to be "an offer you can't refuse" because who would like to have a severely disabled child? An argument against this problem is that we do not know whether the child would prefer to be aborted or live with a severe disability. Summing up, in Denmark, however, prenatal diagnosis is accepted as an open possibility for parents, in particular risk groups and older women. However, as we can see from the critical points in the debate, risk evaluation should receive greater attention.

To answer the question which actors (politicians, religious actors, disabled groups) support which position with which arguments in the context of the debate on prenatal diagnosis, we see that this form of diagnosis has been introduced by pragmatic concerns of science to improve fertility treatment. In particular, this has been supported by the general population, patient groups, and medical science. Conservative religious groups have been more critical toward these new technologies, and there has also been skepticism from the part of critical humanities, philosophy, and theology.

The cultural and religious context of the Danish debate on bioethics, which influences this debate on the status of the human embryo, embryo research, cloning, and prenatal diagnosis, is a confrontation between some very traditional worldviews. Here, we can first of all mention a traditional pietistic concept of Christianity (Indre Mission) that was totally absorbed during the last century. This austere tradition stands in contrast to a much more lively and communitarian concept of Christianity, based on the theologian and philosopher F.N.C. Grundtvig who was very important for the people's movement in the nineteenth and twentieth centuries. Grundtvig's philosophy is a focus on the people and community based on the living spirit of Christianity. In contrast to this philosophy, but also sometimes considered as a leftist radicalization of Grundtvig, we find the so-called cultural radicals (de kulturradikale), who are the leftist communitarian Christians combined with Marxist and left-wing Kantians. They criticize in particular the pietistic approach of the strong Christians. Furthermore, we can mention a conservative Christian movement (Tidehverv) which started as very radical Christians following Kierkegaard and Nietzsche proposing a sort of dialectical concept of Christianity, based on individual belief. However, soon they became rather right wing, as the ideological basis of the right-wing movement in defense of Danish nationality. In fact, as recently argued, all these movements had in common the search for something negative to define what Denmark is not. We can say that today we are facing the situation of post-Denmark, where a radical other should be defined. We can say that we live in the reality of a neoliberal competition state where these movements have partly dissolved into utilitarian pragmatism and neoliberal ideas of biopolitical self-discipline.

We can see this if we go further into the legal and political context of the Danish bioethics debate. Biopolitics and bioethics in Denmark can be characterized by the confrontation between utilitarian, different Protestant Christian, and egalitarian approaches. At the same time, bioethics has challenged the Danish constitutional framework. This challenge deals with issues of foundations of legal dogmatism and the legal system. Denmark has an old constitution from 1848 where the concepts of human rights and human dignity do not play a significant role. A strong legal positivism and realism has been present. This is illustrated, for example, by the fact that the legal scholar Alf Ross refused medical ethics as a source of law.

At the same time, there has been an institutionalization of bioethics in the welfare state. The modern debates on bioethics started with the abortion debates in the 1960s and 1970s. Danish abortion legislation is quite liberal. Today patient autonomy is a strong concept of health law in the welfare state—a bill of patients' rights has been institutionalized. Bioethics has today been institutionalized within different institutions: the National Board of Health, the Danish Medicines Agency, the Central Scientific Ethical Committee, and the Patient's Complaints Board. In particular, the Danish Council on Ethics (with 17 members) has been important for the institutionalization of bioethics.

The foundational research on bioethics in Denmark was initiated during the 1990s when in 1993 two research projects on bioethics were funded by the National Research Councils. The first project GRAN discussed the foundations and applications of bioethics. This project was headed by professor of theology, Svend Andersen,who combines Løgstrup and Lutheran theology and analytic philosophy. The project was pluralistic since utilitarian philosophers and family lawyers were also involved. The second project focused on bioethics and law with the Center for Ethics and Law, headed by the theologian and philosopher Peter Kemp who worked together with Jacob Dahl Rendtorff and Mette Lebech. The research of the Center for Ethics and Law lead to the formulation of basic ethical principles, which has become fundamental for the European approach to bioethics and biolaw.

This book was a result of a European research project focusing on the basic ethical principles for Europe. Here, the foundations of bioethics and biolaw were found in the phenomenology of the body: people as fragile and vulnerable beings. The tension between vulnerability and autonomy was considered as a central problem of bioethics: the formation of narrative identity. Human beings are defined as singular individuals with existential choices in vulnerable bodies. In the book by Rendtorff and Kemp entitled *Basic Ethical Principles in European Bioethics and Biolaw* (Rendtorff and Kemp 2000; Rendtorff 2002, 2015), these basic ethical principles were defined in the following manner:

8.1 Autonomy

Five important meanings of autonomy can be put forward: (1) autonomy as a capacity of creation of ideas and goals for life; (2) autonomy as a capacity of moral insight, "self-legislation," and privacy; (3) autonomy as a capacity of decision and action with a lack of outer constraint; (4) autonomy as a capacity of political involvement and personal responsibility; and (5) autonomy as a capacity of informed consent (Rendtorff and Kemp 2000, p. 25). Autonomy should be considered as a principle of the self-legislation of rational human beings taking part in the same human life world. This does not exclude the recognition of pluralism as a political fact of modern society. But it is necessary to work with a more comprehensive idea of autonomy, recognizing the tensions between different conceptions of the good. The republican sense of autonomy is based on the vision of "the good life with and for others in just institutions" (Ricœur 1990, p. 202). This vision is put forward as the basis for privacy, confidentiality, and informed consent.

8.2 Dignity

Human dignity has the following meanings as an intersubjective concept: (1) It expresses the intrinsic value of the human being in a community or society. (2) It includes respect for the moral agency of the human subject. (3) It means that every human being must be considered as not having a price and being unable to be commercialized. (4) This includes that human dignity refers to the indeterminate position of human beings in the universe. (5) Self-esteem, to be proud, shame, feeling of inferiority, and degradation are essentially matters of human dignity expressed in the intersubjective relations between individuals. (6) Dignity can establish restrictions on interventions in human beings in taboo situations, because of the necessity of human civilized behavior. (7) Finally, dignity relates to metaphysical experiences of human beings in existential limit by degrading treatment. But the relation between rights and dignity is also essential. In that context human dignity expresses the intrinsic worth and fundamental equality of all human beings.

8.3 Integrity

The principle of integrity may be said to refer to the totality of life in that it should not be destroyed. Integrity is a coherence that in a certain sense must not be touched. This coherence, or rather "Lebenzusammenhang," is the narrative coherence of a person's life (the life story) or the narrative (historical) unity of human culture. On this basis, integrity has four meanings. (1) integrity as a narrative totality, wholeness, and completeness; (2) integrity as a personal sphere of self-determination; (3) integrity as a virtue of uncorrupted character, expressing uprightness, honesty, and good character; and (4) integrity as a legal notion, where it expresses the moral coherence of the legal or medical system. In bioethics and biolaw, the idea of integrity as an untouchable core, the personal sphere, which should not be subject to external intervention, is the most important. The personal body must be considered from a phenomenological perspective of the self-mastery of the body. Integrity expresses bodily completeness in a private sphere. In medicine it is indispensable for trust between physician and patient (Engelhardt 1986). This is because there is a close link between respect for identity and respect for integrity. Integrity includes a personal narrative that expresses the identity and life context of the individual.

8.4 Vulnerability

Vulnerability of mental and corporeal life is closely linked to integrity, but it expresses more characteristics of the human condition. Protection of vulnerability is considered as the bridging factor between moral strangers in a pluralistic society, and therefore respect for vulnerability is essential to policy making in the modern welfare state. Vulnerability should be considered as a universal expression of the human condition. Moreover, it appeals to the protection of both animals and the teleological auto-organization of the world. However, vulnerability has been largely misunderstood in modern society, which has been guided by a so-called vulnerability-reducing agenda, which aims to eliminate all vulnerability, i.e., suffering, abnormality, deafness, and disability, in order to create perfect human beings. Respect for vulnerability must find the right balance between this logic of the struggle for immortality and the finitude of the earthly presence of human suffering. As an expression of the destiny of finitude, the moral receptivity of vulnerability, i.e., the vulnerability for the vulnerability of the other, is the foundation of ethics in our time (Rendtorff and Kemp 2000; Rendtorff 2002, 2011).

Thus, what is the role of religion in the public debate about bioethics in Denmark (Rendtorff 2003)? Here we can emphasize that Protestant theology and religion have in general been strong defenders of human autonomy, dignity, integrity, and autonomy (Lebech et al. 1997; Kemp et al. 2000; Kemp and Rendtorff 2009). This is reflected in the debates about stem cell research, cloning, and prenatal diagnosis. Protestantism is, however, also very pragmatic, liberal, and pluralistic in relation to reproductive technologies and bioethics. Therefore, the development has been rather liberal in Denmark. Protestantism recognizes the importance of rational ethics in addition to personal belief and personal choice based on existential commitment.

If we should summarize the positions in Denmark related to the discussion of reproductive technologies, the status of the human embryo, embryo research, cloning, and prenatal diagnosis, we can put emphasis on the following basic positions of the debate on reproductive technologies in Denmark: (1) an autonomy-based view arguing for personal self-determination as a major value; (2) a humanistic "Samaritan" position seeing the need to care for the vulnerable and weak as the most important concern; (3) an individualistic, consequentialist position based on a utilitarian positive attitude toward medical science and technology; (4) a community-based communitarian vision of the relation between human beings in society; (5) a Christian conservative position arguing for the protection and respect of the dignity of human life; and (6) a position of the basic ethical principles in European bioethics and biolaw.

This concern for human autonomy, dignity, integrity, and vulnerability can be interpreted as an expression of the Danish ethical tradition of respect for the embodied person. However, this position is presently challenged by a strong utilitarian, consequentialist approach where utilitarian philosophers focus on equality and rights to reproduction. Moreover, this is combined with a strong presence of neoliberal biopolitical rationalities focusing on personal responsibility for health and self-discipline in relation to the use and abuse of the human body. Accordingly, there is an ongoing discussion of the necessary respect for the human embryo in the context of biopolitical neoliberalism in Denmark. Here, the confrontation between principalism and utilitarianism is not over, and there is still a strong disagreement present in the Danish debates about bioethics and biolaw.

References

Arnoux, Irma. 1995. *Le droit de l'être humain avant la mort et après la naissance*. Bordeaux: Presses Universitaires de Bordeaux.

Coughlan, Michael J. 1988. *The Vatican, the law and the human embryo*. London: Macmillan.

Dunstan, G.R., and Mary J. Seller, eds. 1988. *The status of the human embryo*. London: King Edward's Hospital Fund.

Dyson, Anthony, and John Harris. 1990. *Experiments on embryos*. London: Routledge.

Engelhardt, Tristam. 1986. *The foundations of bioethics*. Oxford: Oxford University Press.

Eser, Albin. 1988. Rechtsvergleichende Aspekte der Embryonenforschung. In *Respekt vor dem werdende Leben: Ein Presseseminar der Max Planck Gesellschaft zum Thema Embryonenforschung*. Freiburg: Max Planck Institut für Strafrecht.

Etiske Råd, Denmark. 2001. *Etiske problemer vedrørende kunstig befrugtning, 1. del. Ret til børn?* Ret til hjælp til at få børn? Redegørelse. Copenhagen: Det Etiske Råd (The Danish Council of Ethics).

———. 2002. *Etiske problemer vedrørende kunstig befrugtning, 2. del. Anonymitet og selektion i forbindelse med sæddontion. Redegørelse*. Copenhagen: Det Etiske Råd.

———. 2003a. *Menneskeligt livs begyndelse og fosteranlægs etiske status, Et debatoplæg*. Copenhagen: Det Etiske Råd.

———. 2003b. *Etiske problemer vedrørende kunstig befrugtning, 3. del. Mikroinsemination og præimplantationsdiagnostik*. Copenhagen: Det Etiske Råd.

———. 2005a. *Etisk Forum for Unge 2005. Den tjekkede tvivl. Om stamceller og etik*. Copenhagen: Det Etiske Råd.

———. 2005b. *Conference report and summaries: The ethics of patenting human genes and stem cells*. Copenhagen: Det Etiske Råd.

———. 2010. *Etiske aspekter ved nye typer af stamceller og befrugtningsteknikker*. Copenhagen: Det Etiske Råd.

———. 2013a. *Det Etiske Råds udtalelse om kompensation for ægdonation*. Copenhagen: Det Etiske Råd.

———. 2013b. *International handel med menneskelige æg*. In *rugemoderskab og organer. En redegørelse fra Det Etiske Råd*. Copenhagen: Det Etiske Råd.

———. 2014. *Det Etiske Råds udtalelse om embryo- og dobbeltdonation*. Copenhagen: Det Etiske Råd.

———. 2015. *Udtalelse om opbevaring og brug af befrugtede æg*. Copenhagen: Det Etiske Råd.

Etiske Råd, Denmark, and Det dyreetiske Råd. 2007. *Mand eller mus? Etiske aspekter ved kimæreforskningen*. Copenhagen: Det Etiske Råd.

Ford, Norman. 1988. *When did I begin? Conception of the human individual in history, philosophy and science*. Cambridge: Cambridge University Press.

Hare, Richard. 1992. *Essays in bioethics*. Oxford: Oxford University Press.

Harris, John. 1983. In vitro fertilization: The ethical issues (I). *The Philosophical Quarterly* 33 (132): 217–237.

———. 1985. *The value of life*. London: Routledge.

———. 1993. *Wonderwoman and Superman, the ethics of human biotechnology*. Oxford: Oxford University Press.

Kemp, Peter, and Jacob Dahl Rendtorff. 2009. The Barcelona declaration. Towards an integrated approach to basic ethical principles. *Synthesis Philosophica* 46 (2): 239–251.

Kemp, Peter, Jacob Rendtorff, and Niels Mattsson Johansen, eds. 2000. *Bioethics and biolaw*. Vol. I–II. Copenhagen: Rhodos International Publishers.

Lebech, Mette, Jacob Rendtorff, and Peter Kemp. 1997. *Den bioetiske vending. En grundbog i bioetik*. Copenhagen: Spektrum, Gyldendal.

Rendtorff, Jacob Dahl. 1997. Legislation, bioethics, judgement. In *Rechtstheorie*, Beiheft 19, 211–222. Berlin: Duncker & Humblot.

———. 1998. The Second International Conference about bioethics and biolaw: European principles in bioethics and biolaw. *Medicine, Health Care and Philosophy* 1 (3): 271–274.
———. 1999. *Bioetik og ret. Kroppen mellem person og ting*. Copenhagen: Gyldendal.
———. 2001. Biobanks and the rights to the human body. In *The use of human bioBanks, ethical, social, economical and legal aspects*, ed. Mats G. Hansson, 5–61. Uppsala: Uppsala University.
———. 2002. Basic ethical principles in European bioethics and biolaw: Autonomy, dignity, integrity and vulnerability—Towards a foundation of bioethics and biolaw. *Medicine, Health Care and Philosophy* 5 (3): 235–244.
———. 2003. Bioethics in Denmark. In *The annals of bioethics. Regional perspectives in bioethics*, ed. John F. Peppin and Mark J. Cherry, 209–224. Lisse: Swets & Zeitlinger Publishers.
———. 2004. Basic ethical principles for a common European culture of science and research. In *The common European space of education, science and culture*, ed. G.V. Makovich, 151–157. Chelyabinsk: Southern Ural State University.
———. 2008. The limitations and accomplishments of autonomy as a basic principle in bioethics and biolaw. In *Autonomy and human rights in health care: An international perspective*, ed. David N. Weisstub and Guillermo Díaz Pintos Pintos, vol. 36, 75–87. Doordrecht: Springer.
———. 2011. La notion de vulnérabilité comme principe fondamental dans la bioéthique et dans le biodroit. In *Une société de soins: santé travail philosophie politique*, ed. Paulette Guinchard and Jean-Francois Petit, 57–61. Ivry-sur-Seine: Les éditions de l'Atelier.
———. 2014. European perspectives. In *Handbook of global bioethics*, ed. Henk A.M.J. ten Have and Bert Gordijn, 293–310. New York: Springer.
———. 2015. Update of European bioethics: Basic ethical principles in European bioethics and biolaw. *Bioethics Update* 1 (2): 113–129.
Rendtorff, Jacob Dahl, and Peter Kemp. 2000. *Basic ethical principles in European bioethics and biolaw, autonomy, dignity, integrity and vulnerability* (Vol. I–II). Barcelona, Copenhagen.
Ricoeur, Paul. 1990. *Soi-même comme un autre*. Paris: Le Seuil.
Singer, Peter. 1993. *Practical ethics*. Cambridge: Cambridge University Press.
Steinbock, Bonnie. 1992. *Life before birth: The moral and legal status of embryos and fetuses*. Oxford: Oxford University Press.
The Catholic Church. 1997. *Congregation for the doctrine of the faith, instruction of respect for human life in its origin and on the dignity of procreation*. London: Libreria Editrice Vaticana.
Thévoz, J.M. 1990. *Entre nos mains, l'embryon, recherche bioéthique*. Geneva: Ed. du Cerf.
Tooley, Michael. 1983. *Abortion and Infanticide*. Oxford: Oxford University Press.
Trappenburg, Margot. 1993. *Soorten van gelijk: Medisch-ethische discussies in Nederland*. Zwolle: Tjeenk Willink.
van der Burg, Wibren. 1995. Embryo experiments in the Netherlands. In *Archiv für Staats- und Rechtsphilosophie*.
Warnock, Mary. 1983. In vitro fertilization: The ethical issues (II). *The Philosophical Quarterly* 33 (132): 238–249.
———. 1995. *A question of life: The warnock report on human fertilization and embryology*. Oxford: Basil Blackwell.

Chapter 10
Religion and Biopolitics in Sweden

Göran Hermerén and Mats Johansson

Abstract After a brief sketch of the historical background, the current political situation in Sweden is outlined. Since 1985 the National Council on Medical Ethics (Smer) has played an important role in the biopolitical debate in Sweden, and its activities and influence are described. Religious teachings are often combined with moral precepts. But what makes a statement religious is not easy to say, especially in societies where religious concerns have been intertwined with humanitarian common sense morality for a long time. Summing up, it can be said that the debate in Sweden on biopolitical issues has not shown many signs of the influence of religious arguments, though some religious actors are active in this debate.

1 Brief Historical Background

Historically, Sweden has been a Lutheran country. In some ways it continues to be, but official ties between the state and the Lutheran Church were abolished in the year 2000, and in Sweden today, unlike its neighbour Norway, there is no longer an official state religion. Numbers of formal members of the Church of Sweden have dropped every year over recent decades, and although the majority of Swedes (some 60%) are still members, Sweden is now a highly secularized country, with only a small fraction of the population attending church services.

In previous centuries, the Lutheran State Church influenced political life in a number of ways; bishops and priests in general had a socially elevated position and were respected. In the seventeenth, eighteenth and nineteenth centuries, this was very evident; but that was long before the separation of state and church and before the emergence of biopolitics. Biopolitical issues became prominent in Sweden only comparatively late. In the 1980s and 1990s, a number of issues were debated and dealt with in governmental commissions, such as criteria of death, transplantation

G. Hermerén (✉) · M. Johansson
Medical Ethics, Department of Clinical Sciences, Biomedical Center (BMC), Lund University, Lund, Sweden
e-mail: goran.hermeren@med.lu.se

M. Weiberg-Salzmann, U. Willems (eds.), *Religion and Biopolitics*,
https://doi.org/10.1007/978-3-030-14580-4_10

and organ donation and priority setting in healthcare. Abortion had also been a hotly debated topic where a political compromise was finally reached.

In 2000 however, the ties between the state and the Lutheran church were abandoned, as already indicated; so in this short transition period—the last decades of the twentieth century—the church still had some influence, but it did not last. Beyond the Lutheran Church of Sweden, there are several other Christian denominations, including Roman Catholic, Baptist, Methodist and Salvation Army churches. There are also a significant number of Muslims in Sweden. In 2010 this number was estimated at 451,000 (or 4.9% of the population), and since then the number has increased as a result of immigration.

2 The Current Political Situation in Sweden

The political landscape in Sweden has changed dramatically over the last 5 years. The once important divide between the Social Democratic Party (*Sveriges Socialdemokratiska arbetarparti*) and the conservative Moderate Party (*Moderaterna*) has lost much of its relevance. Both parties are smaller than they were, and their political agendas are more similar. Based on several polls, the Sweden Democrats (*Sverigedemokraterna*)—a party other parties are reluctant to collaborate with owing to its somewhat dubious and racist past—is now the second largest party in Sweden. The extent to which immigration should be controlled has been a controversial issue. The Liberal Party (*Liberalerna*), the Centre Party (*Centerpartiet*), the Left Party (*Vänsterpartiet*) and the Christian Democrats (*Kristdemokraterna*) are the other parties in the parliament, and they bring with them a range of attitudes to social and religious issues. It is likely that the next election in 2018 will result in a coalition government.

Support for the Christian Democrats, the only party with an explicitly religious foundation, is currently diminishing. Indeed the party is in danger of losing its seats in the Swedish parliament. If they and the Green Party (*Miljöpartiet de Gröna*) have no seats in parliament after the 2018 election, the political landscape in Sweden will be radically changed.

3 Scientific and Demographic Background

Biotechnology has played one important role in the economies of industrialized countries. Breakthroughs in molecular biology have opened up new avenues for diagnostics and therapy in medicine. Research activities in these areas have been rapid and intensive, both in Sweden and globally. The mapping of human DNA, the discovery of epigenetics and the cloning of Dolly are other important milestones in developments which have presented new challenges to both religious and secular societies.

In addition to scientific and technological discoveries, and the digital revolution as well as the explosion of computerized collection and analysis of information, several other factors have played an important role in biopolitical debate in Sweden. Today more people are living longer, and those over 75 usually have several diagnoses. This has changed the disease panorama and created new challenges for the healthcare system. In recent years Sweden, like many other European countries, has received a large number of refugees from non-European countries, which has led to some cultural and lifestyle clashes. Active 24-h media and other information available in the Internet have also helped to raise expectations about healthcare systems.

4 The Preparation of Biopolitical Regulations

In Sweden there is an established system of government and parliamentary commissions through which reports are prepared. In the past these have dealt with topics like criteria of death, transplantation, organ donation, biobanking, genetic integrity and so on. Responding to such a report, the government may propose legislation to the parliament, which may vote to accept or reject the proposal—that is how Swedish biopolitical and other laws have been, and still are, created.

Since 1985 the Swedish National Council on Medical Ethics (*Smer*) has played an important role in the bioethical and biopolitical debate. The Council advises the Swedish government and parliament. It is composed of representatives from political parties, which is unique among other national councils, as well as individuals representing the National Board of Health and Welfare, the healthcare union Swedish Association of Health Professionals (*Vårdförbundet*), the ethics delegation of the Swedish Medical Association and the Ministry of Social Affairs, as well as a number of independent ethical and legal experts. In addition to the chair, the Council has eight representatives of political parties and ten experts. Its mission is to protect and promote respect for human dignity and integrity, particularly in the context of medical research, diagnostics and treatment. The Council cooperates with similar bodies in other countries and takes part regularly in international activities. Each year *Smer* publishes reports and opinions on a number of issues and organizes hearings and public debates. In the past the Council issued publications in a series[1] which included debate books on controversial topics such as euthanasia (1992), the beginning of human life (2000) and genetic screening (2002), as well as a widely read introduction to ethics (with several reprints and editions) (Smer 1989, 2008).

Recent reports include booklets on ethical aspects of wearable technologies and e-health (2017), ethical problems on the borderline between medical research and healthcare (2016), ethical challenges raised by attention deficit hyperactivity

[1] *Smer*, Etiska vägmärken (a series of books on ethical declarations, euthanasia, beginning of human life, genetic screening, etc.)

disorder (ADHD) (2015), the analysis of foetal DNA in the blood of the woman—ethical problems raised by non-invasive prenatal testing (2015)—and robots and surveillance in the care of the elderly (2015). Earlier reports examined among other things ethical aspects of co-payment and the replacement of mitochondria in cases of serious inherited diseases.[2] The working methods of the Council have changed in some respects over the more than 30 years in which it has operated. These changes are described and discussed in some of the contributions to a book published to celebrate the Council's thirtieth anniversary (Smer 2015).

5 *Smer*'s Influence

The influence of the National Council on Medical Ethics is illustrated by the debate on assisted reproduction, and in particular egg donation. In 1995 the Council published a report on assisted reproduction (Smer 1995). This served as the basis for a government legislative proposal to the parliament, Swedish government (2001).[3]

In the 1995 report, the Council proposed that law 1988:711 on IVF (in vitro fertilization) should be amended so that donated sperm could be used in connection with IVF. The Council did not see any relevant difference between allowing donated sperm to be used in the context of insemination, on the one hand, and using it for IVF, on the other, and the former had already been legalized in Sweden in earlier legislation (1984:1140) on insemination. The Council submitted that infertility in men and in women should be regarded in the same way. It pointed out that just as some men cannot produce functional sperm, some women cannot produce functional eggs; and it argued that therefore IVF with egg donation and sperm from a husband/partner should be allowed, provided there was a medical indication, for women of fertile age. A genetic connection with one of the parents is preserved in both cases. *Smer* also argued that a child born with the assistance of these techniques should have the opportunity at maturity to discover the identity of the donor.

But the Council was not unanimous. Some members considered that the law was being extended too far, others not far enough. One member of the Council dissented from the proposal to allow donation of germ cells (sperm and eggs). In his view, any donation of germ cells should be unlawful. He also argued against allowing storage of fertilized eggs in deep freezers. Moreover, one of the experts of the Council wrote a special statement on the proposal to allow IVF with egg donation and the sperm of the woman's partner or husband. Other experts dissented from the view that IVF with both egg donation and donor sperm should be outlawed in the rare cases where

---

[2]These reports can be downloaded free of charge from the *Smer* website: www.smer.se/publications.

[3]Swedish government. Behandling av ofrivillig barnlöshet. Proposition to the parliament 2001/02:89.

both the man and the woman in a couple lack functional germ cells. Nevertheless, the main argument for egg donation in the 1995 report was that sperm donation was already allowed and practised and that egg donation is relevantly similar to sperm donation in that it can give rise to parenthood while preserving a genetic connection with one of the parents. The main concern raised by the donation of fertilized eggs, according to the majority in the Council at the time, was the perceived risk that eggs and sperm would be regarded as freely available objects which could be used to create human beings and that this might promote an unacceptably technological view of human beings.

In 2003 the Swedish parliament essentially followed *Smer*'s proposals. As might be expected, in the parliamentary debate, there was a minority who opposed these proposals; but there was little public debate about the decision taken by the parliament. The same year, the donation of unfertilized eggs was permitted. Besides this direct kind of influence on biopolitics, it is reasonable to assume that *Smer* has had a more indirect effect on Swedish regulation. After all, *Smer* is firmly anchored in the political parties, and it is also frequently asked to comment on ethical aspects of the proposals of government and parliamentary commissions.

6 New Challenges

Many new ethical challenges arise with the opportunities and possibilities created by novel technologies, and in particular following breakthroughs in genetics. Prenatal genetic testing, the sequencing of the human genome and gene editing belong to this category. Other challenges are connected with the extension and interpretation of principles like the right to information, to consent, and to self-determination, and to the ideals of justice and fairness, and other ethical concepts. These challenges can arise in combination, and when they do, they may intensify the issues. Clearly, in debates over euthanasia and physician-assisted suicide, the issue is not new technologies. It is whether patients should be allowed to have a say about their own destiny at the end of life; and if so, under what conditions. This is similar to the case of abortion in that the problem is not driven mainly by new technologies—although, of course, the foetus cannot speak for itself.

A third type of challenge—not unrelated to the others—is created by the growing role of economic models to manage the increasing demands that are being placed on the healthcare system. Clashes between ethical principles and concerns, on the one hand, and the underlying assumptions of economic models and methods for measuring the effects of medical interventions, including the use of QALYs (quality-adjusted life years), on the other, have been the focus of recent debates in Sweden. Moreover, there has been vociferous criticism of New Public Management and the role of economists in healthcare management. The most recent addition to this set of issues, VBV (värdebaserad vård, value-based healthcare), practised at the *Nya Karolinska* hospital in Stockholm, is currently the subject of lively critical debate (Akner and Järhult 2016:113:DXUT; Järhult 2017:114:EFHM).

There are some fairly recent examples of clashes between religious beliefs and healthcare work. A woman who applied for a job as a midwife made it clear to the employer that she would refuse to carry out abortions because they were inconsistent with her religious beliefs. As a consequence, she was not given the job. The case went to court, where the employer argued that abortions were an established part of a midwife's work in Sweden. She lost the case and she also had to pay the costs of the trial, but she was supported financially by a pro-life organization. The midwife has now moved to Norway. The case caused much debate in Sweden (cf. Malmgren 2014), and more is likely to follow as the verdict was appealed against and was forwarded to the European Court of Human Rights.

7 Embryo Research and Its Underpinnings

In Sweden research on fertilized human eggs up to 14 days after fertilization is permitted by law. Once the research has been carried out, the fertilized egg must not be implanted in the womb of a woman. The controversies over embryo research are obviously relevant also to human embryonic stem cell research. In the Swedish debate, three positions can be discerned, all of which have been advocated (cf. Smer 2000).

1. Human life begins when the process of fertilization is complete, and the fertilized egg has an absolute right to life. This means that abortion and research on human embryos, including human embryonic stem cell research, are impermissible.
2. A fertilized human egg is just like any group of human cells, ethically speaking, and it does not need any special protection. This obviously gives a green light to any kind of research on human embryos.
3. A fertilized egg is not just any collection of human cells. It deserves protection, and it must not be used for arbitrary or meaningless experiments. The more human-like the embryo becomes, the more protection it deserves.

The third position is generally accepted in Sweden, both by the National Council on Medical Ethics and in the general debate, but it is on the whole rejected by Roman Catholics, who typically accept position (1). The first position draws a clear line, but it makes abortion and research on human embryos, including human embryonic stem cell research, morally unacceptable. The third position is admittedly somewhat vague and arbitrary: Why up to 14 days? Why not 15 or 13? However, it does help to protect a woman's right to decide about her own body, and it makes important research possible. The third position is often supported by the observation that the beginning of life is a process in which certain stages, like implantation, are important. Before implantation somewhere in the region of 50% of fertilized eggs are lost, and yet nobody would entertain the idea of arranging a funeral for any of them. They have the potential to become human beings, but this outcome is conditional on many factors. A presidential candidate is a potential president, but does not have the rights of a president. Similarly, it has been argued, a potential human being does not necessarily have the rights of a human being. It does not, for example, have an absolute right to life, and to freedom of speech and movement, and such.

Others have stressed the importance of gastrulation. This is a process early in embryonic development during which the single-layered blastula is reorganized into a multilayered structure. It is argued that an essential requirement of being human is having consciousness. This requires certain neurological capacities which, before gastrulation, the blastula lacks. Human identity is important ethically. Before gastrulation, it is not possible to know whether the fertilized egg will develop into one or several individuals. This debate on the conditions of embryo research, and the pros and cons, is generally carried out on the basis of general philosophical and humanitarian arguments. However, in the reasoning of those who defend the first line of thought, an echo of religious views may be discerned—perhaps an echo of the notion of *imago Dei*, the idea that humans are created in the image of God and therefore have an absolute right of life, plays a role here.

We are aware of two relevant opinion polls concerning research on fertilized human eggs in order to get stem cell lines or, more generally, to carry out human embryonic stem cell research. According to a survey in 2002, presented in the newspaper *Dagen* (Wärn, 20.04.2002), 39% of those asked in Sweden would like to see such research prohibited. 22% of the respondents disagreed with this, and the rest were uncertain or did not have an opinion. According to a later survey, published in NEJM 2011, a different picture emerged (Blendon et al. 2011). Now 67% of the respondents in Sweden did not think that embryonic stem cell research should be forbidden. This is a rather significant change in numbers, given that less than 10 years separated the surveys. The difference between the pictures emerging may depend on the selection of respondents, on what the respondents knew about human embryonic stem cell research and alternative approaches, as well as on how the questions were phrased. It is one thing to ask a general question whether research of a certain kind should be permitted or forbidden and another thing to ask if this research should be permitted, if the outcome of the research could result in new therapies—or in therapies that could cure so far incurable diseases. Nonetheless, the 2011 survey also indicated that in both the United States and in Europe as a whole, 'people with the least religious involvement are less likely than their compatriots to believe that embryonic stem-cell research should be forbidden. Those who are more involved religiously are more likely to support restrictions'.

8 Assisted Death: An Embattled Question

In December 2016 the daily newspaper *Dagen*, an independent but Christian paper, published a debate article signed by representatives of many Christian organizations arguing very strongly against all forms of euthanasia (Arborelius 2016). These organizations are very critical of developments in which healthcare institutions and staff are expected to help patients to die.

It should be noted that no representative of the Church of Sweden, Svenska kyrkan, signed the article, and in fact to date, the Church of Sweden has not expressly taken a position on the euthanasia issue. Proposals to the general assembly of the church, 'kyrkomötet', that the church should prepare a position on this issue have been rejected twice, in 2008 and 2015. However, the church apparently considers the issue to be important from an ethical and theological perspective and is likely to continue to discuss and debate it.

The Clapham Institute, a Christian foundation, active in the debates in Sweden, has recently presented a report in which it critically examines the model of assisted dying applied in Oregon, focusing particularly on the need to protect patients with psychiatric diagnoses and problems (Clapham Institute 2017). The report has been criticized by many researchers in the ensuing debate.

9 Religious Influences on Swedish Biopolitics

Has religion had any significant influence on Swedish biopolicies? To address this question, it is necessary to ask what such influence might amount to. Although it is safe to say that openly religious beliefs have played little or no role in Swedish politics for a very long time, it is much more difficult to tell whether such beliefs (or the values associated with them) have played a more indirect role—for instance, in research on animals or euthanasia. Another question is whether religious groups, through public debate and lobbying, have achieved their biopolitical goals? What is of interest here, however, is whether or not religious arguments have influenced the debate.

What, then, makes an argument, statement or value religious? This may be a matter of debate, especially in societies where religious concerns have been mixed and intertwined with humanitarian common sense morality for a long time. 'Religious' and 'religion' are not well-defined concepts; they are vague, value-loaded and contested. Religions vary, as is well known, in a number of respects. For instance, they differ over the possibility of reincarnation, whether there is one god or more than one and whether this god or these gods can appear in human form. Some religions maintain a clear distinction between religious and political ambitions; others do not. Religions also differ in their views of the value or importance of the life and rights of humans as against those of other animals.

For the purposes of this chapter, it will suffice to distinguish three types of religious arguments. Thus we shall say that an argument belongs to religious statements of type 1 if it can be derived directly from a scripture regarded by believers as holy—for example, the Bible, the Koran, the Torah and the Bhagavadgita. These, we might say, are clearly religious statements. We shall say that an argument belongs to religious statements of type 2 if it can *also* be found, and made sense of, in non-religious sources—for instance, in philosophical literature. These, then, are not exclusively religious arguments. Finally, if these other philosophical or literary texts predate and are in that sense independent of religious scripture, we shall say that their arguments belong to religious statements of type 3. Arguments and statements referring to the resurrection of Christ are thus religious in the strong type 1 sense, while arguments about the equal and infinite value of all human beings belong to the last group, type 3. The idea that all humans have equal value originated with the Stoic philosophers. It is not a Christian invention, as has been pointed out by humanists and other secular groups in Sweden. Religious arguments of type 2 or type 3 can be presented as humanitarian arguments without reference to religious or theological doctrines. Religious arguments of type 1 cannot.

10 Religious Actors in the Debate

Clashes between conservative religious groups and others are most likely to occur over the earliest and last phases of human life. Abortion, preimplantation and prenatal testing, as well as various forms of euthanasia and physician-assisted suicide, are all likely to be controversial, as already indicated. In this respect Sweden is unexceptional, even if the groups supporting religious arguments in the strong type 1 sense are a clear minority in the political life in Sweden.

The Christian cultural heritage has been applied to biopolitical issues in Sweden to a limited extent. There is no large Christian party in Sweden, as already mentioned. From the beginning the National Council on Medical Ethics had as members a bishop and a Jesuit pater. However, they took part in the debate on the same terms as the other members. The current Council has no members with religious affiliations representing churches, although like other parties in the parliament, the Christian Democratic Party has a representative on the Council. Today the Roman Catholic Church in Sweden has some 116,000 members, and it estimates that there are about 150,000 Catholic believers in Sweden, a clear minority in a country of more than 10 million inhabitants. But Roman Catholics are active in the biopolitical debate, with a website and a journal, *Signum*. They organize seminars and present the views of the Catholic Church on topics like abortion, euthanasia, IVF and genetic testing. Their views on these issues have had no influence on legislation in Sweden, however.

Pulling these threads together, it can be said that the debate in Sweden on bioethical and biopolitical issues has not shown many signs of the influence of religious arguments in the strong type 1 sense, although some religious arguments of type 3 are used in the debate. A few exceptions exist—for instance, when the focus has been on issues at the beginning and end of life, particularly when the issues have involved the protection of the unborn. Here religious arguments of type 3 (the weak sense) have been used, and Christian members of the Council on Medical Ethics have been active in discussions of issues like euthanasia and abortion.

Moreover, some Christian democrats have been, and still are, involved in governmental commissions and in making proposals to the parliament. Thus, in the governmental commission SOU 2004:20 on genetic integrity and issues raised by genetic diagnostics, gene therapy and cloning, there are some separate statements and expressions of dissent from the main proposal of the commission concerning the use of embryos for research and preimplantation genetic diagnostics (PGD). The Christian Democrats' dissent here refers to the Oviedo convention, but it is also backed up by underlying religious propositions (Landgren SOU 2004: 20). Some additions to the constitutional law, *Regeringsformen* 1 kap 2 para, are proposed in which the inviolability of human dignity is declared to be an overarching principle in all societal activities, and the author proposes that preimplantation genetic screening and PGD/HLA should not be permitted because there is a risk that this method will be used to create 'designer babies'. These attempts to block the proposed law were unsuccessful.

This governmental report also discussed the conditions under which insurance companies could have access to the results of genetic tests. The insurance industry wanted to avoid legislation and preferred more flexible arrangements, such as voluntary agreements between the state and the insurers. This debate did not involve religious arguments. Rather, it revolved around commercial interests and the concern that some members of society would not be able to obtain health insurance at all and that others would be offered insurance only at a price they could not afford. In their proposal to the parliament 2016/17:3305, several Christian Democrats argued that the parliament should support their objections to assisted reproduction for single persons and to assisted reproduction by donated germ cells (Henriksson et al. 2016/17:3305). They also argued that the parliament should support their proposal to provide support for those who had conscientious objections to certain medical interventions (suggestions 56–59 in the text of the proposal).

11 Knowledge and Values in the Biopolitical Debate

Controversial issues in the present debate in Sweden have included the donation of fertilized human eggs; the length of time such eggs can be preserved cryogenically; various forms of genetic testing, screening and sequencing; euthanasia, physician-assisted suicide and palliative sedation; human embryonic stem cell research; biobanking and open data in repositories; the extent to which patients are to be allowed to pay the whole, or part, of the additional cost for more expensive hearing aids, multifocal lenses and other equipment they might need or want; co-payment, priority setting and the mission of healthcare; and, until quite recently, surrogate motherhood and the use of electronic devices to measure physiological variables (wearables and e-health).

In Sweden, as in other countries, the arguments in biopolitical debates are essentially of two kinds: knowledge-based and value-related. Ethical and humanitarian values are commonly cited, and it is only rarely that religious arguments and values are explicitly referred to. Important values in the biopolitical debate include justice, the right to self-determination, human dignity, and solidarity. All of these values are vague and ambiguous. This becomes obvious when they are applied to specific, concrete issues. Human dignity is a complex concept. Its ambiguity is illustrated by the fact that it is appealed to both by those who are for and by those who are against euthanasia. 'Death with dignity' can be interpreted in several ways, as can be seen in the Swedish debate in the Journal of the Swedish Medical Association (Läkartidningen) which took place in 2017 (Koivisto 2017;114:ELLZ; Bergström and Jersild 2015). In a different sense, and related to the idea of non-discrimination on the basis of social and economic status, gender, age and religious or political persuasion, 'human dignity' is applied within the priority-setting debate. In this sense, it has played a role—together with principles of need and solidarity—in the debate over how limited and scarce resources for healthcare are to be allocated in Sweden.

'Human dignity' has religious overtones, but these are conspicuously absent in the Swedish debate. Here a Kantian, rather than religious, interpretation has been influential, including Kant's prohibition on using a person merely as a means. Religious interpretations include those based on the idea of *imago Dei*: that human beings were created in the image of God and for that reason have infinite value from the beginning of their lives. But as already mentioned, the idea that all humans have equal value originated with the Stoic philosophers. Some think that certain values and moral principles that are embraced in society are to be understood as a legacy from the time when religion played a central role and that they lack philosophical backing. Interesting as this idea is, it is a possibility that will not be pursued in this chapter.

12 Ethics and Politics in Sweden

Ethical disputes, like political debates, are based on information about trends and possibilities, as well as on perceptions of them. Values and ideologies play a crucial role in both ethics and politics. There are certain states of affairs we want to achieve and others we want to avoid. Such choices are based on values. In a pluralistic society like Sweden, people take different views both on these values and on their relative importance. In ethical debates there is no need to compromise: certain values are interpreted and applied consistently to situations when people are faced with a choice between alternatives. This holds both for the ordinary person and the ethicist. Consistency and the exploration of the consequences of proposed policy options provide conditions for a lively debate. Consistency is important in policymaking: the hierarchy of values that forms the basis for decision-making should be consistent; double standards are criticized in ethics. In both politics and practical ethics, it is important to understand the concerns of the public at large and to be able to anticipate the opportunities and threats created by new scientific developments. A difference between politics and ethics is that those who want to be successful in politics need to have extensive networks. In particular, they have to be able to form alliances and coalitions and to give and take—to make compromises, in other words. Sweden is no exception. In Swedish political life, there has been a long tradition of compromise, and this has benefitted society at a whole.

At the present time, in contrast with past centuries, the Lutheran Church is not involved in the law-making process. This is not to say that all of the members of parliament are agnostics or atheists. They are not. But the church as an institution plays an insignificant role. Of course, this is not to deny that Christian members of parliament sometimes express religiously coloured views in the debates. They do, as has been illustrated above. The Swedish Lutheran Church sees itself as part of a pluralistic and democratic society. The present archbishop, Antje Jackelén, has attempted in a book,[4] and in newspaper articles, to promote dialogue between secularists and the church on bioethical and biopolitical issues. She has expressed concern about the fear of contamination that prevents non-believers from engaging in serious debates with believers. However, so far this has not resulted in much progress, and it remains the case that few news items and opinions on biopolitical issues are officially published by the Lutheran Church. When religious statements in the strong type 1 sense are published, they have little impact on the decisions being taken, as already mentioned.

References

Akner, Gunnar and Bengt Järhult. 2016. *Värdebaserad vård införs snabbt: värdig efterträdare till NPM? Läkartidningen*. 113:DXUT.

Arborelius, E. 2016. Vi kristna är kallade att stå på livets sida. Dagen. December 9.

Bergström, Staffan and P.C. Jersild. 2015. Att ta på allvar: det självvalda livsslutet. Läkartidningen. 112:DL4D.

Blendon, R., M.K. Kim, and J.M. Benson. 2011. The public, political parties, and stem-cell research. *New England Journal of Medicine* 365: 1853–1856.

Clapham Institute. 2017. *Hotet mot värdigheten* (The threat against dignity).

Henriksson, Emma, et al. 2016. *Proposal to the Swedish parliament 2016*. Vol. 17, 3305.

Järhult, Bengt. 2017. *Ska värdebaserad vård införas av konsulter utan vetenskaplig evidens? Läkartidningen*. 114:EFHM.

Koivisto, Maj-Lis. 2017. *Kritik mot Oregonmodellen i ny rapport*. Läkartidningen. 114:ELLZ.

Landgren, Per. 2004. Expression of dissent. *Government commission on genetic integrity*, SOU 2004:20.

Malmgren, Bengt. 2014. *Stötande att inte ta barnmorskans motiv på allvar*. Läkartidningen.

Smer (Statens Medicinsk-Etiska Råd). Since 1989. Skriftserien Etiska vägmärken. Stockholm: Smer.

———. 1995. *Assisterad befruktning: synpunkter på vissa frågor i samband med befruktning utanför kroppen*. Stockholm: Smer.

———. 2000. *Om livets början*. Stockholm: Fritzes.

———. 2008. Etik: en introduktion, 3 uppl. Stockholm: Fritzes.

———. 2015. *Tankar om Medicinsk etik*. Stockholm: Smer.

Swedish Government. 2001. Behandling av ofrivillig barnlöshet. Proposition to the parliament. 2001/02:89.

Wärn, Daniel. 2002. Dagen, April 20.

[4]Antje Jackelén (2016). *Samlas kring hoppet*. Stockholm: Verbum, 2016.

Chapter 11
Consensus vs. Confrontation: Negotiating Embryo Politics in Norway and Italy

Mirjam Weiberg-Salzmann and Massimiliano Passerini

Abstract Many countries have imposed legislation on biotechnological procedures. However, the way the debates have developed, as well as the degree to which biotechnologies have been restricted, differs widely from country to country. While some countries, such as Italy, were until 2004 incapable of passing a law on biotechnological procedures, others such as Norway had already been imposing restrictive regulations since the 1980s. One reason for this was the different historical relationship between state and church and the different involvement of religious actors in the debate. In Norway, a cooperative approach prevails. The Norwegian Church sees itself as part of a pluralistic society/the public and intends to convince with religious and secular arguments. It therefore explicitly seeks accordance with the argumentation and a consensus in the positions with humanists and non-Christian ethics. In Italy, church and state have chosen a confrontational approach rather than a cooperative one. Overall, the Catholic Church regarded itself as an agent and defender of rational, universal and non-negotiable values and principles, which all reasonable humans could share. In contrast, particular secular and laical forces found that the Catholic Church has gone beyond its constitutional rights by trying to illegitimately impose its world view and ethical-religious values on society as a whole.

1 Introduction: The Development of Biopolitics and the Role of Religious Actors

A number of spectacular biotechnological innovations emerging since the 1960s—like the transfer of DNA from one species to another, prenatal diagnosis procedures or artificial insemination—were highly influential on biopolitics. From a technological viewpoint, this was only the beginning of a development with endless potential. However, questions concerning its societal use and its moral-ethical

M. Weiberg-Salzmann · M. Passerini (✉)
University of Münster, Münster, Germany
e-mail: mirjam.weiberg-salzmann@uni-muenster.de; massimiliano.passerini@uni-muenster.de

M. Weiberg-Salzmann, U. Willems (eds.), *Religion and Biopolitics*,
https://doi.org/10.1007/978-3-030-14580-4_11

justification had not been resolved and political and legal regulation became an increasingly important topic for discussion. Problems to be addressed included: for example, which procedures were deemed acceptable under which conditions, who would gain access to the new biotechnological resources and what ought to be publicly funded. In addition, moral and ethical issues arose, e.g. what was considered socially acceptable and morally appropriate (Lembke 2008). Thus, biopolitical issues have been part of the political agenda in almost all industrialised countries since the 1990s. Many countries have legislated on biotechnological procedures. However, the way the debates have developed, as well as the degree to which biotechnologies have been restricted, differs widely from country to country. While some countries, like Italy, were not able to pass a law until 2004, others like Norway had already imposed strict regulations since the 1980s.

Religion played an important role in the debate on the new biotechnologies. Since the life sciences were a highly specialised field, its regulation required the consultation of a number of experts from different disciplines who had systematic knowledge about “life”. Today, experts from all fields are involved in the discussion. But at least in the beginning, biopolitics were in large part the domain of religious actors, who brought their understanding of how to lead a good and righteous life and of the status of human beings to the discussion. Nonetheless, religion was a double-edged sword within the debate: on the one hand, it was assumed to confer expertise on moral and ethical issues and in finding solutions to borderline situations and in handling contingency. As these skills were needed for solving many bioethical issues, religion was an acceptable participant. On the other hand, there emerged, as early as the 1960s, a counter trend that was set to eliminate any religious argumentation. Secular specialists started taking over the field, and many theologians and religious ethicists acceded to society’s demand for secularisation. Religious actors that wanted to appeal to a wider audience had to secularise their arguments. Moreover, the formal separation of church and state made it difficult for them to argue “religiously”, at least as far as the institutionalised forms of the bioethical discourse were concerned. In parliament, as well as in bioethics commissions, secular language was used. In order to appear in papers and reports, non-religious argumentation and a language intelligible to all were often required (Myskja 2009; Campell 1990). Today, the scientific community and economists alike demand that biopolitical decisions are to be made without reference to religious beliefs (Lauritzen 2005; Schlieter 2005). From a theoretical point of view, the influence of religious actors as well as religious argumentation is viewed critically too. Political science considers the separation of religion and politics as an important (although not sufficient) prerequisite for a stable, democratic state. Although religion was also accorded a positive and useable potential in political opinion and decision-making processes, this demonstrated most benefits when religion remained within the spheres of private life or civil society. Political decisions appeared most legitimate if they were based on secular argumentation that is accessible and understandable to all civilians.

As previously mentioned, in the development of biopolitical debates, the role of religious participants and the use of religious arguments differ widely from country to country. For example, in Italy a legislative agreement was deeply affected by the conflict between religious and secular actors, while in Norway a more or less smooth debate between church and state led to mutual consensus. Our objective is to explain these differences by comparing the debates on in vitro fertilisation (IVF), pre-implantation genetic diagnosis (PGD), human embryonic stem cell research and reproductive/therapeutic cloning. We assume that differences are caused by the historical relationship of religion and politics, the degree of secularisation and the characteristics of the respective hegemonic religious tradition (Weiberg-Salzmann and Willems 2019). It is of special interest how religious actors engage and are integrated into the discussion, and what form and content the arguments used have.

2 Norway: Applying the Christian Cultural Heritage to Biopolitics

2.1 The State–Church Relationship

The current state and church system in Norway can be traced back to the Reformation of the country in the sixteenth century. Ever since, the evangelical protestant church (with a short interruption) has been the state church of Norway. Until 1814 Norway was part of the Danish monarchy, but it was administered by a small Norwegian bourgeois élite out of which priests of the Protestant state church also recruited themselves. Since the indigenous nobility was almost extinct, societal structures were quite homogenous—or egalitarian, as the Norwegians would put it. This has barely changed until today. Democratisation and social and political mobilisation were closely linked to Christian revivalist movements since the seventeenth century.[1]

Initiated by land-holding farmers, the revivalist movements gradually affected the whole society. The Protestant version of Christianity was transformed into a civil theology that focused on individual freedom and cooperation. This civil religion was a mix of Christian "common sense", democratic representation and nationalistic elements. The religious and sociocultural associations that resulted out of this have set the paradigm for political parties since the beginning of the nineteenth century, as well. Up until today, most parties in Norway are closely linked to certain social groups. There is, however, no large Christian party.[2] Norway is a consensus-based democracy characterised by negotiation. Characteristic for this form of government

---

[1]The much revised constitution of 1814 justifies a constitutional hereditary monarchy on a parliamentary-democratic basis. Freedom of religion had not at first been granted by the constitution, but then there had never been an obligation to publicly profess one's commitment to the Protestant state church (Frandsen 2009).

[2]The Christian People's Party, for instance, originated from the Christian laymen's movement and other religious associations from West and South Norway (Stenius 1997).

are as follows: including a broad spectrum of interests in the political decision-making process, pragmatically oriented competition between parties and a sociopolitical consensus. In general, Norwegians place great trust in political regulations based on solutions that incorporate the whole society; decisions that are left to special interest groups gain little trust (Christensen 2007; Petersson 1989, p. 16; Christiansen et al. 2010).

The constitution establishes a strong state and church system—the Protestant-Lutheran religion is the "official state religion" and the "Norwegian-Lutheran church" is the official state church, while the substance of the constitution is strongly shaped by Christian principles. The king is the head of the Protestant church and has to belong to the state church, as does at least half of the government cabinet. Since the king signs laws and as he is commander-in-chief of the armed forces, this gives religion—at least formally—the opportunity to exercise great influence on politics. On the other hand, the church is controlled by means of the state: laws and financial issues concerning the church have to be authorised both by the parliament and by the royal ministry for renewals, administration and church (fornyings-, administrasjons- og kirkedepartement); furthermore, bishops and priests get appointed by the king (Groß and Rothholz 1999; Von Unruh 1989). In practice, however, the state has only seldom intervened in ecclesiastical issues and, if so, only at the request of the clergy and in respect to administrative issues but never to challenge a religious doctrine. In conflicts (for instance, concerning the ordination of women and homosexual priests), the government regularly supported the liberal forces of the church against the conservative wing. From the beginning of the 1990s, a reform process has been initiated in order to transfer more institutional autonomy to the church. Since 2006, discussions have been led to abolish the privileged status of the church (Danielsen 1995; Hiroth 1997; Schmidt 2010; Stortingsmelding[3] 2008). Yet the majority both of the clergy and of the population continued to oppose a complete separation of church and state and prefer to retain a system of church privilege. Since other religious communities benefited from this system too, the regulations made have not been questioned by them so far.[4] However, in 2012, the Norwegian parliament passed a constitutional amendment which meant the separation of the Lutheran church and the state. This ended a liaison of 475 years. Remarkably, the church not only approved this decision; it even turned out to be the driving force. What appears to be at first sight a reduction of the rights of the church and resembles a displacement of religion from the public sphere is actually a win-win situation. Different religions and confessions are from now on of equal status. Certainly, the

---

[3]Act of parliament.

[4]Two secular associations constituted an exception: the Human Ethics Association (Human Etisk Forbund) with 50,000 members and the Pagan Association (Det norske Hedningsamfunn) with 300 members have heavily criticised the system in the past (US State Department 2007, 2008).

Protestant-Lutheran church loses its privileged position, but this is outweighed by a gain in autonomy and by a strengthening of its social credibility.[5]

Conflicts between political-secular forces and the church were prominent in the 1920s and 1930s when the Labour Party opposed the influence of religion in general and that of the state church in particular. Up to today, however, the overall image of the church is quite positive. In recent history, its image is closely connected to the German occupation period: here, the church played an important role with the Resistance against nazification and the Norwegian collaborationist regime. Its image today is shaped by actions such as standing up against legal arbitrariness and violence and for the respect of human dignity (Austad 2009). Religious leaders are met with respect and their opinion on moral-ethical topics is broadly accepted. At the same time, the influence of the church on politics is less evident. It is mainly shaped by single persons or through Christian-oriented politicians. The church does not possess a strong lobby within the Norwegian parliament, but a number of Norwegian parties with Christian roots and attitudes do exist: the Christian National Party (Kristelige Folkeparti) led a coalition government under Bondevik 1997–2000 (I) and 2001–2005 (II).[6] The coalition partners, the Conservative Party of Norway (Høyre, H) and the Liberal Party (Venstre, V), are not explicitly wed to a Christian programme but have a positive attitude towards religious values. Even the Labour Party follows a rather moderate course today. Atheism is an unaddressed subject in the Norwegian party landscape, and as far as the issue of the separation of church and state is concerned, for a long time a "wait-and-see" strategy was being followed. Arguments about moral-ethical conflicts tend to be based on general human rights (in part derived from the Christian image of humanity).

Even though Norway has had an official state church, religious-funded parties were marginal for a long time. Since the contrast between workers, capitalists, farmers and industrial employers was dominant and Christian democratic parties did not emerge from a church–state conflict, they never achieved the same status as broad-based parties. The reformed Christian faith, however, entered a special union with the rural community. This pattern was upheld in spite of secularisation. Today, most Norwegians regard religion as a private matter. In 2009, eight out of ten ethnic Norwegians belonged to the state church, while in 1980 almost 88% were recorded.

[5]The Protestant-Lutheran church is no longer the official state religion. The Norwegian church, however, is still to be the "Norwegian National Church" (Norges Folkekirke). Article 2 no longer states: "*The Protestant-Lutheran religion is the state religion*" but: "*The basic value is our Christian and humanistic heritage*". The church now appoints priests and bishops on its own. The king must still profess his belief in the Protestant-Lutheran religion; ministers, however, are not obliged to do. The Norwegian church will continue to be financed by the state. Other religious communities are also supported by the state, depending on the budget. Forssman (2012).

[6](1) Coalition: Kristelige Folkeparti (KrF), Senterparti (Sp), Venstre (V), (2) Coalition: Kristelige Folkeparti, Höyre (H), Venstre.

Religion still plays an important role, yet in general it is not expressed by means of the active religious participation in organised communities.[7]

2.2 *The Biopolitical Debate in Norway*

The debate on biotechnologies can be divided into three phases: the discussion in the 1970s (risks of biotechnology), the one at the beginning of the 1980s (scientific and ecological paths) and the one starting at the end of the 1980s (ethical-moral aspects) (Bleiklie 2004, p. 210; Brekke 1999). Predominately, with reference to Christian values, the Christian Democratic Party (*Kristelig Folkeparti*, KrF) has consequently spoken up against abortion, in vitro fertilisation and stem cell research (Madeley 1994). The Socialist Left Party (*Sosialistisk Venstreparti*, SV), on the other hand, justified its rejection with general scepticism against the technologisation of human beings and society as well as with the fearing of a society based on selection. The Labour Party (*Arbeiterparti*, Ap) offered liberal and permissive suggestions. It must be taken into account however that although the political parties represent different societal viewpoints, they follow a general, underlying national consensus that is not questioned by anyone—the exception is the Progress Party (*Fremskrittspartiet*, FrP). Consequently, with regard to concrete policies, they are often not so different from each other.

The first Norwegian child conceived by in vitro fertilisation was born in Trondheim in 1984. Thereupon, the health and social sector asked the government to develop a framework of legal regulation. Starting from 1985, legislation was developed for in vitro fertilisation and prenatal diagnostics, first under the KrF (in government from 1981 to 1986, although from 1983 in coalition with a right-wing party and the Centre Party (*Senterpartiet*, Sp)), then, from 1986, under the Labour Party, and passed in 1987.[8] It approved of in vitro fertilisation under very restrictive conditions. Accordingly, only married (heterosexual) couples were taken into account,[9] only the husband's semen could be used, with donated semen only allowed if the man was impotent or the carrier of a serious hereditary disease. Also, IVF could only be employed if the mother was proven to be infertile. Embryo

[7]Only 10% take part in services or other Christian meetings more than once a month (Botvar 2010; Statistisk sentralbyrå 2015). Other confessions and religious communities only make up a minor part in public life.

[8]Lov om kunstig befruktning No. 68, 12 June 1987, Decision. O. No.61 (1986–1987). In contrast to other countries, the debate in Norway focused on research done on inseminated egg cells; therefore, there was initially no distinction made between embryonic stem cell research, PGD and human cloning. It was not until more recent regulations that the medical fields and methods were differentiated from one another.

[9]Since 2009, same-sex marriages are also legal in Norway. Consequently, married women or those living in a stable relationship are allowed to undergo in vitro fertilisation. Homosexual men, however, remain excluded since the donation of ova or surrogacy is still prohibited.

research was prohibited, the freezing of ova came to be regulated and all treatments had to be carried out at institutions approved by the health ministry.[10] Even though the law proved to be more restrictive than had actually been discussed in the debates of previous years, KrF and Sp—now the opposition—voted against the proposals.

In order to clearly state ethical guidelines, the ethics panel *(etikkutvalget)* was founded in 1988. In the final 1991 report, "Human beings and bio-technology", a number of biomedical methods, amongst others in vitro fertilisation and embryo research, are also dealt with (Norge Sosialdepartementet 1991). While the panel supported in vitro fertilisation, it rejected PGD and the selection of human life that it implies. There was a wide set of arguments used that were based on the UN human rights declaration and the Helsinki declaration, as well as on the Christian and humanist idea of man. Wherever the board supported its claims on the basis of religious arguments, it also cited equally important argumentation from an ethical standpoint. Shared values and shared goals (a legal provision) were seen as more important than argumentation strategies (religious or secular). The report did consider aspects that are important to the Norwegian Church subject to discussion (marriage, family, the unity of procreation), but it did not explicitly make reference to the church's position. If at all, the church's general approval of bioethical laws was emphasised. The Ethics Commission *(Bioteknologinemnda)* which arose out of the ethics panel and which still advises the government today comprises a changing group of experts (inter alia, theologians, scientists, doctors, jurists, social scientists and philosophers) and representatives of different associations (agriculture, environmental protection, disabled persons, industry).[11]

In 1999, 2001 and 2004, the commission published documents with the aim of evaluating the legislation and which rejected embryonic research and cloning. The participants, however, did not agree on all subjects. The arguments remained the same: they made reference to human dignity, the uniqueness of man and Kant (means to an end) and warned of the slippery slope loose regulations might lead to. Once again, the protection of human life found a broad consensus based on a wide set of arguments that could be shared by many Norwegians. With the evaluation of the law of 2011, the Ethics Committee supported the current restrictive PGD regulation, while no unanimous decision could be achieved concerning the research of fertilised ova, cloning and stem cell research (Bioteknologirådet 2011).

At the beginning of the 1990s, the ruling Labour Party once again spoke out for a liberalisation of the law. Despite the criticisms of the Christian National Party, the Centre Party and the Socialist Left Party, the biotechnology law was passed in 1994. It dealt with regulations on methods of in vitro fertilisation, PND [prenatal diagnostics], PGD [pre-implantation genetic diagnosis] and embryo research and proved to

[10]All medical research and treatment of childlessness takes place in the public sector in Norway. There are seven public and four private clinics which carry out IVF (Helsedirektoratet 2011, p. 78).

[11]Although the church cannot demand to participate in the Council, it is traditionally taken into consideration.

be slightly more liberal than the law on in vitro fertilisation of 1987.[12] The purpose of the law was, according to the preamble: "[. . .] *to use biotechnologies in the best interest of humankind in a society in which there is a place for everyone. This is to occur following the principles of respecting human dignity, human rights, and personal integrity. It shall not discriminate against someone's genetic disposition, following ethical norms derived from our Western cultural heritage*" (Besl.O. nr.73 [1993–1994]). While all parties could agree on the reference made to human dignity, the term "our Western cultural heritage" stirred discussion. "Our Christian-humanist cultural heritage" was proposed as an alternative but rejected in the end, although the opposition (KrF, H, Sp and Frp) opted for this term. With the exception of this conflict, the debate made no religious references. Besides emphasising the importance of a society in which every human being is included, all parties highlighted the great preventive and curative potential that these technologies have in cases of severe illness. The opposition, however, raised concerns that opening up research possibilities might endanger human dignity and inevitably lead to a selective society. Overall though, the opinions of the government and the opposition did not differ widely from one another. The following years have seen some small changes, with laws and legislation, for instance, with respect to gender tests with athletes. Without exhaustive debates, reproductive and therapeutic cloning was prohibited in 1998 and in 2002, respectively.[13]

With the 2002 evaluation of the law, the opinions of the parties barely changed. The most important arguments were that every individual has inherent self-worth and inviolable dignity. Nobody is to be discarded and human life is not to be used as a resource for others. All parties used the Bible in some specific instances as a point of reference (the tree of knowledge, human sinfulness, man is made in the image of God). The value of the human being was stressed as a normative basis as well as the autonomy/right to self-determination of the patient and the precautionary principle ("prevention is better than cure"). Also in the context of the evaluation, first of all extensive changes were proposed and then, later on, a new law. The biotechnology law *(bioteknologilov nr. 100 5/12/2003)* which, however, does not differ significantly from the present law abolished the anonymity of semen donors. The research with fertilised ova—and therefore PGD, stem cell research and cloning—remained prohibited. A couple of months later, the case of the 6-year-old Mehmet became public. His parents hoped to cure his blood disorder with a bone marrow donation of an (up to that point unborn) sibling. PGD was therefore discussed again and the law,

---

[12]Law on the medical use of biotechnology (*Lov om medisinsk bruk av bioteknology, nr. 56 5.8 1994*). It relaxed the prerequisites for in vitro fertilisation and extends the storage time for fertilised ova. Embryo research remained prohibited, even though PGD was at least theoretically approved of in case of serious diseases which could not be treated accordingly.

[13]Decision Odelsting (Decision O.) No. 79 (1996–1997) [Ban on sex determination except for medical indications];

Decision O. No. 19 (2000–2001) [Enlightening concerned persons];

Decision O. No. 27 (1997–1998) [Ban on producing individuals of the same genes];

Decision O. No. 21 (2002–2003) [Amongst others, ban on therapeutic cloning].

which only had been just passed, was changed. The law now allowed for a genetic examination of the fertilised ova in case of a serious, hereditary disease without possible proper treatment and if a committee approved of the case; the embryos were, however, not to be genetically modified. Ap and Frp supported PGD in cases of severe genetic disorders. The Frp justified this with reference to mercy. The opponents, however, H and KrF, rejected PGD, arguing that a human being is not a means to an end for others and that PGD systematically discards human life. At the same time, all parties focused on their shared values with even the "liberal" parties demanding only limited, regulated research. In comparison with other countries, this was also restrictive.

In 2005, the Labour Party took over the government in collaboration with SV and Sp and announced that research on fertilised ova—amongst other things stem cell research—as well as PGD would be approved of under specific conditions. In the same year, the Progressive Party made a proposal.[14] It was accepted after extensive discussion within and outside the parliament in 2007.[15] The governmental parties pleaded for regulated research on inseminated egg cells, while the opposition was strictly against this. The main argument of the proponents was that from an ethical viewpoint it did not make sense to destroy a surplus of inseminated egg cells instead of using them in research. The government argued on the basis of the humanitarian view on human dignity, human value, the health-care system, UN guidelines and the recommendations of the ethics committee, which went along with the government's policy. The opposition also emphasised the importance of protecting human dignity and naturalness, but arrived at a different conclusion. They opposed opening up the law by arguing that it is the responsibility of ethics to set limits to technologies. They also argued that embryonic research implies the instrumentalisation of human beings and feared that there would be no place left for the weakest in society. In the end, the parties managed to come to an agreement as the government only wanted to permit embryonic research and PGD under very strict regulations. The law is therefore far from being liberal.

2.3 *Religious Positions and Arguments Within the Biopolitical Debate*

The religious landscape of Norway is mainly restricted to the Norwegian Church of which 80% of all Norwegians are members; Muslims (2.5%), Liberal Humanists

[14]Innstilling Storting (Innst.S. = proposal on part of the committee to Storting) No. 225 (2004–2005) and Odelsting proposisjon (Ot.prp = proposal on part of the government to Odelsting) No. 26 (2006–2007). It allows for research with surplus fertilised ova for reasons of quality assurance and stem cell research, as well as PGD for the determination of the tissue type. Cloning remained to be expressly prohibited.

[15]Decision O. No. 78 (2003–2004).

(1.7%), Catholics and adherents of other faiths do not play a role in the public discussion on biotechnology (Statistisk sentralbyrå 2015). The church was always involved in the law-making process. Members of parliament would have contact with the bishops, the church council, general convention and members of the theological faculty of the University of Oslo and the Norwegian School of Theology. Furthermore, the church actively participated in the ethics commission. At the same time, the Norwegian Church—despite being the state church—does not actually demand to have a say in laws, as it regards this as the affair of the Parliament. Nevertheless, it sees its responsibility in giving advice to society in these matters and to ensure that the protection of the weak is included in legislation as well (Bispemøte 24/77[16]). Outside the church, only few interest groups have commented on biopolitical topics. Commonly known are the pro-life group Menneskeverd (Human Dignity, previously Pro Vita; many conservative Christians and members of the Free Church are committed to the group)[17] and the Association of Liberal Humanists (Human-Etisk Forbund) which supports the idea of free research.[18]

Overall, the involvement of the church related to human biotechnology is mainly concerned with termination of pregnancy. "*All other bio-technological developments, like PGD or vitro fertilisation, have played a much smaller role and are only of interest once abortion is being dealt with*" (Christoffersen, interview, April 2011). The Norwegian Church has voiced its opinion several times on the protection of unborn life since the 1970s and still questions the abortion policy (Kallerud 2004, p. 83). Within this debate, the Bishops' Conference placed the question of the value of the embryo onto the agenda[19] (Bispemøte 1971, 1975, 1977, 1978, 1979). The discussions have stressed the Christian idea of man and the moral responsibility of the workers in pastoral and health care. On the one hand, criticism is exerted on the "abortion mentality" and on the "abortion pressure" of society, and on the other hand (Bispemøte 1971, paragraph 1), sympathy is also shown for the situation of the women. In all documents, the Bishops' Conference stands up against abortion since man was created in the image of God and therefore had an infinite value right from the beginning. The debate on abortion, human stem cell research, PGD, PND and cloning is closely intertwined in Norway. The general question here is how the embryo is being treated. Since the termination of pregnancy, however, has now been accepted by society, the church remains silent on "subcategories" like stem cell

[16]Bispemøte (BM = report of the bishops conference) 1977, Sak (= Sache) 24/77.

[17]In 2007, the group was disappointed by the decision to open the biotechnology law for the research of embryos as it requires a grading of human values due to development, functionality and characteristics (Menneskeverd 2007, p. 10).

[18]Despite the ethical problems of the human embryonic stem cell research, they view them as justifiable as long as surplus ova are concerned. Høringssvaret om Utkast til endringer i lov om humanmedisinsk bruk av bioteknologi med mer (Helsedirektoratet 2006).

[19]Since 1975, Norway has got a deadline regulation which implies that pregnancy can be terminated within the first 12 weeks. The woman is free to decide on her own whether she is capable of carrying out her pregnancy or not. After a period of 12 weeks, the pregnancy can only be terminated under specific conditions and after 18 weeks only in case of serious reasons.

research, PGD, cloning, etc., to a greater extent. Nevertheless, since the 1980s, genetic counselling and prenatal diagnostics have been discussed. The recommendation "Human image and human value" of the General Synod of 1985 (Kirkemøtet 1985), which was sent to all parishes, starts off with several Bible quotations and a strongly religiously oriented foundation on human dignity: God as the Creator, the image of God and the resulting inviolable human dignity. Key points on the premises underlying right behaviour are then addressed, amongst other things, "the responsibility to re-build the legal protection of unborn life in our country by campaigning for a new abortion law" and "to contribute to a conscious ethical responsibility as well as the further development of laws for medical research and practice". Even if these are proposals on concrete actions, as is mentioned in the preface of the documents, the premises are kept very general and only provide a general overview of the Christian idea of the man of the church. In the context of the programme, "Challenges of modern technology: Bio-technology, environmental issues and new technology" scheduled for the years 1988 until 1991 by the Council of Churches (Kirkerådet 1987), a working group was formed which published "More than genes" in 1989 (Kirkerådet 1989). In the same year, the synod also dealt with the topic, namely under the title "The value of human beings". Besides dealing with abortion and PND, the document, "Bio-technology and human dignity", also focused upon research with fertilised ova for the first time (Kirkemøte 1989). The document provided information on the position of the church vis-à-vis children, elders and disabled persons in society and sympathised with concerned families. Not only the role of the church was, however, stressed, but it also asked for action. "More Than Genes" is, on the other hand, a basic information brochure. It provides a historical review of how the beginning of life has been understood. It also introduces other basic ethical principles both for informational and normative reasons, against a Christian-oriented background where it becomes evident that non-Christian ideas are also becoming accepted and respected. At the same time, medical methods are explained.

In the 1990s, little new on biopolitics was officially published by the church. With regard to the engagement of the church in the issue of the termination of pregnancy, the Bishops' Conference only very briefly commented on the current state of affairs. More up-to-date statements were released in 2006 in the context of the elaboration of the new biotechnology law. In this context, the synod, the Council of Churches and individual bishops commented during extensive hearings.[20] The 14-page document was addressed to the government and provided a general overview of the situation and the general position of the Norwegian Church and set out the dilemmas in the regulation of biotechnology. Documents of the Ethics Commission or the Parliament were also often quoted. Thus, it seeks to persuade on the basis of well-founded arguments without appearing to be missionising. In 2008, the Bishops' Conference once again released a short statement arguing that human dignity is independent of its level of functioning. The Norwegian Church then commented extensively on the

[20]The most extensive document is Kirkerådet (2006).

proposed amendment in connection with genetic research in a hearing by the Council of Churches in 2011 (Høring 2011). The topics PGD, stem cell research and cloning once again became current at the beginning of 2012 in relation to debates on bringing forward ultrasounds from the 18th to the 12th week of pregnancy. The Bishops' Conference again argued against an early ultrasound and, as with the documents of the previous centuries, referred to the Christian idea of man as being created in the image of God (Bispemøte 2012 Sak 13/12).

Overall, the argumentation of the Norwegian Church proved to be twin-tracked. On the one hand, in earlier documents ("More than genes", 1989), it already referred to the fact that the Bible does not provide (concrete) statements and answers on biopolitical topics. For this reason, it adds the declaration of human rights as well as ethical concepts to religious references such as God as the Creator, the creation in the image of God with the resulting inviolability of human dignity, the sinful nature of mankind as well as human's responsibility for creation. More than once, the argumentation is not explicitly religious but sometimes placed in a religious context (for instance, protecting the weak, human rights being based upon a Christian idea of man which are, however, also employed today in non-Christian contexts). The Norwegian Church supported stem cell research on adult cells but not on embryonic cells. Although the inseminated egg cells are created in the image of God according to the Norwegian Church, they are not seen as an equal to a human being. They deserve protection, nevertheless, as they are life in the making.[21] Biotechnology is not refused as such; the main argument against research with embryos is related to the developmental potential of embryos and to the exploitation of human beings. In this respect, Kant plays a role, as does the argument of the slippery slope, social issues (amongst others distributive justice), protection of the weak and the value of life. While Christian understandings often mark the starting point, other argumentation strategies were seen as equally acceptable and adequate. With the help of this approach, the church's position and arguments also became reasonable to non-Christians. In particular with regard to the wording of laws, the church finds it necessary to provide commonly understandable arguments that are in accordance with many members of parliament (Christoffersen, interview, April 2011). Papers that are more directed towards their own followers, like "Biotechnologies and human dignity" (1989), emphasise the infinite value of human life, since humans are created in God's image and accepted by Him. There are even direct references to biblical sources. Although they mix religious and secular arguments, the overall argumentation strategy remains religious. In general, the Christian/religious ideological basis can be seen more clearly here than in "More Than Genes". This is mainly due to the intended audience (believers), which shares the same convictions, making it unnecessary to secularise the stated arguments (Kirkemøte 1989, pp. 5, 123, 127). In its hearing from 2006 the Church Council once again underlined the sanctity of human life. It, however, admits that there is not always a clear right and wrong to be found in a pluralistic society. Furthermore, the church rejects

[21]Mer enn gener, p. 4, 35, 48f, 58, 102.

embryonic research for the pure objective of doing research. It did, however, support the restricted use of PGD, while simultaneously voicing concerns.

Since 2006, the church has approved of research with fertilised surplus ova, arguing that research might improve the methods of in vitro fertilisation and thus long-term surplus embryos (and their killing) could be avoided. In order to bolster its position, the church makes reference to the image of God and the implied value of life. Apart from this, there were barely any religious arguments made in the last document. At the same time, the ethics committee, the Parliament and the Norwegian Association for the Disabled were quoted in order to ascribe equal value to all humans, regardless of capability. Furthermore, Kant's Categorical Imperative, the slippery slope argument and (the already implemented) IVF are cited as arguments. In the conclusion, the church mentioned that it sees its own position as a contribution to the discussions and as a suggestion for society as a whole (Kirkerådet 2006, pp. 5–9, 14).

Since the church sees itself as part of an egalitarian society and a pluralistic society, it lacks the Catholic Church's claim to universality. The Norwegian Church does not view itself as an authoritarian body but as one amongst many others. Thus, the church refers to statements of other institutions and quotes them in its own documents without reservation. Furthermore, the church does not explicitly want to have a hand in the decision-making process. This position can easily be claimed by the church since all participants share a similar viewpoint.

3 Italy: Catholic Intervention into a Fragmented Biopolitical Discourse

3.1 The State–Church Relationship

Church and state in Italy have always maintained a relationship of a varying nature. Since the foundation of the Italian state 150 years ago, there has been an ongoing conflict between church and state, which varied in intensity throughout the years. In spite of the constitutional separation of church and state, Catholicism has always been not only a religious and social institution but also a political actor (Livi 2016; Pollard 2008). The proclamation of the Kingdom of Italy in 1861 and the annexation of the Papal State Territories and the city of Rome in 1870 opened up the "Roman Question" which started a conflict between the new Italian state and Catholic institutions. The papal bull, "Non expedit" marked this confrontation by calling on all Catholics to abstain from political activities in the new nation state. Moreover, it demonstrated the Pope's political influence on society. This climate of conflict lasted until the Lateran agreements in 1929. They finally recognised the territorial sovereignty of the Vatican, agreed on compensations for the territorial losses during the Risorgimento, introduced Catholic religion in school curricula and established Catholicism as the official state religion. With the new agreement of 1984, the Italian

Catholic Church officially renounced its status as state church and accepted a new system of financing through the "eight per thousand" taxation.[22] At the parliamentary level, the Italian Catholics gathered in a popular mass party with Christian inspiration (*Democrazia Cristiana*, DC) after the end of World War II.[23] The DC represented the political unity of all Italian Catholics and dominated the political establishment until the 1990s. After the crisis of the "first republic", the DC split into several smaller political parties and factions. The governments of the following "second republic" were formed by volatile centre-right and centre-left coalitions (Sassoon 2013; Köppl 2011).

Although Italy is overwhelmingly Catholic in population and culture, the process of secularisation, as in many other Western countries, generated a gradual erosion of Christian values and a decrease in membership and in the feeling of belonging to the traditional church. One indicator was the promulgation of the divorce law and the failure of a referendum for the abrogation of the law in the 1970s. It revealed the relatively weak position of the church, not only in parliament but also in society. In 1981, during the referendum to abrogate "law 194" on abortion, the appeals of the Catholic hierarchy again failed and so demonstrated the irreversibility of the process of secularisation. The church failed to sway the conscience of the Italians even on such an important issue as the life of the unborn (Garelli 2007). According to a leading article in the Jesuit magazine *La Civilta Cattolica*, Italy could no longer be seen a Catholic nation (La civiltà cattolica 1983).

Since the centrality of the DC was increasingly declining since the 1980s, the church began to rethink its spiritual and social role. Under the pontificate of John Paul II (1978–2005), the Italian Church tried to recover the strength and popular power of Catholicism by finding "common criteria of orientation between believers of different political orientations" and by encouraging and spreading the laity's engagement in public institutions. The church's activities followed a double strategy: on the one hand, a reflective attitude on the part of bishops and on the other hand a rather active and visible presence of the papacy in society. The meeting at Loreto in 1985 characterised this contrast between a "Catholicism of mediation" and a "Catholicism of presence".[24] It established a strategy of "new evangelisation", embodied not only in parishes but also in Catholic associations and movements, demanding to be more involved and to regain an "effective leading role on the path

[22]A new arrangement replaced the old Lateran Treaties in 1984 under the Craxi government (1983–86, 1986–87, Socialist Party). With the Concordat, Catholicism officially lost status as the state religion and other religions were finally able to receive tax revenues from the state. However, the Catholic Church de facto already ceased to be state church in 1948 with the new Italian constitution, which recognised the full right to worship of other denominations and religions as well. Ferrari (1995).

[23]The first Catholic party, Partito Popolare, was founded in 1919 by Don Luigi Sturzo and disbanded in 1926 under fascist rule (Altgeld and Lill 2005; Masala 2004).

[24]"Catholicism of mediation" describes the attempt to bring traditional Catholic values and secular world views together in a modern world. "Catholicism of presence" describes the active participation of Catholic lay organisations in society.

towards the future".[25] Although the process of secularisation had significantly reduced active participation in Church life, Catholic identity was still claimed by the majority of Italians and the percentage of practising believers was still higher than in other European countries (Cartocci 2011; Bolzonar 2016).

In 1991, Cardinal Camillo Ruini became president of the Italian Bishops' Conference (*Conferenza Episcopale Italiana*, CEI) and immediately began to fight for the political unity of Catholics and to defend "several fundamental ethical and anthropological precepts" (Garelli 2006). The political unity of Catholics was no longer seen in one political party or in the orientation of the voters to a confessional party, but as an invitation to all politicians and Catholic voters to act for the defence and affirmation of Christian ethical principles in society (beyond party lines). In 1995, during the third ecclesial meeting in Palermo, the Catholic Church started the "*progetto culturale*" (cultural project). Initiated by Cardinal Ruini and the CEI, the cultural project promoted a Christian anthropological ideology. The project's purpose was to reshape the relationship of faith and culture, as well as to mediate between Christian values and new issues that have arisen from globalisation and modernisation. Catholics were asked to engage more actively in social, philosophical, economic, ecological and bioethical questions. The cultural project promoted Christian ethical values like human dignity, the centrality of family life and the significance of the Catholic community. The self-perception of the church was that of a mediator between different Catholic actors. While the church claimed to accept religious and cultural pluralism, at the same time it felt responsible for the defence of Catholic-inspired human values and for the protection of a Catholic Italy. By dint of this new approach, the Catholic Church was able to generate a lot of media attention and to strengthen Catholicism (Garelli 2007).[26]

3.2 *The Bioethical Debate in Italy*

The birth of Louise Brown in 1978, the first IVF baby, marked the start of the bioethical debate in Italy.[27] Immediate reactions to the birth were numerous and almost exclusively negative. The type of arguments made about the issues referred to, inter alia, human interference in the balance of natural processes, the temptation to implement eugenic plans, the separation of sexuality and the creation of life and the exclusion of women from the creative process. While some journalists praised it as a "great step for science", medical professionals and scientists—regardless of their religious or political affiliation—expressed deep concerns with respect to the safety

[25]"Un ruolo guida e un'efficacia trainante, nel cammino verso il futuro" (Ruini 2004).

[26]The church engaged in traditionally Christian fields, like bioethical issues, Church taxes and the freedom and financing of Catholic schools.

[27]The first attempts to legislate on this field were already in the 1950s, but never achieved any results.

and usefulness of the new method, due to its low success rate.[28] Despite criticism, IVF spread rapidly and technical improvements meant that many more children were born healthy. Sterility and childlessness were seen more and more as a curable disease and not a fate.

In Italy, the first IVF child was born in 1984. The same year, the Catholic Church accepted GIFT (gamete intrafallopian transfer) as a method for artificial insemination. From the very beginning, Catholic bioethics were divided into two fractions: on the one hand, there were the critics of the technique, who denounced the exploitation of the newborn, used as guinea pigs for researches. Even feminists condemned the expropriation of their "power to generate". On the other hand, there were the optimists, who valued IVF's ability to alleviate the suffering of infertile couples and to promote new births. Also in 1984, the Pontifical Academy for Sciences organised a conference dedicated to the study of the moral problems raised by IVF. The positions of the Catholics were very open and receptive to the admission of IVF. One exception was Bishop Caffarra—an expert on bioethical issues and since 1980 leader of the papal institute "Pontifical John Paul II Institute for Studies on Marriage and Family" at the Pontifical Lateran University—who rejected IVF because it allowed the conception outside the constraints of marital love (Flamigni and Mori 2005, p. 22). However, the decisions of the conference have never been published.

In the meantime, bioethics became highly topical and the then Minister of Health, Costante Degan, appointed the Santosuosso Commission in 1984 to discuss the moral and legal issues of homologous fertilisation and to prepare a draft law. The document was presented in 1986, but met with criticism from both secularists and Catholics, and never became law.[29] In 1985, a decree of health minister Degan forbade the donation of gametes (for heterologous insemination) in public institutions. This was also one of the results of the Santosuosso commission. The decision to restrict the donation of gametes shifted the monopoly of IVF to the private sector, leaving it without public control. The private clinics underwent barely any inspection and the sector was regulated almost exclusively on the basis of individual administrative decisions and rulings (Ramjoué and Kloti 2004). In 1987, the Catholic Church issued *Donum Vitae* (DV), publishing its official position on the matter. The instruction defined the embryo as a human person and forbade all forms of IVF treatment because the act of marital love should never be separated from procreation as the encyclopaedia *Humanae Vitae* (HV) had laid out previously. Medicinal and scientific experts were disappointed by this restrictive viewpoint. For Catholic politicians especially, DV was of utmost importance, as it provided them with clear instructions and argumentation for the biopolitical debate. Moreover, DV

[28]LXI Congresso della società italiana di ostetricia e ginecologia, 31.05.1982. Compare Flamigni and Mori (2005).

[29]Santosuosso Commission (1984): Its president Fernando Santosuosso has been judge at the Federal Court (Corte di Cassazione) and the Constitutional Court. Compare Santosuosso (1984) and Betta (1995).

sparked the interest of the Catholic world for the nascent discipline of bioethics. Catholic bioethics, which exercised a foundational role within the new discipline, received a further boost. Secular bioethics was established more slowly in Italy. But all in all, the pluralism of bioethical debate was remarkable. Within and amongst the parties, different positions and law proposals emerged. Worthy of note are two bills of 1988, which represent the opposing positions, Bill S. 1126 and Bill C. 3490. The first bill pleaded to restrict the access to IVF to couples (homologous), while the latter—introduced by a parliamentarian of the ultraliberal Radical Party (*Partito Radicale*, PR)—granted access also to unmarried women and to widows with the semen of the deceased husband.

In 1990, Italy finally established the National Bioethics Committee (*Comitato Nazionale per la Bioetica*, CNB). The CNB was intended to give advice to politicians on the possibilities, perspectives and risks regarding all bioethical issues. The committee included experts from different scientific disciplines and religious denominations to represent the various ethical positions within Italian society. But from the very beginning, politicians, intellectuals and non-Catholic actors repeatedly accused the CNB of over-representing Catholic ethics at the expense of religious and political pluralism.[30] The political and institutional crisis that swept Italy in the early 1990s brought the bioethical debate (IVF and its applications) to a standstill until 1994. But after the demand for IVF and its regulation had increased, the new health minister Raffaele Costa (*Forza Italia*, FI) appointed a new commission of experts, chaired by Elio Guzzanti to produce a set of guidelines to allow the practice.[31] Its objective was to impose safety and quality regulations on IVF. This plan was blocked by the political crisis of the first Berlusconi government. The CNB published a number of documents on IVF and related topics; the most important documents for IVF and the status of the embryo are as follows: *Prenatal Diagnostics* from 1992, *Opinion of CNB on artificial insemination processes. Synthesis and Conclusions* from 1994, *Artificial Insemination* from 1995 and *Identity and Status of*

[30]Maurizio Mori—professor for bioethics at the University of Turin—confirmed in an interview that the CNB is supposed to be open to all viewpoints and arguments. Even without the direct participation of the Catholic Church, the CNB displayed considerable Catholic influence. The lists of the committee members, for instance, were checked by the Vatican. Furthermore, he discovered that organisations like “Scienza e Vita” are controlled directly by the church and that Pro-Life organisations are normally Catholic. Carlo Flamigni—professor for fertility at the University of Bologna and CNB member—explained in an interview that the CNB always concludes its work by a vote and that the minority opinion is always added in the appendix of the document. However, only the final document—without the minority opinion—makes it to parliament. Furthermore, Francesco D’Agostino—professor of philosophy of law at the University of Rome II and former president of the CNB—always states his position prior to the vote and generally the CNB members will adopt his viewpoint.

[31]Eugenio Lecaldano—professor of moral philosophy at the University La Sapienza Rome and member of the CNB—reports in an interview that he resigned from the commission after Berlusconi came to power. During this time the composition of the commission changed, because Berlusconi dismissed a number of persons, amongst these Carlo Viano and other liberals/secularists. The commission was now almost exclusively Catholic and advocated Catholic morals.

the Human Embryo from 1996 (Comitato Nazionale per la Bioetica 1992, 1994, 1995a, b, 1996, 1997, 2000). One important point within the CNB discussion was the meta-discourse on the legitimacy of the church's intervention in public debates and in normative processes. The second focus was on the status of the human embryo ("What is an embryo, a human or a person? What about its dignity and its judicial rights? Does an embryo have the right to legal protection like other humans? What about the problem of eugenics, which is becoming a real threat due to embryo research?"). The Catholic position in the CNB was generally conservative: the embryo was seen as a person from the moment of conception. Consequently, any form of experimentation on and selection of embryos was strictly rejected. IVF was described as a technology that dehumanises the act of procreation. Since the embryo is not only a "potential human" but also a "present individual" (both medically and philosophically), it can demand the same fundamental individual rights, especially the right to life. Most of the positions were legitimated by referring to the use of human reason.[32] Overall, the overwhelming majority of CNB documents refrained from using explicitly religious arguments, although the viewpoints presented conformed to those of the Catholic Church. While conservative positions dominated amongst the Catholics, secular and liberal members of the CNB assessed the status of the embryo differently. They cast doubt on the view that the embryo had to be treated as a person from the moment of conception, but rather hold the view of its gradual development and that it gained full rights to protection at a later point. Although the embryo could not be reduced to a mere object and is always worthy of protection, its rights and well-being are not equal to an already born human being. Therefore, they tend towards accepting and ethically justifying the production, experimentation and selection of embryos for research. Moreover, the secularists' central objective was to uphold the absolute separation of religion, ethics and law.[33]

In 1995, *Evangelium Vitae* was enacted, dedicated to the problems of bioethics. It established the doctrine of the "sanctity of life". In the same year, a new commission was appointed, chaired by Francis D. Businelli, to prepare a new draft law. As before

[32]Compare F. D'Agostino: "*The separation of the unitive dimension from the procreative dimension (to use a wording that has been often repeated, criticized by many, and mostly misunderstood) does not create ethical problems because it constitutes a violation of the highest divine laws or natural laws (whose existence and content cannot be discussed at this point), but rather because it depersonalizes human procreation*" (Comitato Nazionale per la Bioetica 1995a, b, p. 133). For a similar position, compare CNB member Mons. Kard. Elio Sgreccia, a bioethicist, founder and ex-president of the Center for Bioethics at the Catholic University of Sacro Cuore.

[33]Arguments were amongst others that "[...] *juridical norms should never be derived from moral evaluations and that political discussions should never be subject to ethical evaluations. There could be no objection against heterologous or homologous IVF for anyone, if it is used to create a new life with the consent of the parents, under legislation which does not discriminate against IVF-children and whereby all technical methods are aiming at the protection and health of the unborn child. Furthermore, the metaphysical, ontological and religious discussion on the status of the embryo should not influence the legislation of a liberal state because of the different philosophical and religious orientations of its citizens.*" E. Lecaldano, see: Comitato Nazionale per la Bioetica (1995a, b) p. 193.

the report never reached the parliament. With the spread of the news about the first animal clone, the sheep Dolly, a new discussion concerning artificial reproduction (IVF, stem cell research, cloning) started and health minister Rosy Bindi issued a decree banning any cloning practice in 1997.[34] Additionally, a couple of committees dealing with legislative proposals on IVF and the status of the embryo were founded. The commission on social affairs of the chamber of representatives began to write a legislative draft to regulate IVF. At the end of January 1998, a restrictive law, the "project Bolognesi", was presented. The bill, introduced by Marida Bolognesi, a delegate of Democrats of the Left (*Democratici di Sinitra*, DS), would grant access to IVF only to heterosexual, married couples or stable partnerships with fertility problems. The bill intended to summarise 18 previous bills on IVF to find a common solution that would include different political and ethical positions. The law presented IVF as a therapeutic method to combat sterility and infertility as opposed to a mere alternative way of procreation (Buzzi and Tassi 2011, pp.77–78). Since the bill included a number of liberal exceptions—like the possibility of heterologous donations of gametes and cryopreservation of embryos—it faced severe opposition from Catholic representatives in parliament, who defended the view of the church. The important catholic newspapers *Avvenire* and *L'Osservatore Romano* criticised the bill harshly and the Catholic hierarchy clashed with the Catholic majority government from the Italian Peoples Party (*Partito Popolare Italiano*, PPI).[35] Subsequently, the "bill Bolognesi" underwent drastic changes, due to an alternative plan designed by Catholics of different parties with the objective of representing Catholic ethics. The new bill introduced the rights of the embryo ("the unborn child") and banned heterologous IVF ("proposal Cé").[36] Following a series of parliamentary disputes between the majority and the opposition, and between Catholics and secularists of both parties, the bill failed in the XIII legislature.[37]

[34]"*The watershed year for Italian bioethics was 1997. The birth of Dolly the sheep in Scotland at the end of February 1997 caused such cultural turmoil that bioethics was given a new social status. Before Dolly's birth, bioethics was already a growing discipline that increasingly attracted attention from educated people, but it was still a field restricted to a few specialists*" (Mori 2003, p. 163). From this point on, many works on several bioethical questions have been translated and published.

[35]The newspaper of the CEI, Avvenire, published, amongst others, a number of articles and reports on the new bill, which criticised the "double judgement" (*doppio giudizio*) criteria. Catholic politicians used the criteria to make a distinction between the ethically wrong and the juridically acceptable, in order to back the proposed law. The bishops insinuated that this criterion was unacceptable because it "[. . .] *takes rational arguments out of religious vision* [. . .]". The newspaper of the Vatican, *L'Osservatore Romano*, published an article with the title "Is a Catholic allowed to damage to God's project over mankind?" It argued that human law is not to be superior over divine law, as the latter is not only part of the Ten Commandments but also of the conscience of every human being. 4 June 1998.

[36]Changes addressed, for instance, the rights of the unborn child, which were added to a law for the first time, as well as the abolition of prenatal adoption of surplus embryos.

[37]It is noteworthy that centre-right Catholics (FI, Alleanza Nazionale, Unione dei Democratici Cristiani e Democratici di Centro) first agreed on the text proposed by the centre-left PPI. Maurizio

Already in 2000 health minister Umberto Veronesi appointed a special commission (Dulbecco Commission) to prepare guidelines for stem cell research in September 2000. The Committee was headed by Renato Dulbecco, Nobel laureate for medicine, and was to prepare a text for the end of the year. Dulbecco proposed allowing the creation of totipotent stem cells by cloning and the use of already cryopreserved embryos for research purposes. The opposition, mostly Catholic, prevented the vote on the joint text and its discussion in the chamber of representatives (Scheffino et al. 2009).

In 2001, a centre-right coalition won the elections thanks to the Catholic vote. Immediately, the parliament began discussions on the draft law on IVF of 1999 and approved the text in June 2002 with a large majority. The opposition of the Catholic centre-left (DS, *La Margherita*, *Comunisti Italiani*) voted in favour of the law, too. Following an official meeting of government representatives at the Vatican in February 2003, and a speech of the Pope in May, the centre-right government accelerated the parliamentary process. The Senate debated the law in September and approved it in December. The text was finally passed in February 2004. The new law (*Law 40*) regulated IVF, banned embryo stem cell research by prohibiting the production of supernumerary embryos and their cryopreservation and prohibited all forms of human cloning.[38] Although the law was restrictive, it did not prohibit the research on stem cell lines imported from abroad. According to the law, Italian researchers could use embryonic stem cells produced in other countries (Grißler et al. 2008, pp. 81–96; Metzler 2007).

Until then the debate did not resonate far outside parliament. But soon after the approval of the law, a sour public debate started with heated reactions from politicians, intellectuals and the media against the law calling it "confessional" and "obscurantist".[39] Catholic politicians of the centre-left who had passed the law by a large majority were criticised and the Catholic Church was accused of improperly meddling in the legislative processes of a secular state and of imposing a sectarian ethic on the whole society. The main actor of this criticism was the RP, which immediately suggested the collection of signatures for a referendum to abrogate the law. Meanwhile, both the centre-right and centre-left began working on a new draft law to amend the recently passed Law 40. On the other hand, the church refused any amendment of the law, because it considered the law to already constitute a

---

Mori suggests that "[...] *right-wing Catholics senators did not show up in order to prevent the left-wing Catholic party from gaining the Church's gratitude from the new law*" (Mori 2003, p. 171).

[38]Law 40: Only heterosexual married couples received access to artificial procreation; only three embryos were allowed to be produced per insemination cycle and all had to be implanted into the womb (Boggia 2005).

[39]Obscurantist means anti-Enlightenment/anti-progressive/anti-educational, metaphysical or religious. One example of such harsh criticism can be found in Corriere della Sera, 22.07.2004. "*Today as back then, we fight with the hope to win. Because if we win, democracy wins. With the legalization of divorce, families were allowed to lead marriages founded on love, and not on fear of the police. Today, we must defend the right to scientific research. This is a fight against Taliban obscurantism.*"

maximum of possible concessions. In 2005, the RP succeeded in collecting the necessary signatures for a referendum and the constitutional court admitted four referendum questions for the abrogation of the law. From this moment on, the Catholics began to defend the law, because they feared a new law could be even more liberal. The CEI began a massive campaign to raise awareness on the "issue of life" and to call for an abstention from voting in order to prevent the referendum from reaching the necessary quorum. Avvenire, supported by other lay organisations, began a media campaign in favour of Law 40 and an ad hoc committee was formed (*Scienza e Vita* = science and life) for the promotion of Catholic ethics and the defence of life against the modern medical and biological sciences. In the end, the referendum failed to reach the quorum, with a 26% turnout of voters. This failure was also a result of the successful campaign of the Italian Church. Law 40 is still operative today, but some significant amendments have been made by court of law. Amongst others, the courts decided to permit PGD, the production of more than three embryos per cycle and the cryopreservation of embryos (Italian Constitutional Court 2009, 2014). In public, the discussion of IVF and embryo research has faded. Other bioethical issues like euthanasia, the treatment of end of life or the admission of RU486 (the abortion pill) have monopolised political and public attention.

3.3 Religious Actors in the Italian Bioethical Debate

The religious landscape in Italy is homogenous; about 80% are Roman Catholic, although only a smaller number describe themselves as active practising believers. Other religions and denominations make up about 5%, amongst them a couple of Protestant denominations (the most important is the Waldensian Church), Muslims and Jews (Cesareo et al. 1995). The Catholic Church in Italy intervenes in public debates occasionally and usually through the lectures or speeches of the President of the CEI. Direct intervention in public debates is left to Catholic lay organisations and individual religious actors. Since the 1990s, a close cooperation between the church, lay activists and scientists opened up a new perspective. Different organisations like *Communione e Liberazione, Azione Cattolica* or *Movimento per la Vita* ensure a Catholic orientation on a political and social level. Since 2004/05, *Scienza e Vita* has been the main lay organisation to fight for certain "non-negotiable values".[40]

Generally, the Catholic Church emphasises the internal rationality of Catholic morality. Through this rational nature, its validity could be extended universally to all people: "*The objective norms governing the right action are accessible to reason,*

[40]These lay organisations are officially recognised by the church and pursue different areas and goals. *Comunione e liberazione* is a global movement that represents the political, ethical and religious principles of Catholicism. *Movimento per la Vita* was originally founded with the objective to protect unborn life against abortion and advocates the protection of the embryo today, too. *Scienza e Vita* is a committee of scientists that is exclusively responsible for bioethical issues.

prescinding from the content of revelation".[41] These are rules which, "[...] *may be deduced from the structure of man himself, as a good which is supreme, unavailable for any transaction. Only by referring to man in his entirety, with irreducible rights, and safeguarded before any further political determination, does one grasp the basic code, one that gains the status of objective rational basis, common to all peoples*".[42] From the church's point of view, the rationality of Catholic moral law legitimises the intervention of the church into politics to safeguard those values it defines as "not negotiable". These values apply to every person, regardless of personal belief and religion. The moral law allows the establishment of ethical positions through philosophical and scientific arguments without necessarily resorting to theological ones.[43]

Concerning bioethical issues, the positions of the Italian Catholic Church reflect those of the magisterium. The nature of the arguments used to justify and defend their positions in bioethics is, without significant change over time, expressed in several encyclicals, documents and speeches of the Vatican on the topics of the family, life and the relationship between faith and reason. The Italian Catholic Church is opposed to IVF, PGD, stem cell research and cloning. Arguments include human dignity, the creation in the image of God, life and its value as a goal in itself and not as a means to another end (Kant). Human and human life (born and unborn) are values in themselves, and cannot be subordinated to any end, and every medical or technical intervention which puts its health at risk is neither ethically acceptable nor legally permissible. According to the concept of "ontologically founded personalism", formulated by Bishop Sgreccia—the former president of the Academia Pro-Vita—the human being is a person from conception until death; i.e. it is an inseparable complex of body and soul which can never be reduced either to its biological or to its intellectual or spiritual dimensions (Sgreccia 2007). IVF and contraception should be forbidden because they represent a human interference in the natural processes of procreation. According to the church, the experimentation on human embryos, the production of embryonic stem cells and therapeutic cloning mean the suppression of human lives, which should not be available to man in any way. Life begins with conception, by the merging of an oocyte and a sperm cell, and from that moment, it has to be protected and defended without gradations. The

[41]Meeting with the representative of British Society, including the diplomatic corps, politicians, academics and business leaders. Address of his Holiness Benedict XVI Westminster Hall, Friday, 17 September 2010.

[42]Sono regole desumibili dalla struttura dell'uomo stesso, quale bene che sta al vertice, indisponibile per qualunque transazione. Solo indicando l'uomo nella sua integralità, dotato di diritti incomprimibili, e salvaguardato prima di ogni ulteriore determinazione politica, si ha il codice basilare, quello che acquista il valore di fondamento razionale oggettivo comune a tutti i popoli Lecture of the president of CEI, 8/11/2010, §4.

[43]This argumentation is used in religious contexts with respect to natural law, too. One example is the opening speech of the president of the CEI (08.11.2010, §4) in which he establishes that natural law is firmly embedded in human nature. The personal rights that derive from this and human dignity exist regardless of circumstances, of "more or less enlightened conventions", or of the culture. They are rather a deep-seated truth, or a "value par excellence".

embryo is seen as a person with dignity and rights and has the right to be born and not be manipulated or exploited for research purposes. The Catholic Church is opposed to PGD, too, because it could serve eugenic plans by reducing the dignity of human life to its genetic health.

Although "Law 40" did not represent the Catholic morality on these issues, the Italian Catholic Church supported and defended it, because at least the law introduces limits on the production and implantation of embryos and prohibits any form of experimentation and economisation of embryonic life. First of all, Law 40 introduces and sets out the rights of the embryo, elevating the embryo to the status of a legally recognised person. Within the discussion the church was using an ostensible secular language. The use of any religious references or arguments was explicitly avoided.[44] In particular, the Catholic lay organisations like Scienza e Vita, which were led by medicinal and humanist experts, focused on human dignity, whereby humans were deliberately not seen as God's creatures but as the highest rational beings on earth. Embryos, children, adults, healthy and sick individuals have to be regarded as humans. Thus, it is prohibited to interfere in the reproduction process of whoever is considered human on an ontological level. Simultaneously, these individuals were considered humans on an ethical level, which means that their bodily integrity and dignity must be protected. On a juridical level, a juridical person is guaranteed individual rights.[45] The implementation of this view on humans has been a part of the so-called Questione Antropologica (anthropological question), a new programme, or rather mission, of the church, whose objective was to support and protect human dignity, value and human nature.[46] Overall, the Catholic Church has only rarely made use of religious arguments or of religious teachings in the

[44]Compare the opening speech of Cardinal Ruini at the forum Progetto Culturale, 3 November 2011. In an interview Domenico Delle Foglie—journalist, former vice director of Avvenire and president of Copercom, the network of the communication agencies of Catholic organisations—describes the dissemination of the Catholic view as "*work through concentric circles*". He states that the Catholic Church tried to anchor Catholic ethics into the Catholic world and into Italian politics and civil society. Their new communication strategy consisted of the foundation of Catholic networks like Forum delle familiglie and Scienza e vita as a biopolitical network and Rete in Opera as a social network. Already existing organisations like Movimento per la Vita were included. A dialogue with the political elite and lobbying activities that address all political parties (left, right or centre) were initiated as well. The objective was to ban artificial insemination by law. When a referendum challenged the law, the Catholic Church supported the "compromise law" to prohibit a further liberalisation. Throughout the discussion, it was a "*fundamental decision*" to use secular language and only to argue on the basis of reason. This way one could never be "[. . .] *driven into a corner on the basis of one's faith*".

[45]For example, the arguments regarding the protection of the embryo and the restriction of IVF are of ontological-philosophical ("the embryo is a human being"), ethical ("the embryo has dignity and is a person"), juridical ("the embryo has the legal status of a person") and psycho-sociological nature ("the interference in the life of the couple, the donation of third-party persons and the manipulation of genes can yield significant consequences for the psyche of the people involved and for society").

[46]For the secular communication strategy, Casini (2006).

discussion on embryo research, but instead defends—in its opinion—objective moral principles that are reasonable and universal to all people.

The second biggest church in Italy, the Waldensian Church, holds a completely different position. While only 1% of the population are members of this church, it has a strong network in politics and science (Rubboli 2007). The Waldensian Church has always been very active in bioethics by publishing documents on various issues and forming a *Working Group on ethical Problems posed by Science* and, later, a *Bioethics Committee of the Waldensian Board*. Despite their many internal differences, the Waldensian Church has always assumed liberal positions, defending the freedom of conscience and supporting a clear separation between religious ethics and state law (Commissione per i problemi posti dalla scienza 2005). At the beginning, the Waldensians were concerned about the new bioethical methods like artificial insemination or embryonic stem cell research, because of their medicinal risks. Because their position became very liberal over time, they were regarded as an arbiter of secular bioethics. The Waldensians do not consider the embryo equal to a person, but assigned it a graduated protection of life.[47] Although the embryo does not have the same rights as a person, it must be treated with respect. All prenatal procedures or PGD should be permitted as long as they serve important medical purposes. Moreover, the Waldensians have criticised the Catholics' interference in political debates. Although the Gospels are seen as the source of all morals, the Waldensians do not demand their adoption into state law. This would be, in their opinion, "fundamentalist" (Chiesa Evangelica Valdese, Commissione bioetica 1995).

4 Conclusion: Negotiating Biopolitics in Norway and Italy

The relationship between church and state in Norway was influenced by the Reformation and the founding of the Lutheran state church. The separation of church and state was not the result of a conflict as in other European countries, but rather a reciprocal agreement. The Norwegian state model can possibly be best described as secular Lutheranism. The laws, norms as well as the moral and legislative code of the state are identical to those of the church. The Norwegian Church does not compete with the state; on the contrary, loyalty and obedience to the authorities is part of the belief system. This tradition made way for a culture characterised by compromises and moderate negotiation. The (Lutheran and governmental) ideal is a unified, inseparable social entity with a consistent set of moral values. Respecting human

Komitee Scienza e Vita: AA.VV. I Quaderni di Scienza e Vita, Roma 2007–2010; Idid: AA.VV. Venire al mondo (Quaderno nr. 3), Roma, I quaderni di scienza e vita, 2007; AA.VV. La legge 40, sei anni dopo (Quaderno nr.7), Roma, I Quaderni di Scienza e Vita, 2010.

[47]"*Although many Protestants say the embryo is already a person at the point of conception, on the basis of its genetic composition it is clear to us that the blastocyst shows no characteristics, by which it could be defined as a human being.*" (Chiesa Evangelica Valdese 2009).

dignity, protecting the weak and avoiding the formation of a selective society were the main points of all parties. Explicitly religious arguments were only used in rare cases, while ethical and humanitarian values that are compatible with Christian morals were frequently referenced. More restrictive political parties, like the Christian People's Party and the Conservatives, made use of an argumentation structure that is in accordance with the Christian idea of man in many respects. However, the Christian background was rarely mentioned, and if so, it was done in a way that the arguments would also hold true from a humanist viewpoint. The somewhat more liberal parties like the People's Party and the Left Party rarely ever made use of religious arguments. However, since all the parties shared common ground with respect to cultural values and all could agree on restrictive policies, the underlying ideology of the arguments used barely played a role. Throughout the debate, the importance of respecting the opinions of others was always highlighted. The delegates often mentioned the good collaborative work they shared with the ethics committee. There were, however, no references made to the official contributions of the church (neither positive nor negative).

The debate was also within a wide spectrum in the ethics committee. The main ethical argument mentioned in all the reports is the means-to-an-end formula. Depending on the author, either a Christian or a humanitarian view is used to justify this stance. Religious arguments and motives do not serve hereby as universal guidelines but rather as additional arguments to evaluate positions. All papers highlight the fact that collaboration is possible even with different opinions, as all could agree on the outcome. This proves once again that the relative homogeneity of the Norwegian population favours finding a consensual solution, as almost nobody deviates too far from the universalist norm (in this case, human dignity and protecting human life).

Throughout the discussion, the Norwegian Church has seen its role in advising both the government and the public in bioethical issues. It also tried to influence respective laws. The main focus lay hereby on protecting the embryo, human dignity and heterosexual marriage. The status of the embryo is, however, disputed within the church. In principle, the church sees its responsibility as speaking for those who are left unprotected by society. The inseminated egg cell represents this category. The Norwegian Church sees itself as part of a pluralistic society and of the public and seeks to convince with religious and secular arguments. It therefore explicitly seeks accordance with the argumentation of and a consensus with the positions of humanist and non-Christian ethics. The first documents stress the Christian actions of Church representatives, a fact which is, however, dropped in later documents. They were based upon religious sources, but they also display other ethical models. Besides the direct reference to religious sources, authorities and biblical passages, a Christian ethics of solicitude and responsibility is used which does not argue in terms of religion but which wants to be seen against the background of Christian traditions. The desire to convince non-Christians as well, so as to come to a consensual agreement with society as a whole, is highlighted. The values mentioned by the church can be found in a secularised version within the law.

In Italy, church and state have chosen a confrontational approach rather than a cooperative one. The Roman Catholic Church has deeply informed Italian history. Since the church lost its territories to the newly founded Italian nation state in 1861, both institutions came into conflict with each other repeatedly. In spite of the secularisation processes and the constitutional separation of church and state, the church maintained an important political and social actor. The reform of family law that legalised divorce (1970) and abortion (1978) nearly came close to a clash of cultures between Catholic institutions and a secularised society. This is due to the fact that national and religious identity is closely interlinked and that the Holy See is located on Italian soil. Furthermore, as the Catholic Church is widely present in parishes and through numerous lay organisations, a broad Catholic subculture has emerged. The Catholic-oriented DC functioned as a political contact point for the Catholic Church up until the 1990s. Against all odds, the Catholic Church succeeded in influencing all parties from left to the right after the political crisis. All political debates oscillate between Catholic values and secular principles. In the debate on AHR (assisted human reproduction) procedures, the meta-debate on the separation of church and state also played an important part. Catholic and conservative actors considered the intervention of the church as legitimate and its arguments as rational, while liberal and socialist actors found the meddling of religious principles in secular state politics inappropriate.

In this context, it turned out to be difficult to establish legal regulations, as it was impossible to mediate between the different positions. The absolute value of human life as the highest rational being, whose life is inviolable in all stages of development, was fundamental to Catholic and conservative views. Abortion, IVF and embryonic stem cell research were seen as dangerous and illegitimate interference with natural processes. From their point of view, the embryo was a person—on an ontological, juridical and ethical level—and any form of intervention posed a threat to its life and integrity. The arguments were predominantly not explicitly religious but taken from natural law, metaphysics, ethics and legal doctrine. At the same time, they were in line with Catholic tradition, or rather Catholic natural law. This posed the question of whether or not this constituted a neutral point of reference for the biopolitical discussion. Proponents of the secular side argued that an embryo is not equal to a person, or a born human being. According to their opinion, an embryo is, ontologically speaking, not yet a human, ethically, not yet a person, nor does it share the same rights as already-born people on a juridical level. Consequently, the support of embryonic research to ensure human development and the healing of illness was acceptable.

The Italian Parliament issued numerous committees, special commissions and, in 1990, a permanent National Committee of Bioethics to advise, inform and present legal propositions on bioethical matters. The Parliamentarian commissions, however, never led to concrete solutions, with the exception of Law 40. Before 2004, discussion rather than decision was the order of the day. After detailed, and often heated, discussion between secular and Catholic-oriented politicians, all proposals had been dismissed. Only Law 40 regulated IVF, PGD, human stem cell research and cloning. The law was conservative and met Catholic principles. While the

Catholic Church would preferably have banned IVF completely, it regards this law as the lesser evil compared to more liberal versions.

The reports of the CNB gave an overall picture of the arguments used in the media and in the parliamentary debates, too. The documents, written by experts from various disciplines, almost exclusively contain the results of the discussion. The CNB supports IVF, although only under very strong restrictions, like the ban of heterologous insemination or the necessity of implanting all the produced embryos. The CNB assigns the status of a person to the embryo and thus grants it ethical inviolability and the right to legal protection. PGD, experimentation and overproduction for research purposes were rejected. In the end, this was the position of the majority reflecting the Catholic-conservative composition of the CNB. A minority that did not ascribe the embryo the status of a person at the moment of conception mainly remained invisible within the papers. In spite of that, Catholic-oriented experts assessed the results as a compromise, because they fell short of a complete ban. Their secular opponents, however, view the CNB as an agent of Catholic ethics and the results as a unilateral assessment of Catholic-oriented scientists. The arguments used were not religious, but rather philosophical, juridical, medicinal, etc. At the same time, they correlate with Catholic natural law.

Overall, the Catholic Church regarded itself as an agent and defender of rational, universal and non-negotiable values and principles, which all reasonable humans could share. In contrast, particular secular and lay forces found that the Catholic Church had gone beyond its constitutional rights by trying to illegitimately impose its world view and ethical-religious values on society as a whole. This conflict led to Catholic participants developing an argumentation strategy that can only be defined as religious in the broadest sense. While it is in accordance with the religious traditions and positions of the Catholic Church, the arguments are based on Catholic natural law and other disciplines from humanities and natural sciences.

References

Altgeld, Wolfgang, and Rudolf Lill. 2005. *Kleine italienische Geschichte.* Bonn: Bundeszentrale für politische Bildung.

Austad, Torleiv. 2009. *Die Kirche als Fürsprecher des Rechtes. Kirche und Staat in Norwegen in der Besatzungszeit 1940-1945 vor dem europäischen Hintergrund.* Vortrag im Rahmen der Ringvorlesung: Die europäische Integration und die Kirchen—Akteure und Rezipienten, Institut für Europäische Geschichte der Johannes Gutenberg-Universität, 14.12.2009.

Betta, Michela. 1995. *Embryonenforschung und Familie: Zur Politik der Reproduktion in Großbritannien, Italien und der Bundesrepublik.* Frankfurt a. M: Peter Lang.

Bioteknologirådet. 2011. Innspill til evalueringen av bioteknologiloven. http://www.bion.no/filarkiv/2011/12/bioteknologiloven_evaluering_Bioteknologinemnda.pdf. Accessed 10.01.2012.

Bispemøte. 1971. Sak 13/71.

———. 1975. Sak 13/75.

———. 1977. Sak 24/77.

———. 1978. BM = Minutes of the Bishops Conference.

———. 1979. Sak 69/79.

———. 2012. Sak 13/12.
Bleiklie, Ivar. 2004. Legislation for protection. Why Norway designed restrictive policies in the field of ART. In *Comparative biomedical policy. Governing assisted reproductive technologies*, ed. Ivar Bleiklie et al., 209–227. London: Routledge.
Boggia, Andrea. 2005. Italy enacts new law on medically assisted human reproduction. *Human Reproduction* 29 (5): 1153–1157.
Bolzonar, Fabio. 2016. A Christian democratization of politics? The new influence of Catholicism on Italian politics since the demise of the Democrazia Cristiana. *Journal of Modern Italian Studies* 21 (3): 445–463.
Botvar, Pål Ketil. 2010. Endringer i nordmenns religiøse liv. In *Religion i dagens Norge. Mellom sekularisering og sakralisering*, ed. Pål Ketil Botvar and Ulla Schmidt, 11–24. Oslo: Univeritetsforlaget.
Brekke, Ole Andreas. 1999. Bioteknologi og politikk—en rapport fra Bergen. In *Bioteknologiens etiske, juridiske og sosiale aspekt. Rapport fra et seminar arrangert av, Grunnleggende bioteknologi og Etikkprogrammet (Norges Forskningsråd) 23th-24th March 1998*, ed. Thorvald Sirnes and Per Sandberg. Oslo: Innsatsområdet Etikk (Universitetet i Oslo).
Buzzi, Fabio, and Giacomo Tassi. 2011. *La procreazione medicalmente assistita: normativa, giurisprudenza e aspetti medico legali*. Milan: Giuffré.
Campell, Courtney S. 1990. Religion and moral meaning in bioethics. *The Hastings Center Report* 20 (4): 4–10.
Cartocci, Roberto. 2011. *Geografia dell'Italia cattolica*. Bologna: Il Mulino.
Casini, Carlo. 2006. *Le cinque prove dell'esistenza dell'uomo*. Milano: San Paolo.
Cesareo, Vincenzo, et al. 1995. *La religiosità in Italia*. Milano: Mondadori.
Chiesa Evangelica Valdese, Commissione bioetica. 1995. *Bioetica—Ricerca e Orientamenti*. §12–86. https://www.chiesavaldese.org/documents/bioetica.pdf. Accessed September 2011.
———. 2009. Cellule staminali. Aspetti scientifici e questioni etiche. Documento nr.14, §4.3. https://www.chiesavaldese.org/documents/staminali_doc14.pdf. Accessed 08.08.2012.
Christensen, Tom. 2007. The Norwegian state transformed? In *Norway in transition. Transforming a stable democracy*, ed. Oyvind Osterud, 16–35. London: Routledge.
Christiansen, Peter, et al. 2010. Varieties of democracy: Interest groups and corporatist committees in Scandinavian policy making. *Voluntas: International Journal of Voluntary and Nonprofit Organizations* 21: 232–240.
Comitato Nazionale per la Bioetica. 1992. *Diagnosi Prenatali*. http://bioetica.governo.it/media/170634/p9_1992_diagnosi-prenatali_it.pdf. Accessed 24.11.2011.
———. 1994. *Parere del CNB sulle tecniche di procreazione assistita. Sintesi e conclusioni*. http://bioetica.governo.it/media/170636/p16_1994_tecniche-procreazione-assistita_it.pdf. Accessed 24.11.2011.
———. 1995a. La fecondazione assistita. http://bioetica.governo.it/media/170121/p17_1995_fecondazione-assistita_abs_it.pdf. Accessed 24.11.2011.
———. 1995b. *La fecondazione assistita*. Rome: Presidenza del Consiglio dei Ministri.
———. 1996. Identità e statuto dell'embrione umano. http://bioetica.governo.it/media/170647/p24_1996_identit%C3%A0-embrione_it.pdf. Accessed 24.11.2011.
———. 1997. Parere del CNB su "La clonazione come problema bioetico". In CNB, La clonazione. http://bioetica.governo.it/media/172053/p30_1997_clonazione_it.pdf. Accessed 24.11.2011.
———. 2000. Parere del Comitato Nazionale per Bioetica sull'impiego terapeutico delle cellule staminali. http://bioetica.governo.it/media/171397/6-pareri-26-ottobre-2007.pdf. Accessed 24.11.2011.
Commissione per i problemi posti dalla scienza. 2005. Documento nr. 8. https://www.chiesavaldese.org/documents/postReferendum3.pdf. Accessed 08.08.2012.
Danielsen, Rolf. 1995. *Norway. A history from the Vikings to our own times*. Oslo: Scandinavian University Press.
Ferrari, Silvio. 1995. Staat und Kirche in Italien. In *Staat und Kirche in der Europäischen Union (2005)*, ed. Gerhard Robbers, 229–252. Baden-Baden: Nomos.

Flamigni, Carlo, and Maurizio Mori. 2005. *La legge sulla procreazione medicalmente assistita. Paradigmi a confronto*. Milan: Net.

Frandsen, Karl-Eric. 2009. The conglomerate state of Denmark-Norway-Schleswig-Holstein 1523-1864 under the rulers of the house of Oldenburg: Integration, preservation and dissolution. In *Die Reiche Mitteleuropas in der Neuzeit: Integration und Herrschaft: Liber memoralis Jan Pirozynski*, ed. Adam Perlakowski, Robert Bartczak, and Anton Schindling, 83–96. Kraków: Księgarnia Akademicka.

Forssman, Berthold. 21 May 2012. Glücklich geschieden. Staat und Kirche in Norwegen trennen sich. https://www.domradio.de/nachrichten/2012-05-21/staat-und-kirche-norwegen-trennen-sich. Accessed 21 May 2012.

Garelli, Franco. 2006. *L'Italia cattolica nell'epoca del pluralismo*. Bologna: Il Mulino.

———. 2007. *La chiesa in Italia*. Bologna: Il Mulino.

Grißler, Eric, et al. 2008. *Stammzellen und Embryonenschutz. Status quo, Rechtsvergleich und öffentliche Debatte am Beispiel ausgewählter europäischer Staaten*. Vienna: Institut für Höhere Studien.

Groß, Herman, and Walter Rothholz. 1999. Das politische System Norwegens. In *Die politischen Systeme Westeuropas*, ed. Wolfgang Ismayr, 129–162. Opladen: Leske+Budrich.

Helsedirektoratet. 2006. Bioteknologiloven. http://www.human.no/Livssynspolitikk/Bioteknologi/. Accessed 04.10.2011.

———. 2011. Evaluering av bioteknologiloven. https://helsedirektoratet.no/Lists/Publikasjoner/Attachments/199/Evaluering-av-bioteknologiloven-status-og-utvikling-pa-fagomradene-som-reguleres-av-loven-IS-1897.pdf. Accessed September 2011.

Hiroth, Finngeir. 1997. *Church and state in Norway*. Oslo.

Høring. 2011. *Forslag til forskrift om endringer i forskrift om genetisk masseundersøkelse i forbindelse med utvidet tilbud om nyfødtscreening 21th March 2011*. Accessed 04.10.2011.

Interview with Svein Aage Christoffersen, 14 April 2011, Oslo.

Italian Constitutional Court. 2009. Sentenza No. 151/2009. http://www.cortecostituzionale.it/actionSchedaPronuncia.do?anno=2009&numero=151. Accessed 15.03.2015.

———. 2014. Sentenza 99/2014. http://www.altalex.com/index.php?idnot=67193. Accessed 15.03.2015.

Kallerud, Egil. 2004. The ambiguity of progress: Biotechnology in Norway. In *Mediating public concern in Biotechnology: A map of sites, actors and issues in Denmark, Finland, Norway and Sweden*, eds. Marja Häyrinen-Alestalo and Egil Kallerud, 83–112. Oslo: Norwegian Institute for Studies in Research and Higher Education Rapport 2/2004.

Kirkemøtet. 1985. Menneskesyn og menneskeverd. En anbefaling fra Kirkemøtet 1985, KM 3/85.

Kirkemøte. 1989. *(KM) 12/89 Bioteknologi og menneskeverd*.

Kirkerådet. 1987. *Kirkerådets rammeprogramm for 1988-1991. Prosjekt No. E3/87: Utfordringer fra moderne teknologi. Bioteknologi, miljøspørsmål og ny teknologi*. Oslo: Den Norske Kirke.

———. 1989. *Mer enn gener. Utredning om bioteknologi og menneskeverd*. Oslo: Den Norske Kirke.

———. 2006. *(KR) 30/06, 15.-16.06.2006. Høringsuttalelse til Utkast til endringer i lov om humanmedisinsk bruk av bioteknologi m.m. (bioteknologiloven)*. Oslo: Den Norske Kirke.

Köppl, Stefan. 2011. *Politik in Italien: Vom Kartell zum Wettbewerb? Parteien-Parlament-Regierung*. Baden-Baden: Nomos.

La civiltà cattolica. 1983. Anno 134, Vol. 4: 4–5.

Lauritzen, Paul. 2005. Stem cells, and human rights. Implications for a posthuman future. *The Hastings Center Report* 35 (2): 25–33.

Lembke, Thomas. 2008. Eine Analytik der Biopolitik. Überlegungen zur Geschichte und Gegenwart eines umstrittenen Begriffs. *Behemoth: A Journal on Civilisation* 1 (1): 72–89.

Livi, Massimiliano. 2016. The Ruini system and Berlusconismo: Synergy and transformation between the Catholic Church and Italian politics in the Second Republic. *Journal of Modern Italian Studies* 21 (3): 399–418.

Madeley, John. 1994. The antinomies of Lutheran politics: The case of Norway's Christian People's Party. In *Christian Democracy in Europe. A comparative perspective*, ed. David Hanley, 142–154. London: Continuum International Publishing Group.
Masala, Carlo. 2004. Born for government: The Democracia Cristiana in Italy. In *Christian democracy in Europe since 1945*, ed. Michael Gehler and Wolfram Kaise, 88–102. London: Routledge.
Menneskeverd. 2007. *Vern om livet: Tide netter den nye bioloven*, No. 02.
Metzler, Ingrid. 2007. Nationalizing embryos: The politics of human embryonic stem cell research in Italy. *BioSocieties* 2: 413–442.
Mori, Maurizio. 2003. Bioethics in Italy since 1997. In *The annals of bioethics. Regional perspectives in bioethics*, ed. John F. Peppin and Mark J. Cherry, 163–192. London: Routledge.
Myskja, Bjørn K. 2009. Rationality and religion in the public debate on embryo stem cell research and prenatal diagnostics. *Medical Health Care and Philosophy* 12: 213–224.
Norge Sosialdepartementet. 1991. *Norges offentlige utredninger: 6 Mennesker og bioteknologi*. Oslo: Statens forvaltningstjeneste, Seksjon Statens trykning.
Petersson, Olof. 1989. *Die politischen Systeme Nordeuropas. Eine Einführung*. Baden-Baden: Nomos.
Pollard, John. 2008. *Catholicism in Modern Italy: Religion, society and politics since 1861*. New York: Routledge.
Ramjoué, Celina, and Ulrich Kloti. 2004. ART policy in Italy: Explaining the lack of comprehensive regulation. In *Comparative biomedical policy. Governing assisted reproductive technologies*, ed. Ivar Bleiklie, Malcolm L. Goggin, and Christine Rothmayr, 42–63. London: Routledge.
Rubboli, Massimo. 2007. *I protestanti*. Bologna: Il Mulino.
Ruini, Camillo. 2004. La chiesa in Italia.*Vita e Pensiero* Nr. 6.
Santosuosso, Fernando. 1984. *La fecondazione artificiale umana*. Milano: Giuffrè.
Sassoon, Donald. 2013. *Contemporary Italy: Politics, economy and society since 1945*. London: Routledge.
Scheffino, Nathalie, Celina Ramjoue, and Frederic Varone. 2009. Biomedical policies in Belgium and Italy: From regulatory reluctance to policy changes. *West European Politics* 32 (3): 559–585.
Schlieter, Jens. 2005. Bioethik, Religion und Kultur aus komparativer Sicht. Zur Bedeutung und Methodik einer neuen Fragestellung. *Polylog. Zeitschrift für interkulturelles Philosophieren* 13: 7–14.
Schmidt, Ulla. 2010. Norge: Et pluralistisk Samfunn? In *Religion i dagens Norge. Mellom sekularisering og sakralsisering*, ed. Pål Ketil Botvar and Ulla Schmidt, 25–43. Oslo: Universitetsforl.
Sgreccia, Elio. 2007. *Manuale di bioetica, Vol. 1. Fondamenti ed etica biomedica*. Milano: Vita e Pensiero.
Statistisk sentralbyrå. 2015. Minifakta om Norge 2015. www.ssb.no/emner/00/minifakta/no. Accessed 13.02.2016.
Stenius, Hendrik. 1997. The good life is a life of conformity: The impact of the Lutheran tradition on Nordic political culture. In *The cultural construction of Norden*, ed. Øystein Sørensen and Bo Strath, 161–171. Oslo: Scandinavian University Press.
Stortingsmelding. 2008. No. 17 (2007-2008), Staten og Den norske kirke. https://www.regjeringen.no/no/dokumenter/stmeld-nr-17-2007-2008-/id507168/sec1. Accessed 24.11.2011.
U.S. State Department. 2007. *International Religious Freedom Report, Norway*.
———. 2008. *International Religious Freedom Report, Norway*.
Von Unruh, Georg-Christoph. 1989. Die Eigenart der Verfassung des Königreichs Norwegen. *Jahrbuch des öffentlichen Rechts* n.F. 38: 277–298.
Weiberg-Salzmann, Mirjam, and Ulrich Willems. 2019. Challenging the political: Religious actors and religious arguments in liberal democracies. In *Religion and biopolitics*, ed. Mirjam Weiberg-Salzmann and Ulrich Willems. Cham: Springer.

Part V
Mixed-Confession States with High Level Religious Pluralism

Chapter 12
The Role of Religion in the Political Debate on Embryo Research in the Netherlands

Wybo J. Dondorp and Guido M. W. R. de Wert

Abstract Until the late twentieth century, there were three main political currents in the Netherlands: Christian, Labor, and Liberal, giving Christian party politics a stronger position than in European countries with a binary division between conservative and progressive. The history of the debate about embryo research coincides with the end of this period. Whereas in the 1980s the Christian Democrat party (CDA) still had strong religiously motivated views about embryo protection, it has since lost both the power and the drive to pursue this. However, due to strategic opportunities arising in the margin of Dutch coalition politics, smaller parties inspired by Christian belief (CU, SGP) are still quite influential on issues of medical ethics in the highly fragmented political landscape of the present.

1 Introduction

The political debate on embryo research in the Netherlands started in the 1980s and in 2017 is still ongoing, 15 years after the Embryos Act came into force. The hot potato remains whether a new government should take steps to lift the present ban on research for which human embryos would have to be especially created. This has been proposed in order to enable important medical research that cannot be carried out by other means including the use of "leftover" IVF embryos. As a small Christian party may be needed for a stable coalition with a majority in both houses of Parliament, and as this party has strong religiously inspired views on the equal value of all human life from conception, it seems unlikely that the ban will indeed be lifted in the next period, however much this is desired by some of the other parties in the envisaged coalition. The news that views about embryos held by orthodox Christians might be a stumbling block for the formation of a new government was brought by the main media with a mixture of disbelief and fascination (Anonymous

W. J. Dondorp (✉) · G. M. W. R. de Wert
Department of Health, Ethics and Society, Research School CAPHRI for Public Health and Primary Care, Research School GROW for Oncology and Developmental Biology, Maastricht University, Maastricht, The Netherlands
e-mail: w.dondorp@maastrichtuniversity.nl

M. Weiberg-Salzmann, U. Willems (eds.), *Religion and Biopolitics*,
https://doi.org/10.1007/978-3-030-14580-4_12

2017). This is a nice illustration both of the extent to which Dutch society has secularized in the past decades and of the fact that Christian party politics is still able to determine the course of the debate on ethically sensitive issues, such as embryo research.

In this chapter, we will summarize the history of the political debate in the Netherlands about whether (and if so under what conditions) human embryos may be used for scientific research. Our main focus will be on the role of religion and religiously inspired actors. We will conclude that because of the need to find common ground with secular parties within the small margins of coalition politics, explicit religious reasoning did not play a role in this debate. But political parties inspired by Christian belief were, and still are, quite influential. Although they did not manage to block all embryo research, they are still in a position to determine its scope in the coming period. Religious parties inspired by other beliefs have until now not been part of the Dutch political landscape, nor have views of other religions played a role in the debate on embryo research. For the purpose of this chapter, "the role of religion" can therefore be understood as referring to the Christian religion in its different forms and shades.

2 Background: Dutch Politics as the Art of Compromise

The Netherlands is a constitutional monarchy (since 1848), with a parliamentary system based on proportional representation (since 1918) (Netherlands Institute for Multiparty Democracy 2008). Political parties emerged in the late nineteenth and early twentieth centuries with the aim of representing the political interests of specific parts of the electorate. The first such party was the neo-Calvinist "Anti-Revolutionary Party" (ARP), founded in 1879 by the leader of an orthodox movement in the Dutch Reformed Church, with the specific aim of securing funding for religious schools on an equal basis as state-run public schools. Segments of the population with different worldviews soon had their own political parties, including the Roman Catholics, social democrats, and liberals. Until the last decades of the twentieth century, those parties were the political expression of a strong vertical segregation of Dutch society along denominational lines (so-called societal pillars), where segments of the population with different worldviews also had their own schools, hospitals, youth and labor organizations, newspapers, etc. Given that under the system of proportional representation no single political party could ever win an absolute majority, coalition governments working on the basis of political compromise always have been (and still are) the rule in Dutch politics (Aerts et al. 2013).

2.1 Christian Party Politics

As a result of the policy of "antithesis" between Christian and secular that was explicitly sought by the leadership of the protestant and Roman Catholic parties in the early twentieth century, the political landscape in the Netherlands did not evolve into the binary divide between conservative and progressive parties as seen in other European countries (Meijering 2012). Until late in the century, there were three main currents: Christian, Labor, and Liberal. In this situation, the main Christian parties (Protestant and Roman Catholic) were in the luxury position of being able either to exclude the other two groups or to choose with whom to govern, leading to coalition governments that always contained a strong religious element and to parliamentary debates in which it was not at all strange to also hear representatives of secular parties quote the Bible or refer to the Christian tradition, if only to point to selective reading, dissent, or hypocrisy on the part of their opponents (Meijering 2012). After seeing a sharp fall in their share of the electorate in the early 1970s (reflecting the onset of cultural changes that led to a dismantling of the "pillar" system and rapid secularization), the main Christian parties merged to form the "Christian Democratic Appeal" (CDA) in 1977. This allowed them to hold on to power as major partner in coalition governments, providing prime ministers from 1977 till 1994 and again from 2002 till 2010 (Aerts et al. 2013).

With a nation-wide presence in all layers of public policy (local, provincial, national), the CDA (and its predecessor parties) has always been a strong governance-oriented party. The famous words allegedly spoken by a CDA politician during the 1980s, "We rule this country," may be apocryphal, but they nicely capture how the party saw itself and how it was regarded by other actors in that period as well. Against this background, it came as quite a blow when, after new losses, the 1994 parliamentary elections led to a coalition of the social democrats (PvdA) with two liberal parties (VVD and D66), pushing the Christian democrats to the opposition benches. The so-called purple coalition of labor and liberals was the first since the beginning of the century in which no Christian party participated (Peters 2015). For the purpose of our chapter, this is an interesting period, as the purple coalition succeeded to get legislation passed on two important medico-ethical issues, notably the Euthanasia Act (2002) and the Embryos Act (2002).

2.2 *How God Disappeared from Dutch Politics (But Not Entirely)*

Although the CDA's founding "Program of Principles" still explicitly referred to the Bible and Christian social ideals as a source of inspiration for practical politics (CDA 1989), the party developed into a center-right popular party in which the link with the Christian tradition is no longer very prominent, akin to, e.g., the German CDU. In 2017, the CDA website only briefly refers to "the Jewish-Christian roots that have made our country one of the best places on earth" (CDA 2017). This seems the inevitable outcome of the party's wish to remain a strong political force in what is now one of the world's most secularized societies. Whereas in the past, Christian party politics has been much stronger in the Netherlands than in other European countries, the cultural revolution of the 1960s and 1970s has effectively "driven God out of Parliament," as the title of a book on Dutch parliamentary history has it (Meijering 2012).

However, two smaller parties with a still explicit Christian profile have more recently obtained some political influence, thanks to the fragmentation of Dutch political landscape. These are the evangelical "Christian Union" (CU) and the orthodox "Reformed Political Party" (SGP), currently the oldest party in Dutch parliament (established 1918). With a culturally conservative but socially progressive message explicitly referring to biblical values, the CU attracts protestant and Roman Catholic voters dissatisfied with the meager Christian profile of the CDA, whereas the SGP is the political voice of a few smaller churches on the utmost right wing of Dutch Protestantism. With only six seats in Parliament (of 150), the CU was needed for a majority coalition government from 2007 to 2010 and very probably again (with five seats) in 2017. In its first period as a coalition party, the CU threatened to bring down the government over allowing or not allowing a wider scope of indications for pre-implantation genetic diagnosis (PGD). And when the 2012–2017 liberal–labor coalition (VVD and PvdA) lacked crucial votes in the Senate, the two small Christian parties made clear that in exchange for their possible support, they at least wanted a halt to further steps in medico-ethical dossiers, including with regard to embryo research.

3 The Introduction of IVF in the Netherlands

Since the early 1970s, Rotterdam-based physiologist, Zeilmaker, belonged to the small international group of pioneers who tried to develop a new and revolutionary form of medically assisted reproduction based on fertilization of oocytes in the laboratory, outside the woman's body (Troost 1996). His British colleagues Edwards and Steptoe, whose names will forever be connected with the breakthrough of in vitro fertilization (IVF), were the first to successfully use this technology (Brian 2013). In July 1978, they reported the birth of Louise Brown, the world's first IVF child. Louise's mother had been infertile due to blocked fallopian tubes. Without IVF as a means of circumventing this problem, it was impossible for her to conceive.

In the meantime in Rotterdam, Zeilmaker had partnered with gynecologist Alberda to conduct preclinical experiments with donated oocytes and sperm (Troost 1996). When these proved successful (leading to fertilization in vitro), they asked permission in 1979 from the Board of the Rotterdam Academic Hospital for clinical experiments, in which embryos produced from the gametes of volunteering infertile couples would be transferred to the woman's womb. This request was initially refused. As quoted in a later television documentary about these early days of IVF in the Netherlands, the hospital's advisory ethics committee thought transfer was still too risky: "The committee advises to approve the project up to the moment of embryo-transfer. In this way more experience can be gathered, no false expectations will be generated and there will be occasion for a wider debate about the many aspects of the technologies in question" (Praamsma 2003). Two years later, in 1981, the committee changed its advice from negative to positive, not so much because reassuring evidence about offspring safety had become available, but rather as a result of successful attempts by the Rotterdam pioneers to mobilize patients and influence public opinion (Praamsma 2003). The Hospital Board gave the green light and fertility patients were recruited for the experiment. After several failed attempts, this led to the birth of the first Dutch IVF baby, Stephanie Li, in 1983 (Troost 1996; Zeilmaker et al. 1987).

3.1 Request for a Health Council Report

In 1982, the Minister of Health, Mrs. Gardeniers-Berendsen (CDA), reacted to the rapid developments by requesting the Health Council to draw up an advisory report on IVF and (as was already planned) donor insemination. The Health Council of the Netherlands (Gezondheidsraad, established 1902) is a statutory scientific advisory body whose remit is to provide government and parliament with recommendations on issues of public health and health care (Rigter 1992). The council works through ad hoc committees ideally providing consensus reports based on scientific evidence. The Health Council is politically independent; its membership consists of scientists and other academics appointed by the Crown on the basis of their personal expertise in a relevant field of knowledge. Apart from experts in a wide array of medical disciplines, this also includes health lawyers, social scientists, and medical ethicists (Ten Have 2002). Although several medical ethicists of the first generation were in fact trained as theologians, it was not as theologians and certainly not as representatives of religious perspectives or church views that they were invited to sit on Health Council committees. The approach to doing ethics at the Health Council has always been academic and secular (Ten Have 2002).

With regard to IVF, the minister's questions pertained to safety, indications, and costs, but also to the need for regulations and the acceptability of using any "fertilized cells" not transferred to the womb for research purposes (Gezondheidsraad 1986). At the Health Council, a multidisciplinary committee was installed that would take a comprehensive look at what was referred to at the time as "artificial reproduction," including ethical and psychosocial aspects (final report 1986). Zeilmaker was one of the members. As the pace of developments continued to be high, with other hospitals wanting to step in, the committee felt urged to come up with an interim report on IVF (1984) in order to avoid the unregulated use of this highly sensitive new technology. The interim report recommended concentrating expertise in academic centers, using the introductory phase for gathering more data on all relevant aspects including the health of any children born after IVF and limiting indications for the time being to IVF for blocked fallopian tubes (Gezondheidsraad 1984). These recommendations laid the basis for the licensing system under which IVF in the Netherlands is still being offered, making it perhaps the only country in the world where medically assisted reproduction is not commercially available. A temporary regulation was in place as from July 1985. Reimbursement was initially based on ad hoc subsidies by the medical insurance fund (Van Overbeeke 1992).

3.2 *Cryopreservation of Human Embryos*

One of the new developments of the early 1980s were attempts at cryopreserving IVF embryos. In fact, Rotterdam was the first IVF group worldwide to report an IVF baby grown from a frozen–thawed embryo (Zeilmaker et al. 1984). This new technology permitted the use of all good-quality embryos obtained in a single IVF (hormone stimulation) cycle. Surplus embryos (beyond the maximum number to be responsibly transferred to the woman's womb, set at three in the 1984 Health Council report) can since be cryopreserved and used in a later transfer attempt, thus reducing the number of hormone-stimulation cycles to which women coming for IVF are exposed. This made IVF more efficient, more patient-friendly, and (in view of possible risks of hormone stimulation for the woman) also safer. But the introduction of this important adjuvant technology (now a mainstay of modern IVF) inevitably also meant the accumulation of frozen human embryos, most of which will not be used by the couple for whom they were initially created. The fate of these "spare" or "leftover" embryos gave further impetus to the question of the acceptability of embryo research.

4 Health Council Reports on Human Embryo Research (1984, 1986)

The Health Council's interim report was deliberately brief on ethical issues. Without further discussion, the committee made the following statements: "(1) IVF as a technology to solve fertility problems is ethically acceptable; (2) Fertilizing oocytes outside a woman's body should in principle be directed at helping the woman who provided the oocytes to become pregnant; (3) Attempts at culturing a conceptus beyond the implantation stage with the aim of letting a human individual develop outside the womb or in order to generate parts of an individual (organs) are widely rejected in the international literature. The committee concurs with this view; (4) The committee is of the opinion that under specific, well-defined conditions, a conceptus that has no further use for inducing a pregnancy in the woman in question may be used for research aimed at developing IVF methodology, or for research aimed at furthering basic knowledge; alternatively [such conceptus may also be used] to bring about a pregnancy in another woman" (Gezondheidsraad 1984). The committee added that of course both embryo research and donation were only acceptable with the informed consent of the providers of the gametes; with regard to embryo research, it was important to acknowledge that "this is about experiments with conceptus at the pre-implantation stage, where only a small number of undifferentiated cells are present and where there is not yet an embryo in the gynaecological sense of the term (. . .)" (Gezondheidsraad 1984).

4.1 *The Semantics of Embryo Politics*

This phrasing ("not yet an embryo") reflects the same reasoning that a few years later, in the fierce British debate following the 1984 Warnock report, led embryologist Anne McLaren and other proponents of conditionally accepting human embryo research to embrace the term "pre-embryo" as better suited to reflect what was really at stake (McLaren 1986). Although the new term rapidly made its way in the scientific literature and ethical debates all over the world, its use was understood by opponents as defining moral problems away rather than facing them. As a result of this criticism, the new term became ethically suspect and was gradually abandoned again before the mid-1980s. However, it has been suggested that in the British debate the use of this short-lived term paid out for the proponents, in the sense of contributing to a hard-won majority for the liberal legislation (the Human Fertilization and Embryology Act of 1990) proposed by the Warnock committee (Colomer and Pastor 2012). In the Netherlands, there certainly was no such effect, as the term had already eclipsed before a more liberal bill was sent to Parliament in the last half-decade of the century (see Sect. 5.2).

4.2 *Qualifications*

The Health Council's committee's final report (1986) discussed a wider array of issues, including gamete and embryo donation and surrogacy (Gezondheidsraad 1986). With regard to embryo research (or pre-embryo research, as was the new terminology for the moment), the committee reiterated the position of the interim report, but did so in more reserved and qualified language. The committee started from the consideration that the early embryo "has an intrinsic value because it contains the unique genetic information for a (future) human being and is therefore worthy of protection. (. . .) However, this does not yet determine the value of the pre-embryo as compared with full grown human life. [Its] value is on the one hand determined by its potential to grow into an individual (. . .), but on the other hand by the fact that this development has only just begun. It is conceivable that other values and interests outweigh the worth of the pre-embryo. If large interests of very many people are at stake, for instance in important research aimed at scientific knowledge that cannot be obtained through experiments with animal embryonic material or by other ways, it can be considered as an exceptional measure to instrumentally use human pre-embryos. However, precisely because such use is only acceptable in exceptional cases, it must be subject to strict conditions" (Gezondheidsraad 1986).

The conditions given in the report are both procedural and material. Procedurally, the report insists that decisions on embryo research protocols be made on a national level by a committee able to assess the medical aspects and conduct "a weighting of ethical interests." Moreover, the handling and preservation of embryos should be made a prerogative of the medical profession and trading in human embryos must be forbidden and made punishable. Materially, the committee "finds it morally unacceptable to culture a human conceptus more than two weeks after fertilization as that would bring it in the phase of embryogenesis" (the last words of course revealing the "pre-embryo" framing discussed above). At this point, the committee referred to the consensus about a 14-day limit in "most foreign reports." However, it is also clear from the committee's discussion that this consensus was based on pragmatic rather than convincing ethical considerations, and one of the ethicists (Fretz, Delft) explicitly disagreed, proposing the ability to experience pain as a more objective criterion in his view (Gezondheidsraad 1986).

4.3 Creating Embryos for Research?

With regard to the possibility of using IVF to create embryos for research, the committee took the view that this was unacceptable, as "it would amount to a 100% instrumental use of human life and infringe upon the worthiness of protection of unique human life" (Gezondheidsraad 1986). Interestingly, one committee member made a dissenting statement at this point: "Prof. Zeilmaker considers it ill-advised to reject the fertilization of oocytes with the aim of using them for research," as this would block research in fertilization problems as a cause of infertility. One may add here that the preclinical experiments carried out by Zeilmaker and his group prior to receiving the green light for clinical IVF (see Sect. 3) of course also amounted to creating embryos for research. Indeed, without such preclinical research IVF would never have been developed. The report, however, is silent on this.

In a separate ethics sub-report, theologian Kuitert (Vrije Universiteit Amsterdam), one of the ethicists on a committee working group, explained the distinction between using leftover and creating new embryos for research as a matter of intention. Especially the idea of creating research embryos treated them as a mere means from the very beginning, which would be at odds with the dignity of human life, whereas spare embryos were created with the intention of letting them grow into a child. Declaring the use of spare embryos acceptable was justified by the situation: it concerned leftover embryos for which destruction was the only other option, thus leading to a choice between two evils. Kuitert seemed to acknowledge that this reasoning was shaky, as he added that "the best solution would therefore be to apply IVF technology in a way that avoids the occasioning of any 'left-overs'" (Gezondheidsraad 1986). It is indeed not obvious why deliberately creating more embryos than can be responsibly transferred to the womb would be less "instrumentalizing" than creating embryos for research (Devolder 2005). However, this question was not discussed, neither in the report nor in Kuitert's own publications on the issue in the previous years (Kuitert 1983; Kuitert

1985). In one of these publications, Kuitert had made the further restriction that research using spare embryos should be observational only (Kuitert 1983). This would of course have seriously reduced the usefulness of embryo research.

Of the two other ethicists contributing to the sub-report, one (Fretz) took a less principled line on creating embryos for research. His view was that this "seems morally unacceptable for the time being, given the extra dilemmas this implies. It will be necessary to further reflect on this in the future" (Gezondheidsraad 1986). With those "dilemmas," Fretz may have also meant to refer to whether the oocytes needed for creating research embryos can be responsibly obtained, which is indeed an important further issue in this whole debate. The fact that it was not separately discussed in the report may be due to the committee's categorical rejection of the research for which such oocytes would be needed.

Whatever the reasoning, it is obvious that the committee (unlike their counterparts on the British Warnock committee) regarded the creation of embryos for research a bridge too far. Although the report's reasoning on this point met with criticism already at that time (De Wert 1987), and although later Health Council committees took a different view (Gezondheidsraad 1998; COGEM/Gezondheidsraad 2017), the idea that creating embryos for research was ethically far more problematic than using spare embryos has since determined the scope for embryo research in the Netherlands. Already in the same year (1986), when colleagues of Zeilmaker in Maastricht (Geraedts and his team) managed to obtain a grant for a research project that involved creating embryos to study fertilization pathology, the hospital's ethics committee refused to give the green light for this part of the protocol because it regarded the creation of embryos for research as ethically unacceptable (Anonymous 1987). When Geraedts appealed against this decision, a committee of appeal was formed consisting of three experts: ethicist Kuitert, physiologist Zeilmaker, and health lawyer Gevers. This committee confirmed the rejection of the protocol, although one of the members (Zeilmaker) did not share the ethical objections (Health Council 1987). When these events were reported in the media (summer of 1987), MPs of the CDA asked parliamentary questions, urging the minister (also CDA) to prioritize legislation based on the notion that human embryos deserve protection (Tweede Kamer 1987; van Hengel 1987).

5 The Route to Legislation

Whereas in Britain the whole legislative process was completed in 8 years, between the start of the Warnock Committee in 1982 and the passing of the Human Fertilization and Embryology Act in 1990, things took much more time in the Netherlands: from 1982 till 2002 (Kirejczyk 1999). This was partly due to the fact that within the CDA, the main coalition party until 1994, many considered the recommendations of the Health Council as too liberal. In 1988, the scientific bureau of the CDA had issued its own report on assisted reproduction and embryo research. This extensive report, entitled "Meaningful life" (1988), did not leave any room for embryo research whatsoever (Wetenschappelijk Bureau voor het CDA 1988).

5.1 Meaningful Life: Christian-Democratic Report on Embryo Research

Referring to documents and position statements issued by the churches (Roman Catholic Instruction *Donum Vitae*, World Council of Churches Report on Genetic Engineering, reports from the French protestant and German evangelical churches, and from the Church of England), the CDA report observed that there was every reason to critically assess the new medico-technical reproductive methods. Although "Christian-democratic politicians should not try to impose a pattern of values with regard to procreation, that they themselves regard as convincing, upon society," it was clear from the party's "Program of Principles" that the government as God's servant had a responsibility to protect important values and that this perspective should not be abandoned in order to simply follow majority views. "Drawing on biblical sources," this required working for a society based on justice, solidarity, and stewardship, and regulating developments in economy, science, and technology accordingly. With regard to the handling of human embryos, the relevant perspective was respect for and protection of human life, regardless of its stage of development or form of appearance. "Physically or mentally handicapped persons are not of less worth than those who are healthy. If so, it is unclear why making such a distinction would be morally acceptable with regard to embryonic life that has not yet assumed a [recognizable] human form" (Wetenschappelijk Bureau voor het CDA 1988).

Both the destruction of embryos and their use as research material were at odds with these notions. "Destruction of embryos must be rejected, even when balancing this option [the creation of more embryos than can responsibly be transferred to the womb] with the interest of the woman [to avoid the burdens of multiple hormone stimulation cycles]. The physical burdens for the woman do not outweigh the destruction of early life." And although "the aim of certain experiments (improving the knowledge of infertility, genetic disorders, etc.) is not just legitimate but even desirable, this should not be attained through experiments that would infringe upon the potential and the possibility of this life to grow into a complete human being." These considerations led the report to recommend a maximum number of oocytes to be fertilized (to be determined by the government) and the requirement that "all fertilized oocytes will be transferred" (Wetenschappelijk Bureau voor het CDA 1988).

5.2 Drafts

The first draft for legislation (1992) consisted of an addendum to the bill on medical experiments. This was based on the principles of the CDA report, rather than on the recommendations of the Health Council. Except for experiments on (supposedly nonviable) polyploid embryos, it left no room for embryo research that would not benefit the embryo in question (den Hartog-van ter Tholen 1993). This phrasing meant to accommodate the new technology of pre-implantation genetic diagnosis (PGD) as a possible (still experimental) alternative for prenatal diagnosis followed by a possible abortion. However, experiments aimed at developing this technique in embryos not intended for transfer would clearly not be allowed. Moreover, conditions were that the research was "inevitable" and would remain "exceptional." When this bill was sent to the Council of State for review, it was fundamentally criticized for all kinds of inconsistency. The Council of State also pointed out that the bill and its explanatory memorandum failed to convincingly explain why the uncontested view that human embryos have an intrinsic value would lead to an almost absolute rejection of using them for research (Tweede Kamer 1992 No. 23 016B). In response to this criticism, a revised (truncated) version was sent to Parliament in February 1993, mainly consisting of a 2-year moratorium on all forms of research using human embryos (Tweede Kamer 1992 No. 23 016 1–2).

When in the summer of 1994 the "purple coalition" (see Sect. 2.1) took over, one of its first policy acts was a withdrawal of their predecessor's Embryos Bill. In a Memorandum to Parliament, an entirely new attempt was outlined in early 1995 (Tweede Kamer 1994). With regard to the moral status of the human embryo, the new legislation would start from the notion that human embryos, as beginning forms of human life, are worthy of respect, but also that this worthiness is not absolute and leaves scope—under conditions of proportionality and subsidiarity—for research using spare embryos. Creating embryos for research, however, would amount to an unacceptable violation of the embryo's worth and would therefore be forbidden in the envisaged legislation. This position was in line with the earlier Health Council recommendations and also with the views that led to a similar prohibition statement in the so-called Oviedo Convention of the Council of Europe (1997), of which the Netherlands was one of the initial co-signatories: "The creation of human embryos for research purposes is forbidden" (Article 18 part 2).

There were several reasons why drafting of the Dutch Embryos Act took five further years (the bill was presented to Parliament in 2000). Firstly, the pace of developments made it a matter of regulating a moving target. This included the birth of the cloned sheep Dolly in 1997, showing that it would be possible in principle to also create human embryos through cloning rather than fertilization. This required adapting the definition of an embryo. Whereas in the earlier drafts of the 1990s the human embryo was defined as "the result of the fusion of human gametes prior to birth," no such definition would suffice in the post-Dolly era. The drafters settled for "a cell or collection of cells with the capacity to develop into a human being." As has been pointed out, this is not entirely unproblematic either, as it implies that nonviable embryos are not really embryos in the first place, thus by definition falling outside the scope of legal regulation (Health Council 2005). A second reason why drafting took more time was a change of mind on the part of the drafters concerning the very issue of creating embryos for research (or broader: for reasons other than the induction of pregnancy). The need for this was spelled out in a new report from the Health Council (1998).

5.3 New Health Council Report

In view of updating the regulatory system for IVF, the outgoing coalition had asked the Health Council for a further report in February 1994. As this included a request to indicate "possible new ethical or societal issues", a new multidisciplinary committee was formed that in 4 years produced three reports, extensively discussing the science and ethics of a broad range of developments, ranging from IVF indications and treatment quality to new IVF-related technologies (including PGD) and, importantly, the place of embryo research as a necessary element of responsible innovation of new reproductive technologies (Gezondheidsraad 1998).

The committee's standpoint on using spare embryos for research was not dissimilar to that of the earlier Health Council reports or the 1995 Government Memorandum, but with regard to creating embryos, it took a different view. Although the committee did not deny that there might be a moral difference between using spare embryos and especially creating embryos for research purposes that would require creating new ones, it argued that the relevance of any such difference would be limited in the light of the general starting point that pre-implantation embryos have a relatively low moral status. In other words, even if such a difference could be construed, it did not follow that human embryos may not, under conditions, be created for research (Gezondheidsraad 1998).

One of these conditions was that the relevant research cannot also be carried out using spare embryos. As the report made clear, this is often the case with research aimed at preclinically investigating new reproductive technologies. For instance, the safety and effectiveness of technologies such as in vitro maturation (IVM) or cryopreservation of human oocytes could not be investigated using spare embryos as these did not originate from in vitro matured or frozen/thawed oocytes. The committee used these (and other) examples to underline the importance of reconsidering the proposed ban. These new adjuvant technologies were important to make IVF less burdensome for women, but might come with risks, the assessment of which required preclinical research in animals and human embryos. If the proposed ban led to skipping the step of such safety research, this meant putting women and children at risk for the sake of protecting embryos. The alternative of awaiting the results of research carried out in countries with more liberal legislation, such as the UK, was morally insincere (Gezondheidsraad 1998).

As creating research embryos requires mature oocytes, a limiting condition was that these are obtained in a morally sound way. Whereas some committee members thought it was not acceptable to ask reproductively healthy women to donate oocytes for research, others regarded this as not really different from asking healthy volunteers to participate in medical experiments. Asking women in an IVF program to donate part of their oocytes was problematic, unless a high number of oocytes were obtained and a previous cycle had shown a high fertilization rate. Other options were asking consent for using any non-fertilized oocytes or using any mature oocytes from women undergoing a gynecological operation (Gezondheidsraad 1998).

6 The Hybrid Character of the Embryos Act

As becomes clear from the Explanatory Memorandum that was sent to Parliament together with the new Embryos Bill in 2000, the government accepted the reasoning of the new Health Council report regarding the creation of embryos for research. Although it still thought that this was a larger infringement on respect for human life, making this infringement was justifiable in principle in the light of the great moral importance of research that would lead to better medical treatments (Tweede Kamer 2000). At that time (2000) the prospect of "better treatments" also referred to a possible future scenario of personalized cell therapies using human embryonic stem cells obtained from embryos created through somatic cell nuclear transfer, often referred to as "therapeutic cloning" (Health Council 2005; Dondorp and De Wert 2007).

However, the government saw pragmatic reasons for not yet eliminating the envisaged ban from the new act. These were threefold: (1) a presumed lack of public support—this was concluded from a public consultation with professional and societal groups including several churches and other organizations with a religious background; (2) the wish not to step too much out of pace with most other European countries; and (3) a perceived lack of urgency—it was felt that the ban would not immediately lead to significantly undermining important scientific developments (Tweede Kamer 2000). With respect to each of these points, the government expected further developments within a few years that together would lead to sufficient grounds for a decision to lift the ban. With time, a better understanding of the importance of embryo research would lead to greater acceptance for doing away with legal barriers, both nationally and internationally (Tweede Kamer 2000).

As a result, the 2002 Embryos Act has two faces. It still forbids the creation of embryos for other purposes than pregnancy (Article 24a), but it also contains the provision that this ban will be lifted (Article 33, part 2) and specifies the conditions (Article 9 and 11) under which research for which embryos are to be created will then be allowed. In the original version of Article 33 part 2, it was specified that the proposal for the Royal Decree lifting the ban was to be made within 5 years after act came into force (meaning: before 1 September 2007).

Why did the government regard it as necessary to have the lifting of the ban already inscribed in the act? Because at the time the Netherlands still intended to ratify the Oviedo Convention (see Sect. 5.2). The government wanted to avoid that, as a consequence of ratifying this European treaty with its explicit prohibition of creating embryos for research, the Netherlands would deprive itself of the future possibility (short of revoking or trying to change the Convention) to change its legislation on this point. In order to keep this option open, it was necessary (under Article 36 sub 1 of the treaty) to make a formal proviso when ratifying, which had to refer to existing legislation that explicitly diverged from the relevant Article of the Convention.

6.1 *A Perpetual Cliff-hanger?*

Although the general expectation was that the purple coalition would continue after 2002, this did not happen. The dramatic elections (with the murder of Fortuyn and the specter of a breakthrough of populist politics) brought the CDA back to power as the major coalition party and in control of the medico-ethical dossiers at the Ministry of Health (Aerts et al. 2013). Although revoking the Embryos Act was not politically possible, the new government's policy statement declared that, during its office, the ban on creating embryos for purposes other than pregnancy would not be lifted (Anon. 2007).

A new report of the scientific bureau of the CDA (Human Biotechnology, 2003) suggested that there was no need for creating embryos as there were possible alternatives for "therapeutic cloning" that could be considered (Anon. 2007). This ignored the fact that embryonic stem cell-based therapy was only one among the several developments affected by the prohibition. With regard to spare embryos, the new report no longer insisted that only so many oocytes should be fertilized in IVF as could be responsibly transferred in a single cycle. That indeed had been the clear and consistent position of the party's previous report on the matter (see Sect. 5.1). Instead, it was now suggested that still to be developed techniques such as oocyte cryopreservation would allow *reducing* the number of embryos ending up as leftovers. Although the rhetoric sounded familiar: "whoever is serious about the worth of beginning human life will do his utmost to choose embryo-saving policies," the formulations made clear that embryo protection had become a policy ideal rather than a condition sine qua non. Obviously, the political price of insisting on making IVF more burdensome and risky for women in order to avoid instrumentalizing human embryos had become too high. Of note, the report failed to address the need for testing those embryo-saving technologies in preclinical safety studies, studies indeed for which human embryos would have to be created (Gezondheidsraad 1998). A significant difference with the previous (1988) CDA report was also the lack of explicit reference to principles based on religious belief. Instead, the new report referred to vague notions such as "Jewish-Christian cultural sources" (Hoedemaekers 2003).

Although in the first official Evaluation Report of the Embryos Act (2006) it was concluded that in order not to block important research, the intended lifting of the ban should no longer be postponed (Olsthoorn-Heim et al. 2006), what happened was the opposite. Briefly before the 5-year period stipulated in the Act ran out, a new CDA-led coalition (in which for the first time also the Christian Union (CU) participated) decided to simply erase the words with this time frame from the act, thus turning the announced lifting of the ban into a perpetual cliff-hanger, a moratorium without expiration date.

6.2 *Citizens' Perspectives?*

Apart from a small study in the early years of IVF (Holmes and Tymstra 1987) and the sounding consultation with representatives of professional and societal organizations carried out as part of the legislative process, no efforts had until then been made to find out about the views of the public. When the Rathenau Institute, an independent Health Technology Assessment Institute with the official task of advising the Dutch Parliament, announced it would do a survey to fill this lacuna, the

CDA regarded this as undermining government policy and at first tried to prevent the institute from carrying out its plans (Sander and Lucas 2007). But when the report was published, it seemed that any worries at CDA headquarters had been premature. The main results suggested there was quite some support for the decision to keep the ban in place: 49% were in favor of keeping the ban, 35% wanted it to be lifted, and 14% was undecided (Steegers et al. 2008).

However, there were some problems with this study (De Wert and Dondorp 2008). Surveys about public attitudes about policy issues unconnected to peoples' daily lives are notoriously problematic. A crucial condition is that the information given to respondents is correct and balanced. Whether that was the case here could be questioned. For instance, the mistaken view that the ban would mainly affect research into "therapeutic cloning" was not corrected in the information. There was no explanation of the importance of preclinical safety studies as part of the introduction of new reproductive technologies and why spare embryos are unsuited for such research.

A remarkable finding was that participants tended to be more positive about research using specially created embryos if that research was bound by more conditions, such as the absence of alternative research approaches (subsidiarity), a prohibition of reproductive cloning, and rules forbidding the use of embryos for trivial aims. However, they were not told that those conditions were already specified in the Embryos Act. Had they understood this, the support for lifting the ban may have been higher. Finally, correct information telling them that, under up-to-date careful stimulation protocols, the true level of risk for women donating their oocytes for research is very low was also lacking. Interestingly, the presumed high risks of this procedure for candidate donors were the main reason for rejecting research using specially created embryos, rather than concerns about embryos. As far as we are aware, no further studies of attitudes of the Dutch public with regard to embryo research have been conducted.

6.3 Present Situation

In 2012, the second evaluation report of the Embryos Act again recommended lifting the ban (Winter et al. 2012). This report observed that cryopreservation of oocytes was now an accepted technology, but that its premature introduction in several countries should not be taken as a recipe for dealing with further innovation (Dondorp and de Wert 2011; Harper et al. 2012). For instance, one of the technologies on the horizon was reproduction using stem cell-derived gametes, as a means of helping people to have genetically related children who until now are dependent on donor sperm or oocytes. As the safety of this technology could not be taken for granted, it should not be introduced in the clinic without proper preclinical research, including research with embryos created from stem cell-derived gametes.

As the report was published just ahead of the inauguration of a new coalition in which for the first time after the "purple" period Christian parties were not participating, one might have expected some new action in response. However, the recommendation to finally lift the ban was met with virtually the same reaction as from earlier CDA-led governments: no need to consider this for the time being (Tweede Kamer 2013). Politically, this was understandable: as the government had no majority in the Senate, it was dependent on help from varying groups of other parties for getting its policies accepted. Whereas the CDA took a nonconstructive opposition role, the smaller Christian parties (CU and SGP) did not a priori reject the role of providing support in exchange for having some of their issues respected. This may well explain why at least in its early days, the 2012–2017 liberal–labor coalition did not try to shake up any of the smoldering medico-ethical dossiers. This changed in the last year of its period, with government initiatives to take further steps both in the euthanasia dossier (more specifically with the aim of allowing assisted suicide for people who consider their lives to be completed) and in that of the creation of embryos for research.

Referring to data again confirming that the ban does indeed create a problem for important research, the Minister of Health has in May 2016 announced her intention to change the Act in order to allow preclinical studies with specially created embryos for specific technologies (Tweede Kamer 2016). This was limited to technologies that were relatively close to clinical introduction: IVM, mitochondrial replacement technologies, and reproduction using stem cell-derived gametes. Creating embryos for other research, including for fundamental studies, would remain forbidden.

This proposal is more limited than the situation foreseen in the Embryos Act after the lifting of the ban. According to Article 11, it will then (under conditions) be allowed to create embryos for research that can be expected to generate new insights in the fields of infertility, assisted reproduction, genetic or congenital abnormalities, and transplantation medicine. Commentators have also pointed out that the limitation to technologies already close to clinical introduction ignored the need for research that brings a technology to that threshold, including more fundamental studies (Dondorp et al. 2016). In her letter to Parliament, the minister herself strongly underlined the very limited character of her proposal, saying that a complete lifting of the ban was not intended and that "carefulness and respect for human dignity" required this cautious approach (Tweede Kamer 2016).

The letter did not lead to much discussion in the main press, but in the orthodox Roman Catholic journal *Katholiek Nieuwsblad*, former CU politician Wiegman (now director of a religious "pro-life" patient organization) wrote a lengthy piece in which she pointed out that the Rubicon had already been crossed when it was accepted to (create and) use spare embryos. She commented that the feeling of "moral unease" that was still present in the early days of the debate, and of which the ban was an inconsistent but clear reminder, had gradually been eroded. It was "a sign of a sea-change that those who still ascribe some level of value to the embryo are increasingly asked to justify their views" (Wiegman 2016).

6.4 Outlook

At the time of writing (summer 2017), a new coalition is in the making. It is highly likely that apart from the present Health Minister's party (the liberals of the VVD), the new government will include three further parties (D66, CDA, and CU) that all find medico-ethical issues very important, but whose views are very diverse. As the party most connected with the drafting of the Embryos Act, D66 wants the ban lifted. It regards the plans to allow some research with specially created embryos as important but insufficient. CDA and CU, by contrast, reject those plans. Even more so than for the present CDA, embryo research is seen as a battle cause of Christian politics by the CU. As they are needed for a majority (or else at least for political support), it is quite certain that the categorical ban on creating embryos for research will remain in place for the next period.[1]

7 Discussion

Looking back, the political debate on embryo research in the Netherlands can be divided into three periods: an initial decade (1982–1994) in which the CDA owned the dossier and was at the legislative wheel; the "purple" interlude of two parliamentary terms (1994–2002) in which a coalition of three secular parties managed to draft a new bill and see it accepted; and the period since 2002 in which Christian parties either within or without the ruling coalition have successfully blocked the envisaged lifting of the ban on the creation of embryos for research.

In the initial decade, the CDA had strong views about how government, as God's servant, was called to protect the equal value of all human life from conception, and about its own role to ensure that this would also be the basis of the legislation that it was entrusted to draft. However, it had soon to find out that in a plural and rapidly secularizing society, this was wishful thinking. When the "purple coalition" took over the dossier in 1994, it seemed to have realized that conversely, it was important not to lose contact with the part of society for which the pace of developments was high. As we have seen, a supposed lack of societal support was the main reason for stopping short of allowing the creation of embryos for research (Tweede Kamer 2000). It was felt that society needed more time to get used to the idea. From the perspective of the drafters of the act, it may seem unfortunate, in hindsight, not to have moved on when this would have been possible, but one simply had not

[1]On October 10, 2017, it was announced that VVD, CDA, D66, and CU will indeed form the next coalition government (the first four-party coalition since 1977). The parties have committed themselves to a formal Coalition Agreement that also contains a separate section on medico-ethical issues. With regard to embryo research and the debate about keeping or lifting the ban on creating embryos specifically for this purpose, the document states that "this is a very serious issue [that requires] a broader debate also encompassing the ethical and societal aspects. The cabinet will stimulate this discussion while simultaneously looking for alternatives." With these lines, the proposal of the last government to at least partly lift the ban seems effectively relegated to the parliamentary archives, at least for the coming years. The Coalition Agreement: "Vertrouwen in the Toekomst. Regeeraccoord 2017–2021" can be retrieved from the website of the Dutch Parliament at: https://www.tweedekamer.nl/debat_en_vergadering/uitgelicht/kabinetsformatie-2017 (last visited October 11, 2017).

reckoned with the swift return to power of the CDA. Clearly, this return was not so much because explicit Christian politics suddenly again appealed to the electorate, but because in a year of dramatic events that paralyzed the reelection campaign of the purple coalition, the party successfully positioned itself as the solid governance party needed in difficult times (Aerts et al. 2013).

The outcome of the process with the present act that allows research using spare embryos while still containing the embattled ban may be regarded as unsatisfying from the point of view of consistent ethical reasoning on both sides of the debate. Either pre-implantation embryos have a moral status that forbids using them as a mere means or they do not have such a status. In the first case, the only justifiable approach is the one taken in the 1988 CDA report (avoid creating more IVF embryos than can be transferred in the same cycle). In the second case, there is no good reason why human embryos could not also be created for research that cannot be done with spare embryos. It seems the opposing sides have locked the debate in a position that is ethically untenable and that neither can be happy with.

7.1 The Role of Religious Argument

What was the role of religion in the Dutch political debate on embryo research? It depends how we read the question. Did explicit *religious arguments* (referring to Christian creeds and teachings, what the Bible says, etc.) play a role in the debate? For an argument to play a role in a political (or other) debate, the various debating parties must be able to recognize the validity of the argument, at least in principle. As we have seen, religious argument played a widely recognized role in this sense during much of the last century in what was then still a very Christian nation (Meijering 2012). However, this was already something of the past when the embryos debate started in the 1980s. To make themselves heard by other participants to the debate, Christian politicians had already learned to rephrase their arguments in the language of more generally shared values that also nonbelievers might find important such as "human dignity" or "respect for life." But precisely because these notions are not specifically Christian or religious, this means that explicit religious reasoning did not play a role in the debate.

What about the fact that a prominent member of the Health Council's Committee who himself spoke out strongly against creating embryos for research was a professor of systematic theology (Kuitert)? It would certainly be wrong to take this as a sign of religious reasoning in the bosom of the Health Council. Actually, Kuitert is famous for rejecting religious ethics, lecturing his students that one need not believe in God to tell good from bad (Peelen 2016). His role at the Health Council, in later years also as vice-president of the Council's Standing Committee on Medical Ethics and Health Law, would not have been possible had he presented as the spokesperson of a religious perspective. If Kuitert had a specific influence, it must have been the early framing of the debate in terms of the problematic distinction (discarded versus created) that led into the present stalemate. But his reasoning for this (referring to respect for human life) was not based on (explicitly) religious arguments (Kuitert 1983). Interestingly, his successor on the Ethics chair of the Theological Faculty at the Vrije Universiteit (Amsterdam), and briefly also a member of the Health Council (Reinders 1993), wrote a well-received book on the ethics of embryo research, in which he not only took a distinctively religious perspective but also criticized the self-understanding of the Health Council's secular ethics as based on a supposedly worldview free autonomous morality (Reinders 1993).

7.2 Views of the Churches

Did *organized religion* (churches) have a role in the political debate? Not directly, of course, but perhaps in terms of their views? To the extent that they had views on the matter, these were indeed brought into the political arena by the Christian parties. Clearly, the views of the first report of the CDA's scientific bureau (1988) matched to an important extent with those of the Roman Catholic Church as laid down in *Donum Vitae* a year earlier. Of course, it does not follow that these were necessarily also the views of individual members (or clergy) of the Dutch Roman Catholic Church. However, the main Reformed Churches (since fused into the Protestant Church in the Netherlands) never had an official position on embryo research or other biotechnological developments. In 1991, a joint Committee on Biomedical Ethics of these churches issued a report that contained a plurality of views rather than a single position (Boer 1991). Most committee members thought that pre-implantation embryos were not unassailable and could be instrumentally used for important research under conditions akin to those specified by the Health Council. Some of these members went even further and thought that this might also include research with specially created embryos. Only a minority of the committee was against all forms of embryo research. These members concurred with the recommendations of the CDA report aimed at avoiding leftover embryos. Clearly, it is far too simple to suggest a straight relationship between the embryo policies of Christian political parties and the views and positions of churches and church members in the Netherlands.

7.3 Christian Parties

However, if the role of religion is taken to refer to *any organizations on a religious basis* including Christian political parties, then of course "religion" in this sense had and still has a large role in the ongoing Dutch embryos debate. This goes for the CDA in the first part of this story (although its efforts to block embryo research eventually failed) and for the CU (and SGP) up until today. In the second period, the CDA can no longer be regarded as having a religious basis. It has become a conservative party for which a critical position with regard to embryo research is more a matter of underscoring traditional values in a globalizing culture rather than anything else. The two other parties are only small and will remain so, as the traditional, evangelical and orthodox Christians they represent are a small minority in Dutch society. However, as mentioned, their political influence precisely on issues of medical ethics including embryo research is relatively large. This is due to the fragmentation of present-day Dutch politics, where small parties can have greater influence if they manage to strategically position themselves in the arena. Whereas the SGP remains an ideological party that would not get its hands dirty, the CU has shown a willingness to compromise if issues important to its electorate can be moved in the desired direction. For this party, accepting the Embryos Act as a matter of fact certainly means getting its hands dirty. However, it may be the price to pay in order to put itself in a position to effectively keep a new government from doing worse. The margins may be small, but within those margins, Christian parties are still able to significantly influence policy-making.

8 Conclusion

Former senator De Gaay Fortman is known for comparing the practice of late twentieth century Dutch politics with the craftsmanship and eye for detail of ivory workers (De Gaay Fortman 1979). In the end, it has always been a matter of small margins and compromises and of respecting the sensibilities of opponents who may become partners in a next coalition. This was no different for the long and winding process of debate and regulation with regard to embryo research, up until the cautious wording of last year's Minister's letter to Parliament. We have seen that although religious argument was understandably absent from this debate between ideological strangers, Christian parties have, with limited success, tried to block regulations allowing embryo research. Their motivation stemmed from the religiously inspired view that pre-implantation embryos have the same moral status as all human individuals. This position was consistently laid down in the 1988 CDA report "Meaningful life" and is now inspiring the efforts of the Christian Union to prevent the lifting of the ban on creating embryos for research. However, even if the remaining Christian parties manage to prevent this step for the time being, further developments are on the horizon that will make it less obvious what their own position should be. Here we think of developments that will require reflection on how, both scientifically and ethically, the concept of an embryo should be delineated. To the extent that it will become possible to create embryo-like structures (or even "synthetic" embryos) from induced pluripotent stem cells (Pera et al. 2015), would research using such structures be ethically problematic? Clearly this debate is far from over.

References

Aerts, R., Bohl H. De Liagre, P. De Rooy, and Henk te Velde. 2013. *Land van kleine gebaren. Een politieke geschiedenis van Nederland*. Amsterdam: Boom Uitgevers.

Anonymous. 1987. Regels voor onderzoek met (pre-)embryo's? *Graadmeter (Periodical of the Health Council)* 3 (5): 4–5.

———. 2007. Verbod embryokweek remt breed onderzoek ivf. *Trouw*, February 13.

———. 2017. Embryowet wordt grote splijtzwam in de formatie: Nederlandse Omroep Stichting (NOS). June 24. Available from http://nos.nl/artikel/2179841-embryowet-wordt-grote-splijtzwam-in-de-formatie.html. Accessed 15 July 2017.

Boer T. 1991. Mensen in wording. Theologische, ethische en pastorale overwegingen bij nieuwe voortplantingstechnieken en prenataal onderzoek. Een verkenning. Rapport van de commissie "Biomedische Ethiek" van het Deputaatschap en de Raad voor de Zaken van Kerk en Theologie van de Nederlandse Hervormde Kerk en de Gereformeerde Kerken in Nederland. Utrecht: Samen op Weg Kerken.

Brian K. 2013. The amazing story of IVF: 35 years and five million babies later. *The Guardian*, July 12.

Christen Democratisch Appel (CDA). 1989. Program van uitgangspunten. Available from https://www.cda.nl/standpunten/publicaties/program-van-uitgangspunten/. Accessed 15 July 2017.

———. 2017. Waarden en traditie. Available from https://www.cda.nl/standpunten/themas/waarden-en-traditie/. Accessed 15 July 2017.

Colomer, M.F., and L.M. Pastor. 2012. The pre-embryo's short lifetime. The history of a word. *Cuadernos de Bioética* 23: 677–694.

Commissie Genetische Modificatie (COGEM)/Gezondheidsraad. 2017. *Ingrijpen in het DNA van de mens. Morele en maatschappelijke implicaties van kiembaanmodificatie*. Bilthoven: COGEM.

De Gaay Fortman, B. 1979. *De kunst van het ivoordraaien. Handleiding voor het politieke ambacht*. Houten: Spectrum.

De Wert, G.M.W.R. 1987. Het advies van de Gezondheidsraad inzake kunstmatige voortplanting: samenvatting en kritiek. *Nederlands Tijdschrift voor Geneeskunde* 131: 367–371.

De Wert, G.M.W.R., and W.J. Dondorp. 2008. Burgerperspectieven op embryo-onderzoek? Een gemiste kans. *Tijdschrift voor Gezondheidsrecht* 32: 257–262.

den Hartog-van ter Tholen, R.M. 1993. Beleidsvorming over het omgaan met embryo's bij politieke en maatschappelijke dissensus. In *Beleid en ethiek Uitgave in samenwerking met het Centrum voor Bio-ethiek en Gezondheidsrecht*, ed. Brom FWA, B. Van den Berg, and A.K. Huibers, 112–124. Assen: Van Gorcum.

Devolder, K. 2005. Creating and sacrificing embryos for stem cells. *Journal of Medical Ethics* 31 (6): 366–370.

Dondorp, W.J., and G.M.W.R. De Wert. 2007. On waving the embryos banner. Embryonic stem-cell research and moral reasoning in recent reports from the USA, the Netherlands and France. In *Ethics Law & Society*, ed. J. Gunning and S. Holm, vol. III, 7–24. Farnham: Ashgate.

Dondorp, W., and G. de Wert. 2011. Innovative reproductive technologies: risks and responsibilities. *Human Reproduction* 26 (7): 1604–1608.

Dondorp, W.J., M.C. Ploem, and G.M.W.R. de Wert. 2016. Het kabinetsvoorstel tot verruiming van de Embryowet: een halve stap vooruit. *Tijdschrift voor Gezondheidsrecht* 40: 536–244.

Gezondheidsraad. 1984. *Interimadvies inzake in vitro fertilisatie*. Den Haag: Gezondheidsraad.

———. 1986. *Advies inzake kunstmatige voortplanting*. Den Haag: Gezondheidsraad.

———. 1998. *IVF-afrondende advisering. Pre-implantatie diagnostiek; Onderzoek gericht op verbetering van IVF; Onderzoek met menseijke embryo's; Nabeschouwing*. Den Haag: Gezondheidsraad.

Harper, J., M.C. Magli, K. Lundin, C.L. Barratt, and D. Brison. 2012. When and how should new technology be introduced into the IVF laboratory? *Human Reproduction* 27 (2): 303–313.

Health Council of the Netherlands. 2005. *Embryonic stem cells without moral pain?* The Hague: Centre for Ethics & Health.

Hoedemaekers, R.H.M.V. 2003. *Humane Biotechnologie. Een studie naar menswaardige toepassingen van nieuwe biotechnische mogelijkheden*. The Hague: Wtenschappelijk Instituut voor het CDA.

Holmes, H.B., and T. Tymstra. 1987. In vitro fertilization in the Netherlands: experiences and opinions of Dutch women. *Journal of In Vitro Fertilization and Embryo Transfer* 4 (2): 116–123.
Kirejczyk, M. 1999. Parliamentary cultures and human embryos: The Dutch and British debates compared. *Social Studies of Science* 29 (6): 889–912.
Kuitert, H.M. 1983. Experimenten met embryo's. een eerste verkenning. *Medisch Contact* 38: 433–435.
———. 1985. Wanneer is leven menselijk leven? Over de status van het embryo. In: Instituut voor Gezondheidsethiek (IGE). In *Gezondheidsethiek in de huidige samenleving Voordrachten gehouden ter gelegenheid van de opening van het Instituut voor Gezondheidsethiek te Maastricht*, 31–36. Maastricht: IGE.
McLaren, A. 1986. Embryo research. *Nature* 320 (6063): 570.
Meijering, E. 2012. *Hoe God verdween uit de Tweede Kamer. De ondergang van de christelijke politiek*. Amsterdam: Balans.
Netherlands Institute for Multiparty Democracy, Instituut voor Ethiek en Politiek. 2008. The Dutch Political System in a Nutshell Amsterdam: Netherlands Institute for Multiparty Democracy. Available from http://nimd.org/theme-brochures/the-dutch-political-system-in-a-nutshell/. Accessed 15 July 2017.
Olsthoorn-Heim, E.T.M., G.M.W.R. De Wert, H.B. Winter, T.A.M. Te Braake, M.J. Heineman, A. Middelkamp, et al. 2006. *Evaluatie Embryowet*. Den Haag: ZonMw.
Peelen, G.J. 2016. *Spreken over boven. Harry Kuitert, een biografie*. Amsterdam: Vesuvius.
Pera, M.F., G. de Wert, W. Dondorp, R. Lovell-Badge, C.L. Mummery, M. Munsie, et al. 2015. What if stem cells turn into embryos in a dish? *Nature Methods* 12 (10): 917–919.
Peters, K. 2015. *Een doodgewoon kabinet. Acht jaar Paars 1994–2002*. Amsterdam: Boom Uitgevers.
Praamsma H. 2002–2003. Andere Tijden. IVF in Nederland. Hilversum: VPRO, NTR. Available from https://anderetijden.nl/aflevering/504/IVF-in-Nederland. Accessed 15 July 2017.
Reinders, J.S. 1993. *De bescherming van het ongeboren leven. Morele en godsdienstige overwegingen bij experimenten met menselijke embryo's*. Ten Have: Baarn.
Rigter, R.B.M. 1992. *Met Raad en Daad. De geschiedenis van de Gezondheidsraad 1902–1985*. Rotterdam: Erasmus Publishing.
Sander Becker, Nicole Lucas. 2007. CDA verbaasd over embryo-onderzoek van Rathenau-instituut. *Trouw*, 7 April.
Steegers, S., H. Dijstelbloem, and F.W.A. Brom. 2008. *Meer dan status alleen. Burgerperspectieven op embryo-onderzoek*. Rathenau Instituut: Den Haag.
Ten Have, H.A.M.J. 2002. Honderd jaar Gezondheidsraad. V. Ethisch-maatschappelijke vraagstukken. *Nederlands Tijdschrift voor Geneeskunde* 146: 1947–1951.
Troost N. 1996. Veel collega's zagen ivf niet zitten, maar ik geloofde er in. *de Volkskrant*. August 6.
Tweede Kamer der Staten-Generaal. 2013. Evaluatie Embryowet. Brief van de Minister van Volksgezondheid, Welzijn en Sport aan de Voorzitter van de Tweede Kamer der Staten Generaal, juli 11. Vergaderjaar 2012–2013, nr 30 486–5.
———. 2016. Prenatale screening. Brief van de minister van Volksgezondheid, Welzijn en Sport aan de Voorzitter van de Tweede Kamer der Staten Generaal, mei 27. Vergaderjaar 2015–2016, nr 29323-101.
———. 1987. Vragen van de leden Soutendijk-van Appeldoorn, Laning-Boersma en Borgman (allen CDA) over het kweken van embryo's ten behoeve van wetenschappelijk onderzoek. Vergaderjaar 1987–1988, Aanhangsel 99:197–8.
———. 2000. Wet houdende regels inzake handelingen met geslachtscellen en embryo's (Embryowet). Memorie van Toelichting. Vergaderjaar 2000–2001, nr 27:423–3.
———. 1994. Wijziging van de wet inzake medische experimenten in verband met regels inzake handelingen met menselijke embryo's en geslachtscellen, Brief houdende intrekking van het wetsvoorstel en Notitie regelgeving inzake enige handelingen en wetenschappelijk onderzoek met embryo's en foetussen. Handelingen Tweede Kamer vergaderjaar 1994–1995, nr 23:016–7.

———. 1992a. Wijziging van de wet inzake medische experimenten in verband met regels inzake handelingen met menselijke embryo's en geslachtscellen. Advies Raad van State en Nader rapport. Vergaderjaar 1992–1993 nr 23:016 B.

———. 1992b. Wijziging van de wet inzake medische experimenten in verband met regels inzake handelingen met menselijke embryo's en geslachtscellen. Vergaderjaar 1992–1993, nr 23 016:1–2.

van Hengel W. 1987. Kweken van embryo's voor onderzoek ontoelaatbaar. *Reformatorisch Dagblad*, 4 July.

Van Overbeeke, G. 1992. *In-vitro fertilisatie. De introductie van reageerbuisbevruchting in Nederland*. Amsterdam: Centrale huisdrukkerij VU.

Wetenschappelijk Bureau voor het CDA. 1988. *Zinvol leven. Een christen-democratische bijdrage aan de discussie over draagmoederschap, kunstmatige inseminatie, gift en in vitro fertilisatie*. Deventer: Van Loghum Slaterus.

Wiegman E. 2016. Een embryo is leven. *Katholiek Nieuwsblad*. 16 June.

Winter, H.B., W.J. Dondorp, M.C. Ploem, N.O.M. Woestenburg, J. Legemaate, and G.M.W.R. De Wert. 2012. *Evaluatie Embryowet en Wet Donorgegevens Kunstmatige Bevruchting*. ZonMw: Den Haag.

Zeilmaker, G.H., A.T. Alberda, I. van Gent, C.M. Rijkmans, and A.C. Drogendijk. 1984. Two pregnancies following transfer of intact frozen-thawed embryos. *Fertility and Sterility* 42 (2): 293–296.

Zeilmaker, G.H., A.T. Alberda, C.A.M. Jansen, and A. Verhoeff. 1987. Resultaten van het laboratorium voor in vitro-fertilisatie van de Erasmus Universiteit te Rotterdam. *Nederlands Tijdschrift voor Geneeskunde* 131: 2198–2201.

Chapter 13
Moralizing Embryo Politics in Germany: Between Christian-Inspired Values and Historical Constrains

Mirjam Weiberg-Salzmann and Ulrich Willems

Abstract For historical reasons, political debates in Germany tend to stress the separation between religion and politics. At the same time, the church is a well-respected social player and through various institutions interwoven with the state. Concerning the debate on the regulation of biopolitics, religion has played an important role, either through direct interventions by the churches or more indirectly through the religious beliefs of politicians. Germany has passed laws that strongly regulate or prohibit biotechnological procedures. The most important legal result of the debate on human biotechnologies was the Embryo Protection Act (ESchG) in 1990. In the discussion, which was conducted widely in the media and accompanied by public interviews and hearings, the Protestant Church (EKD) and the Catholic Church, deaconries, and associations for the disabled actively intervened and mostly held restrictive positions. Church representatives were also on the official advisory boards. The main topics were freedom of research versus protection of human dignity and historical reservations (National Socialism period/eugenics). Despite some liberalizations (e.g., postponement of the deadline), the law remained restrictive and is today regarded by many actors as in need of revision for various reasons.

1 Introduction

Political conflicts with a religious bias or starting point are not uncommon even in liberal democracies. This is predominantly a question of an appropriate interpretation of freedom of religion, the associated practices, and the presence of religious symbols in public spaces (e.g., wearing a headscarf). A second focus is the entitlement to certain state benefits (e.g., religious education) or state measures of positive discrimination (e.g., exemption from compulsory military service) or state sanctions for deviant behavior (violence). A fourth point concerns religious criticism of certain state policies (e.g., life sciences). The extent to which religious groups follow their religious convictions internally in socially controversial issues is not initially of

M. Weiberg-Salzmann (✉) · U. Willems
University of Münster, Münster, Germany
e-mail: mirjam.weiberg-salzmann@uni-muenster.de; ulrich.willems@uni-muenster.de

M. Weiberg-Salzmann, U. Willems (eds.), *Religion and Biopolitics*,
https://doi.org/10.1007/978-3-030-14580-4_13

relevance for the democratic order (as long as public norms are not violated). Religious convictions are only of interest if they make a public claim to validity and come into conflict with other groups/individuals with differing ideas. Since particular religious positions in a pluralistic-democratic society are not shared by all citizens, they cannot be used as the basis for a general regulation. Moreover, religious positions do not only differ between the different religions and denominations, but there are also considerable differences within the groups. In any case, a secular, ideologically neutral state must not commit itself to a single religious or ideological point of view. Nevertheless, the question remains: What influence should religious groups, their motives, and ideas have on state policy?

In general, this question for liberal concepts of democracy is concerned with the acceptability of religious arguments in state politics. Whether and how religion should be considered in deliberative processes has been answered differently in theory: At the extreme cornerstones of the debate, some authors demand the exclusion of religious and religiously motivated arguments and claims, while others consider the inclusion necessary and justified. A middle position assumes an open discourse, ideally for all, with the focus not on the religious foundation, but on the content they convey. If the content is reciprocally generalizable and could in principle also be formulated in this way, the argument is admissible.

In principle, empirical questions are open to discussion in many conflicts in which religious actors are involved. In the case of stem cell research, for example, it is disputed what benefits the research will have and when results can be expected. These empirical questions—upon which all participants agree—are also treated as empirical questions and explained with the help of science. Thus, medical and biological arguments play an important role when it comes to the ethical and legal evaluation of preimplantation diagnostics, stem cell research, and cloning. The most important reference points are natural embryonic development and the potential of the cells involved. Bible exegesis and religious arguments, as competing empirical knowledge, play no (or only in very rare cases a subordinate) role, even if science cannot give unambiguous answers to the question. Disputes arise over the evaluation of empirical arguments, e.g., their scope and limits. Within the framework of the stem cell debate, the expected medical benefit has been evaluated normatively in relation to other goods. More frequently used than directly religious arguments are the ethical or moral insights gained through the interpretation of religious sources, etc. presented by religious actors.

The interest of religious actors in bioethical questions did not develop in Germany until relatively late in the 1980s. The background was the new developments in medicine and a growing ethical pluralism. In the following debates on the moral, social, and political problems associated with this, initially doctors, philosophers, and theologians became involved and then social scientists, psychologists, and lawyers. In principle, almost all actors—in view of the experiences in the Third Reich—were mistrustful of biomedical practices and their representatives. Many also rejected the theories (pragmatism and utilitarianism) that dominated bioethics in the Anglo-American region, since they were seen as an uncritical and ideological defense of biomedical progress and its economic utilization. The predominant approach was that of Kant and a metaphysical rational foundation of ethics. In the field of medicine, various bodies, e.g., the German Medical Association, established codes of ethics, and sanctioned violations were necessary. At the political level, working groups and expert commissions on biopolitical issues have been commissioned since the mid-1980s. However, Germany was one of the last countries within the European Union (EU) to set up a national ethics council. Germany has passed laws in the various biomedical fields (e.g., euthanasia, abortion, reproductive medicine, embryo research, and cloning), most of which are restrictive and were passed with the public participation of the Catholic and Protestant churches (Schöne-Seifert et al. 1995, 1627–1639).

One major point of contention in the biopolitical debate was the value conflict "embryo protection versus health protection," because the ontological status and claim to protection of prenidative embryos is highly controversial in today's pluralistic society. The advocates of the new technologies, in particular actors from reproductive medicine, demand a liberalization of the laws. The critics, above all the two main German churches, evaluate these demands as being enslaved to progress and going beyond boundaries. They see the limit of research reached where fundamental questions of the understanding of human life and its appreciation are affected. For two decades the Catholic Church has taken the view that the embryo becomes a human being from the moment of conception. From this an absolute protective right can be derived, without any further consideration. In Protestantism, very different assessments of embryo status can be found, ranging from absolute protection to gradual protection (Kreß 2006: 92–99; Kreß 2009: 135–143).

Despite all normative-theoretical reservations toward religious actors and arguments in politics, it must first be asked whether there is at all a specifically religious contribution in the bioethical debate, which positions are extracted from the Christian faith, and how the positions are implemented argumentatively. In the following sections, we will first present the historical framework conditions in Germany, followed by the genesis of the biopolitical debate and the actors involved. Finally, in the fourth part, we will address the positions and arguments of the religious actors.

2 The Historical Dimension: State and Church in Germany

In Germany, the separation of politics and religion is one of the basic democratic principles that have developed over the past centuries. The institutional structure of the state-church relationship in Germany can be described as a hybrid. The Basic Law stipulates a separation of state and church; the state may not favor any religion. On the other hand, the constitution links state and church, e.g., within the framework of church tax and in relation to certain privileges (Robbers 2005: 83–101). For a long time, the relationship between politics and religion was only marginally discussed in Germany. Modernization was apparently inextricably linked with the decline of religion, which continued to exist but was pushed into the private sphere. It was not important for political decisions. German political rhetoric focusses on the separation between church and state. The idea is that religion is considered to work best when it does not interfere with politics. Politicians who invoke religion, use religious rhetoric, or give religious reasons for their arguments are viewed suspiciously.

An overview of the historical development in Germany explains how and why the relationship has developed. Important events can be found in the eighteenth century. In Germany, this was the time of enlightened absolutism in which religion was met with extreme skepticism. This resulted from the observation that the state was dealing with various religious groups that were potentially hostile. This posed a threat to peace and stability in the country. The background to this perception was decisive: First, the religious wars of recent centuries have indicated that religious communities were incapable of resolving their differences peacefully. Secondly, there was much to suggest that the disastrous consequences had resulted from religion having too much influence on politics. Conservative and liberal approaches differed only in their attitude to religion in that the former sought to link the state with a particular religion whose traditional values the state protected. The latter assigned the state a role as an impartial authority between the various groups, combined with a certain distance from everything religious. What was important, however, was the fundamental agreement of both positions with regard to the attitude to religion in general: Primacy was always a protection of the secular, the political from religious dominance, in order to protect the state (directly or indirectly) from the (actual or imagined) danger of religion (Zachhuber 2007: 7–16; Voigt 2007: 17–33).

The relationship between the state and the church in Germany was also significantly influenced by the long process of establishing the state and the nation. Germany is religiously mixed and has been so since the Peace of Westphalia established the principle of the (free) choice of religion by the sovereigns. In the nineteenth century, the belated nation Germany possessed—despite all general reservations against religion—a strong religious component in the process of nation-state formation. Already in the decades before the unification, the idea of a "chosen people" and a pact between God and the Germans existed among the Protestants. After unification, this idea was secularized and instrumentalized by the political elites. Due to religious heterogeneity, however, Prussian Protestantism could not simply be extended to the entire nation as a state religion[1]; at the same time, the elite of the state were predominantly conservative and Protestant. Under Bismarck the state united with the Protestant Evangelical churches (cf. the Lutheran concept of the alliance of state and church) and forced a cultural struggle against the Catholic Church. As a result of the state threat, a Catholic party (Deutsche Zentrumspartei = German Centre Party) was formed, as in France, which remained in existence until the end of the Weimar Republic. The political rift was more pronounced among Catholics and Protestants than between the church and secular

[1]The unification of Germany in 1871 brought together various states and city states in a multi-denominational nation state.

forces. But already in the discussions on the religious regulations of the Constitution of the Weimar Republic, a new line of conflict between religious and secular forces emerged. This then also became apparent in the deliberations on the Basic Law after the Second World War. In general, politics wanted a separation of church and state but ultimately remained dependent on their legitimacy (Minkenberg 2000: 21–48).

After the end of the Second World War, there were far-reaching changes in the relationship between state and church; the symbols of the nation-state were separated from the religious sphere. Also the combination of religion and patriotism awakened negative memories of the era of the Third Reich when people were banished and annihilated because of their religious affiliation; Protestants and Catholics now settled their enmity and jointly founded the Christian Democratic Union (CDU), which was committed to conservative Christian (interdenominational) values. Despite legal separation, various forms of mutual cooperation between church and state developed. Politics and religion respected their mutual independence, and at the same time religion played an important role in many ethical considerations among politicians and voters (Monsma 2000: 81–107). Cooperation between the Catholic and Protestant churches intensified: since the two churches together represented the absolute majority of the German population, it was hardly possible—at least in the early years—theoretically to decide anything against their will. The cooperation of the churches can also be seen from the point of view of the conflict of interests with the social democratic parties and trade unions (Liedhegener 2003: 237–262).[2] In addition, the religious elites, especially in the CDU/CSU (Christian Social Union), together with the Catholic Church—and to a lesser extent also the Protestant Church—fought until the mid-1960s for the program of a "Christianization of society" (Willems 2002: 137–162; Willems 2004: 303–328). Despite a relatively long history of cooperation, some academic voices point to still existing differences between the Catholic and Protestant churches with regard to the theological conception of social order and the sociopolitical activity. On the Protestant side, different positions exist parallel to each other, not least due to the absence of a Magisterium. The internal plurality and differentiation have led to an expansion of sociopolitical activities within Protestantism. By contrast, Catholicism possesses greater potential to voice a more unified position, not least because of the hierarchical structure and the Magisterium. In particular, until the mid-1960s, the Catholic Church was the politically more active and assertive of the two churches. Since the early 1970s, however, in the face of massive social pluralism, the Catholic Church abandoned its campaign of "Christianizing society" and has since confined the political activities to the defense of the main interests and positions of the church (Willems 2007: 316–340; Heinecke 2003: 223–235).[3] Interestingly enough, today there are more theologians in the SPD, Germany's second largest popular party, than in the conservative parties,

[2]Despite continuing competition, there were similarities in many positions (institutional independence, financing, welfare/social services), and they cooperated with the government, but rarely demonstrated this publicly.

[3]Heinecke contrasts a politically active Protestant Church with an apolitical Catholic Church that seeks unity. In general, he paints the picture of a small, passive, withdrawn Catholic Church and a shrinking politically active, critical—in part state/government-loyal—Protestant Church.

because the East German Social Democratic Party of Germany (SPD) was founded in the German Democratic Republic (GDR) by pastors and activists of the Protestant Church.

The attitude of the population toward religion remains divided to this day. On the one hand, there is great skepticism when politicians refer to their personal faith in their actions; this is even viewed critically by the churches. On the other hand, most people consider statements of the churches on certain political questions to be acceptable. The figures show that in Germany there is a widespread trend toward secularization; church attendance and the number of church members are decreasing. Nevertheless, the churches are accorded a high degree of credibility, and the right/obligation to engage politically in certain areas is recognized. This hybrid of separation and cooperation is also reflected in many legal regulations. In Germany, for example, it is compulsory to attend school (in the USA, religious communities can be exempted from compulsory school attendance). In Germany, it is not only the action that is prohibited but also the propagation of hatred against other religions and communities, regardless of the context. In Germany, the state supports ecclesiastical educational institutions (religious instruction at state schools). The protection of life is anchored in the Basic Law (Grundgesetz), but abortion is exempt from punishment. Likewise, the law places marriage under the special protection of the state, and homosexual partnerships are largely legally equal.

3 Debating Embryo Politics in Germany

In Germany the discussion started after new regulations legalized the termination of pregnancy in 1974, a law which was banned by the Federal Constitutional Court just a year later. The law initially defined only the status of the embryo within the womb, but not in vitro. Unlike in the United States of America, in Germany reproduction technologies have been regulated by direct federal legislation. Although IVF was allowed, the legislature adopted numerous restrictions on the creation of embryos (Bundesministerium 1985; Leist 1990). Despite considering the individual circumstances of the mother, in the case of abortion, for a long time a clear preference for the protection of the embryo existed (Eser and Koch 1988; Hoerster: 1995; Kuhlmann 1996). Disability campaign groups, feminists, churches, and members of parliament from CDU, SPD, and the Greens (since 1993 Alliance 90/The Greens) formed a broad alliance against the "exploitation," the "abuse," and the "destruction" of embryos for research. The fierce public discussion at the time ran between the poles of total freedom of reproduction and radical protection of the embryo's life (ESHRE 2007). Their resistance led to a restrictive legislation, which almost prohibits any research with embryos. Violations of the law are punishable by imprisonment for up to 3 or 5 years or fines (Embryo Protection Act = Gesetz zum Schutz von Embryonen (ESchG): 1990). Unlike in the USA where embryonic stem cells are mainly derived from aborted fetuses and embryos surplus to IVF, in Germany the production of human embryonic stem cell lines is prohibited. In particular, the Embryo Protection Act is intended to prevent the misuse of IVF techniques for the production of human embryos and to protect the embryos from use by others, for example, for research purposes. The law defines an embryo as a fertilized, developable egg cell (approximately 24 h after the fusion of the nucleus). This definition is broader than the medical term embryo. The law prohibits, among other things, the transfer of a foreign unfertilized egg, fertilization for a purpose other than pregnancy, and the transfer of more than three embryos within a cycle. The artificial insemination of a surrogate mother or the artificial transfer of an embryo to a surrogate mother is also punishable by law. Also punishable are gender selection (with the exception of serious hereditary diseases) and deliberate artificial alteration of genetic information in embryos intended for fertilization. Reproductive cloning and the creation of chimeras or hybrids are also prohibited and punishable by law. Whether the Embryo Protection Act prohibits "therapeutic cloning" too or whether there is a regulatory gap is today subject of a controversial debate among legal experts. The embryo is also protected under the German Constitution (Grundgesetz). It states that "human dignity is inviolable" and that "everyone has the right to life and inviolability of his person." The German constitution is explicitly focused on respecting and protecting human dignity as a result of Germany's history, in particular the genocide, selective reproduction based on racist theories, and medical experiments of the Third Reich. The concept of human dignity is primarily a result of two sources: the restoration of Christian values in which the holiness of human life is explained by man having been created in the image of God as stated in the Bible and Kant's liberal expression of human dignity, which he derives from man's ability to reason and his autonomy. Germany's particular responsibility to protect human life encompasses the importance of human rights, solidarity with the weak, and acting in the interest of the common good. This broad definition of human rights was used to mark reproductive technologies as anti-democratic, antisocial, and inhuman. Legal intervention by the state seemed necessary to prevent the development of a purely technocratic society. Nonetheless, it also states the freedom to pursue science and research (Eibach 2008: 58–77; Brown 2004: 37–53).

As mentioned above, the debate had become polarized at an early stage and the antagonistic positions left little space for compromise (Damschen and Schönekcer 2003; Ach and Runtenberg 2002). The question of how to regulate the import and use of human embryonic stem cells was a subject of controversial social discussion. Among others, the Commission of the German Bundestag and various bioethics committees of the Federal States, the National Ethics Council, the German Research Foundation, and the German Medical Association contributed to this debate, some of them displaying widely divergent views. While Chancellor Gerhard Schröder, for example, advocated research into stem cells under certain restrictions, as he feared that Germany could suffer economic disadvantages, others, such as President Johannes Rau, vehemently rejected stem cell research. The debate also raised the problem that it is difficult to ban stem cell research if abortion is allowed at the same time.[4] The debate led to the German Stem Cell Act (StZG) of 2002, which was

[4]It was criticised that the ESchG awarded a few hour-old cluster of cells more protection than a fetus in the womb (BMJ 323 7 July 2001).

passed by the Bundestag after a heated debate by a narrow majority. Generally, the bill gave priority to adult stem cells, but the importation of embryonic stem cell lines that were created before 2002 for research purposes was permitted under strict conditions approved by the German parliament. Embryonic stem cells from aborted fetuses may be used for research, as well as foreign-produced stem cell lines that were created before the end of the deadline.[5] The trigger for a renewed debate, and ultimately for amending the Act, came from position papers written by the German Research Foundation (DFG) and the German National Ethics Council in the years 2006 and 2007, respectively. Within a similar heated debate in 2008, the Bundestag again discussed a liberalization of the bill.[6] Supporters and opponents of stem cell research presented their arguments in Berlin. Scientists were quoted as saying: "A lot is happening in Berlin behind the scenes" (Dr Oliver Brüstle). Brüstle and Research Minister Annette Schavan were strongly criticized by the church for their attitude. Four proposals were up for debate: no change, lifting of the import restriction on the date May 2007, lifting of the import restrictions as a whole, and a complete ban on embryonic stem cell research. Minister Schavan (who is considered a conservative Catholic) had previously supported the old regulation and only later advocated a possible postponement of the deadline. The resistance of pro-life groups and the Catholic Church to a change in the law had been enormous in the weeks leading up to its adoption. The political parties had abolished the parliamentary pressure to vote. In the end Minister Schavan voted for a postponement of the deadline; Chancellor Merkel kept a low profile in the run-up to the dispute and then also voted for a postponement of the deadline. 346 voted for a postponement of the deadline, 228 against, and 6 abstained. The proposal to abolish the cutoff date received only 126 votes, a complete ban on stem cell research 118. As a result of pressure from scientists, the act was amended. In April, the German parliament decided to change the stem cell law. The production of stem cells remained prohibited, but the cutoff point moved to 1 May 2007. The usable lines thus increased from 40 to 400 (BMJ 2008a, b). In addition to these criteria, embryonic stem cell lines can only be used for research if they are vital in developing new medical and scientific knowledge. Furthermore, the law created an ethical and regulatory oversight board. Applications to import stem cell lines for research must be approved by the Central Ethics Commission for Stem Cell Research. The German National Ethics Council, set up in 2007, advises and issues opinions to the government and public alike on scientific and medical issues that concern society and human health. The Embryo Protection Act remained as unaffected as the conviction of the majority of the population that

[5]Act ensuring Protection of Embryos in connection with the importation and use of human embryonic stem cells, Stammzellgesetz (StZG –Stem Cell Act), 28 June 2002 at http://bundesrecht.juris.de/stzg/index.html

[6]Act ensuring Protection of Embryos in connection with the importation and use of human embryonic stem cells, Gesetz zur Sicherstellung des Embryonenschutzes im Zusammenhang mit Einfuhr und Verwendung menschlicher embryonaler Stammzellen Stammzellgesetz (StZG—Stem Cell Act), 14 August 2008 at http://bundesrecht.juris.de/stzg/index.html

the state must ensure the protection of the embryo. Cloning is indirectly (through the embryo protection laws) prohibited (Merkel 2002; Robertson 2004: 189–227).

The ethical principles in the law last amended in 2011 remain controversial with regard to their practical consequences. Until a ruling by the Federal Court of Justice in July 2010, it had been assumed that preimplantation diagnostics was banned by the Embryonic Protection Act (Robertson 2004: 189–227; Düwell and Mieht 1998). However, a verdict of the Bundesgerichtshof (6.7.2010)—the Federal Court of Justice of Germany—mentioned that preimplantation genetic diagnosis was not explicitly prohibited by the Act (Krüger and Gollnick 2010: 955–958). Moreover, it stated that when the law was enacted, the legislature did not have in mind the PID developed abroad at that time. This has led to an ongoing political debate on its ban or political regulation, to which the churches have contributed prominently. After the ruling, the Bundestag was forced to enact the Preimplantation Diagnostics Act in 2011 after a long struggle. The new article allowed preimplantation diagnostics (PID) within narrow limits, for example, in the case of a high risk of a serious hereditary disease or a high probability of stillbirth or miscarriage. In addition, the woman had to give her consent in writing after comprehensive clarification and consultation. An interdisciplinary ethics committee and a specially qualified physician in a PID center approved for this had to agree as well.

Many reproductive physicians do not consider the law today any longer up-to-date. They argue that the practice of reproduction is one step ahead of the law. Bioscientists are also in favor of releasing surplus embryos, which have no chance of survival, for research purposes. They would like to get involved in international research in the field of so-called genome editing. This is a technique that will at some point make targeted therapeutic intervention in individual genes possible. Lawyers also criticize the law as anachronistic and vote for a cautious reform in order to adapt it to technical, medical, and biological developments. Embryos should be able to be used at least for high-ranking research purposes. It is also criticized that the law allows sperm donation, but not egg donation. Furthermore, it leads to numerous and risky multiple births. There are various positions within the group of bioethicists: On the one hand, there are still doubts about liberalizing the law, but on the other hand, they state that the law has not fulfilled the legislator's intention—namely, to prevent the creation of surplus embryos. Furthermore, many embryos are now frozen at the pronuclear stage. Although this is legally permissible because they do not yet count as embryos, it is in principle a circumvention of the law. Until today it remains highly controversial, as to whether a liberalization of the Embryo Protection Act is needed (Spiewak 2017; Arp 2017).

4 Religious Positions and Arguments within the German Debate

The particularly difficult ethical question, as mentioned above, concerns the creation or selection of surplus embryos. The moral status of the embryo and the associated worthiness of protection of the prenidative life are questionable and controversial. In the following we will illustrate the reasoning of the Catholic Church and the Protestant Church in Germany in the German biopolitical debate,[7] with the help of the most prominent examples. The churches in Germany have formed a tight and strong alliance advocating restrictive policy recommendations on embryo research, therapeutic, and reproductive cloning and preimplantation diagnostics from the mid-1980s to the debate on stem cell import. They even published a number of joint statements.[8] Before the mid-1980s, at the beginning of the public debate about bio-policy in Germany, the positions on the sides of both churches were quite dissimilar, and in any case there was not yet a complete rejection of IVF or embryonic research for the sake of fundamental progress (Kirchenamt der EKD 1987a, b; Banchoff 2011: 49–50, 101). However, since the Catholic Church committed its Magisterium to a fundamental rejection of embryonic research in Donum Vitae (1987), this position has also been advocated by the Catholic Church in Germany. In 2008 the Catholic Church once again spoke out vehemently against a softening of the stem cell and embryo protection laws. In December a new official teaching letter ("Dignitatis personae. Concerning questions of bioethics") was published which confirms the absolute protection of the embryo. The letter even intensified the "Donum Vitae. On the beginning of human life and the dignity of reproduction" which was published in 1987. The letter described the prenidative embryo as a person fully entitled to the protection of dignity. The earlier version had presented this view in the form of a rhetorical question, leaving very little room for interpretation. Although the Catholic Church took up this position only in modern times (previously the creation of the soul of the embryo was considered to be the 90th day), today it is claimed that the new version has always been represented by the Catholic Church. With this point of view, the Catholic Church justifies a prohibition of PID and embryonic stem cell research. Argumentatively the statement was based on two basic positions: on the one hand, the creationist thesis of the creation of a spiritual soul at the same time as conception (simultaneous instead of gradual ensoulment) and, on the other hand, with the Catholic right of nature,

[7]The "Evangelical Church in Germany" (Evangelische Kirche in Deutschland, EKD) is an umbrella organization formed by 22 Lutheran, Reformed, and United regional churches. The EKD is not itself a "church"—for example, it has no competence in matters of doctrine and no individual members. One of its main joint tasks with which its members have entrusted the EKD is the political representation of German Protestantism.

[8]Kirchenamt der EKD, Sekretariat der deutschen Bischofskonferenz, Gott ist ein Freund des Lebens: Herausforderungen und Aufgaben beim Schutz des Lebens: Gemeinsame Erklärung des Rates der Evangelischen Kirche in Deutschland und der Deutschen Bischofskonferenz, 1989; Kirchenamt der EKD, Sekretariat der deutschen Bischofskonferenz, Wieviel Wissen tut uns gut? Chancen und Risiken der voraussagenden Medizin: Gemeinsames Wort des Rates der Evangelischen Kirche in Deutschland und der Deutschen Bischofskonferenz zur Woche für das Leben 1997, 1997; K. Lehmann, M. Kock, Kirchen bitten um klares Votum für Schutz des Menschen von Anfang an, 2002; W. Huber, K. Lehmann, Für konsequenten Embryonenschutz in Europa: Appell der Vorsitzenden der Deutschen Bischofskonferenz und des Rates der evangelischen Kirche in Deutschland, 2003; W. Huber, K. Lehmann, Klonen ist ein Irrweg: Erklärung des Vorsitzenden der DBK, Kardinal Karl Lehmann, und des Vorsitzenden des Rates der EKD, Bischof Wolfgang Huber, 2004.

according to which a child has the right to be conceived in a natural act within marriage for the sake of his/her dignity.

In its official statements, the Protestant Church now also resolutely rejected research on embryos. The cooperation of both churches ended however with the stem cell debate. The main reason for this lies in the growing plurality of biopolitical positions within the Protestant Church. Since the beginning of the stem cell debate, the various positions have more and more frequently been explicitly named in the declarations by the Protestant Church (Kirchenamt der EKD 2002). In addition a group of protestant theologians has spoken out explicitly against the position of the EKD in the stem cell debate (Anselm 2003b: 197–208). They argued in favor of a limited permission to carry out research on leftover embryos and substantiated their position with an "ethic of healing." They furthermore contested that the argument of the likeness of God allowed deriving statements about the status of the embryo (Anselm 2003a: 47–69). This group of Protestant theologians was also characterized by great skepticism toward the ecumenical cooperation in issues of bio-policy. They particularly criticized the moral absolutism of the Magisterial Catholic bioethics and instead advocated a case-by-case ethics (Kreß 2003: 111–134). In the debate on moving the deadline for the import of stem cells in 2007, the Protestant Church held the position that a single postponement of the deadline was justifiable, while the Catholic Church stood by its rejection of all embryonic research. But in the debate about the statutory regulation of preimplantation diagnostics in 2011, both churches have advocated a ban in separate statements (Pressestelle der EKD 2011; Deutsche Bischofskonferenz 2011). Again, this position was however not undisputed within the EKD (Körtner 2011).

Let us first of all take a look at the way the Protestant Church builds its arguments. The 1987 proclamation by the synod of the Protestant Church in Germany (EKD), entitled "On respect for life—benchmarks for genetic engineering and reproductive medicine" (Kirchenamt der EKD 1987a, b) and published in preparation of the debates for a passage of an embryo protection act, was addressed both to its own church members and to that part of the general public interested in such issues. The proclamation assumes that "the belief in God as Creator, Redeemer and Consummator commits the Church and its members to publicly speak out on this issue and to sharpen people's conscience. The Holy Bible and the belief of Christians form the basis for ethical conclusions even if answers for each and every individual question may not be directly deducible from them. The synod is convinced that not only Christians will be able to agree on these conclusions." In contrast to the discussions on genetic engineering, which above all pointed out possible dangers, the deliberations on embryo research and cloning are based mainly on theological arguments. The central theological argument is that of human dignity substantiated by God's love. This human dignity is also due for "unborn human life starting from the earliest phase of development." As early as with the "unification of ovum and sperm cell, a future person [is] created". The proclamation hence not only advises against the use of in vitro fertilization techniques but also calls both (embryo-destructive) embryo research—even if intended for high-ranking research goals—and cloning as ethically unjustifiable.

The declaration of the EKD Council of 22 May, 2001 "The protection of human embryos must not be limited" (Pressestelle der EKD 2001), i.e., dating to the beginning of the debate about stem cell imports, opposes the "production of human embryonic stem cells for research purposes, the liberalization of the production of embryonic stem cells derived from surplus embryos, but also pre-implantation diagnostics." In this succinct declaration, it is only referred to "human dignity and the human right to life, as anchored in articles 1 and 2 of the German Basic Law," which forbids utilizing human life as a material or a means for other purposes. However, the declaration does refer to the statement "God is a Friend of Life" published jointly by the German Catholic Bishops' Conference and the EKD (together with several other Christian churches) back in 1989 (Kirchenamt der EKD 1989). There, it is explicitly stated in Chap. 4 about "The Particular Dignity of Human Life" that "the likeness of God ... [has become] a central term in the spiritual world of Christendom when describing the special dignity of human life. Also Article 1 paragraph 1 of the German Basic Law stands in this traditional context." However, it is also emphasized there that the content version of the likeness of God also influenced Greek thinking and the philosophy of Immanuel Kant. Greek thinking, so the argument goes, stressed man's rationality, freedom, and individuality, while Kant stressed man's ability for ethical self-determination. But the declaration also makes it clear that the biblical-theological understanding of the likeness of God also contains elements that point beyond Greek thinking and Kant's philosophy. These indications may well be interpreted in such a sense that biblical-theological and philosophical interpretations of human dignity are historically interrelated in such a way that they are mutually connectable. In its most recent intervention in the debate about a legal regulation of preimplantation diagnostics, the Council of the EKD has confirmed its position stated earlier, namely, that this method should be banned (Pressestelle der EKD 2011). The Council justifies this ethical judgment by saying that the selection between worthy and unworthy life linked to PID is not reconcilable with a Christian understanding of human dignity.

Let us now take a look at the argumentation of the Catholic Church. In their word on issues of genetic engineering and biomedicine dated March 2001 (Sekretariat der Deutschen Bischofskonferenz 2001a, b), the German bishops assume that religious traditions offer valuable resources for a bioethical debate because "in them ... [are] contained a wide knowledge and a deep experience of life concerning the ways to handle the world and about how to form them in a way that is conducive to life, which offer ... clear criteria for judgment." They explicitly refer to the encyclical "Fides et ratio," according to which the truths of faith are open to rational reasons, just as those things that are condemned from the point of view of a secular morality are also being condemned by faith. In their statement, they express their position for a ban of preimplantation diagnostics, germline therapy, and therapeutic or reproductive cloning. The basis of their ethical judgment is formed by the biblical view of man according to which the likeness of God constitutes the holiness and non-availability of human life but also its inviolable human dignity due to all, born and unborn, alike. At the same time, human dignity is characterized as the principle which in the sense of an "overlapping consensus" is accepted also by secular rationality as the basis for bioethical reflection. Besides the principle of human dignity, the Catholic bishops however also refer to further arguments like the danger of abuse or the currently still great risks of germline therapy. In their shorter press statements, the Catholic Church uses both theological and secular arguments. So their own positions are always justified by referring to the Biblical-Christian idea of man. But they also always refer to the capacity of discernment of human reason or to constitutional guidelines (Deutsche Bischofskonferenz 2001, 2007). Similar to the EKD Council, the German Bishops' Conference publishes a statement on preimplantation diagnostics, which advocates a clear legislative ban (Deutsche Bischofskonferenz 2011). According to the German Bishops' Conference, from a Christian standpoint, each and every human being is endowed with the status of a personality and human dignity right from the very start. That is why human life is worth protecting without any limitations in each phase of its existence. Moreover, the statement refers to the results of embryo research which according to the statement show that new human life begins by the merging of a human ovum and sperm cell. Altogether, the opposition to human embryonic stem cell research was debated in a much broader sense that was not restricted to the churches' ethical-religious arguments. Instead, part of the debate dealt with defining the general ethical-moral and sociopolitical goals and values of German society. In this discursive environment, the influence of the Christian churches was strong, because they were part of a broad and heterogeneous alliance of groups critical of embryonic stem cell research. While the "rights of the embryo" argument might have immediately split the opposition against stem cell research, the mobilization and combination of a plurality of themes connecting German history with human rights topics discursively created a broad alliance of groups and actors critical of embryonic stem research (Gottweis and Preinsack 2003: 412–432; Gottweis et al. 2009: 105–114).

5 Summary

As in Germany, in almost all Western countries, the separation of religion and politics is regarded as a fundamental principle of democracy. However, for historical reasons, the relationship between the two spheres differs considerably. Political debates in Germany tend to stress the separation. At the same time, the church is a well-respected societal player and through various institutions interwoven with the state. In Germany, the priority has always been the protection of the political from religious dominance. But even in Germany, religion has played an important role in many ethical issues and related public debates, either through direct interventions by the churches or, more indirectly, through the religious beliefs of politicians. The churches still enjoy a high credibility, insisting on their right to intervene in certain public debates, and they are institutionally (e.g., through political parties, governmental commissions) and regularly involved in political decision-making processes.

In Germany, biopolitical debates were held in areas such as reproductive medicine, genetic diagnostics, and stem cell research. In the first wave at the beginning of the 1980s, the opportunities and risks of genetic engineering were discussed and, from the mid-1980s onwards, the discussion was institutionalized in a working group and a commission of inquiry. The second major discussion (on reproductive medicine and embryonic stem cell research) began in 2000. Further ethics committees and commissions of inquiry were set up at the national and the federal level to accompany these discussions. The legal regulations in Germany are restrictive. Germany, including the churches' input, has passed laws that prohibit or regulate biotechnological procedures. At the international level, Germany, together with France was, in 2001–2002, also the driving force behind the UN initiative against reproductive cloning. The most important legal result of the debate on human biotechnologies was the Embryo Protection Act. The provision is a defensive law that follows a guiding principle: The state must set strict limits to fertility medicine from the outset in order to prevent eugenic excesses and the production of designer babies. The law defines the beginning of life—analogous to the teachings of the Christian churches—as the earliest possible time. As soon as egg and sperm cells are fused, sacrosanct human life is created. This regulation meant, among other things, that all embryos conceived in the laboratory must be implanted in the woman, regardless of their ability to survive. In the discussion, which was conducted widely in the media and accompanied by public interviews and hearings, churches, deaconries, and associations for the disabled actively intervened and mostly held restrictive positions. Church representatives were also on the official advisory boards. The main topics were freedom of research versus protection of human dignity and historical reservations (NS period/eugenics). Despite some liberalizations (e.g., postponement of the deadline), the law remained restrictive and is today classified by many actors as in need of revision for various reasons.

Both main churches refer in their arguments to reason and a close connection between faith and reason. The dominant view is that of a faith brightened by reason, which is subjectively assimilated (understood). The German Catholic Church represents the moral teachings of the Catholic hierarchy by using basic arguments of Catholic moral teachings as well as scientific or legally informed nonreligious arguments. Within the Catholic doctrine, the embryo is considered a person with inherent dignity and rights, and to harm the embryo is intrinsically immoral and wrong. Catholic natural law tradition allows Catholic actors to use ostensibly nonreligious arguments. But we have to keep in mind the underlying assumption of a God-ordered world or rather the existence of diachronic and universal law. Due to the cooperation between the Catholic and Protestant Church, the latter adopted some of these positions. Critics within the Protestant group rejected the moral absolutism of the Magisterial and pleaded for a more case-by-case-ethic. Initially the positions were based on religious/theological arguments with reference to God and the Holy Bible. Later documents referred to human dignity and the human right to life (against the background of a Christian inspirational Grundgesetz/Kantian philosophy), too.

The willingness of society to accept religious arguments and actors in public discourse is due to the fact that the church continues to be accorded a social role as a moral authority. Dealing with questions of the meaning of life, solving life problems, and caring for the weak members of the community are seen as historical core competencies of the church. This agreement is independent of the fact that the church is only "salvific important" for a minority of citizens and that religious contents are regarded as irrelevant in most situations. Empirically this leads to a quite high acceptance of church commitments in certain policy fields (human rights, war and peace, euthanasia, marriage and family) as well as a clear support of the "Christian image of man" in the population. The division of roles—on the one hand, an agreement to a sociopolitical positioning of the church and, on the other hand, a desire for a clear separation of tasks between religion and politics—brings with it a number of problems for the churches, which are also reflected in their argumentation structure: Together with other actors, the churches are perceived as part of civil society and can better pursue individual goals in cooperation with it. The problem, however, is that the specific religious/church aspect is in danger of being lost to them, and the churches themselves see themselves less as part of civil society as counterpart and corrective. In view of the separation of religion and politics, it becomes more difficult to promise something specifically Christian on the relevant issues, in the sense of spreading the Christian message. The churches react differently to this challenge: The adaptation of their argumentation structure ranges from a far-reaching secularization over diffuse hybridizations up to religious closure. The Catholic Church has continued its tendency to a selective naturalization of religious arguments, in which certain positions are understood as moral absolutes. Parts of the Protestant Church have adopted these rigid positions without, however, being able to adopt the argumentation structure (because of the different position of the natural law doctrine in Protestantism). Other reformed Protestants have developed hybrid forms in which partly religious and partly secular arguments predominate. That Christianity and its sources do not provide concrete instructions for action for individual questions of genetic engineering is rarely denied by church circles. What is offered are guidelines for leading a good life and examples of correct behavior in morally difficult situations. In the context of the biopolitical debate, the traditions ask how man's relationship to nature and to God is changed by the new technologies and whether fundamental Christian norms are thereby violated. In principle, the churches welcome scientific progress. When it comes to questions of application, however, they see it as their task to ask critical questions and point out the dangers and consequences. To this end, the justification of the aims, the means, and the consequences of genetic engineering are examined.

References

Ach, J., and C. Runtenberg. 2002. *Bioethik: Disziplin und Diskurs: Zur Selbstaufklärung angewandter Ethik*. Frankfurt: Campus.

Anselm, R. 2003a. Die Kunst des Unterscheidens: Theologische Ethik und kirchliche Stellungnahme. In *Streitfall Biomedizin: Urteilsfindung in christlicher Verantwortung*, ed. R. Anselm and U.H.J. Körtner, 47–64. Göttingen: Vandenhoeck & Ruprecht.

———. 2003b. Starre Fronten überwinden. Eine Stellungnahme evangelischer Ethiker zur Debatte um die Embryonenforschung. In *Streitfall Biomedizin: Urteilsfindung in christlicher Verantwortung*, ed. R. Anselm and U.H.J. Körtner, 197–208. Göttingen: Vandenhoeck & Ruprecht.

Arp, D. 2017. Embryonenschutzgesetz: Noch auf der Höhe der Zeit? *Die Zeit*, June 1, 2017.

Banchoff, T. 2011. *Embryo politics: ethics and policy in Atlantic democracies*. Ithaca, NY: Cornell University Press.

BMJ. 2008a. Volume 336, April 19, 2008.

———. 2008b. Volume 336, February 2, 2008.

——— (formerly British Medical Journal). 2001. Volume 323, July 7, 2001.

Brown, E.B. 2004. The dilemmas of German bioethics. *The New Atlantis* 5 (Spring): 37–53.

Bundesministerium für Forschung und Technologie In-Vitro-Fertilisation, Genomanalyse und Gentherapie. 1985. Bericht der gemeinsamen Arbeitsgruppe (Benda-Report), J. Schweitzer. München.

Damschen, G., and D. Schönecker. 2003. *Der moralische Status menschlicher Embryonen*. Berlin: de Gruyter.

Deutsche Bischofskonferenz. 2001. Presseerklärung zu einem Flyer angesichts von "Der Mensch: Sein eigener Schöpfer? Zu Fragen von Gentechnik und Bioethik", March 7, 2001.

———. 2007. Statement des Sekretärs der Deutschen Bischofskonferenz zur Stellungnahme des Nationalen Ethikrates "zur Frage einer Änderung des Stammzellgesetzes", July 16, 2007.

Deutschen Bischofskonferenz. 2011. Stellungnahme der Deutschen Bischofskonferenz zur Präimplantationsdiagnostik (PID). https://www.google.com/url?sa=t&rct=j&q=&esrc=s&source=web&cd=1&ved=2ahUKEwinzqXLi5viAhUOI1AKHUM3BGMQFjAAegQIAhAC&url=https%3A%2F%2Fwww.dbk.de%2Ffileadmin%2Fredaktion%2Fdiverse_downloads%2Fpresse%2F2011-044_Anl_2-PID.pdf&usg=AOvVaw1Q_6yNmbyUcne5u0IXPNaN. Accessed 12 January 2019.

Düwell, M., and D. Mieht. 1998. *Ethik in der Humangenetik: Die neuere Entwicklung der genetischen Frühdiagnostik aus ethischer Perspektive*. Tübingen: Francke.

Eibach, U. 2008. Protection of life and human dignity: The German debate between Christian norms and secular expectations. *Christian Bioethics* 14 (1): 58–77.

Eser, E., and H.G. Koch. 1988. *Schwangerschaftsabbruch im internationalen Vergleich: Rechtliche Regelungen, soziale Rahmenbedingungen, empirische Grunddaten – Europa*. Vol. 1. Baden-Baden: Nomos.

European Society of Human Reproduction and Embryology. 2007. Germany's embryo protection law is 'killing embryos rather than protecting them'. https://www.eurekalert.org/pub_releases/2007-07/esfh-gep070307.php. Accessed 12 January 2019.

Gesetz zum Schutz von Embryonen: Embryonenschutzgesetz. 1990. (ESchG – The Embryo Protection Act). http://www.gesetze-im-internet.de/eschg/BJNR027460990.html. Accessed 13 January 2019.

Gesetz zur Sicherstellung des Embryonenschutzes im Zusammenhang mit Einfuhr und Verwendung menschlicher embryonaler Stammzellen. 2002. Stammzellgesetz (StZG – Stem Cell Act) (Act ensuring protection of Embryos in connection with the importation and use of human embryonic stem cells). http://bundesrecht.juris.de/stzg/index.html. Accessed 14 August 2008.

Gottweis, H., and B. Preinsack. 2003. Religion, Bio-Medizin und Politik. In *Politik und Religion, (Politische Vierteljahresschrift Special Issue 33/2002)*, ed. M. Minkenberg and U. Willems, 412–432. Wiesbaden: Westdeutscher Verlag.

Gottweis, H., B. Salter, and C. Waldby. 2009. *The global politics of human embryonic stem cell science: regenerative medicine in transition*. Houndmills: Palgrave Macmillan.

Heinecke, H. 2003. Konfession und Politik in Deutschland: Ein doppelter Vergleich zwischen evangelischer und katholischer Kirche in der DDR und der Bundesrepublik 1949–1989. In *Religion – Staat – Politik: Zur Rolle der Religion in der nationalen und der internationalen Politik*, ed. M. Brocker, H. Behr, and M. Hildebrandt, 223–235. Wiesbaden: Westdeutscher Verlag.

Hoerster, N. 1995. *Abtreibung im säkularen Staat: Argumente gegen den §218*. Frankfurt: Suhrkamp.

Kirchenamt der Evangelischen Kirche in Deutschland. 1987a. Von der Würde werdenden Lebens: Extrakorporale Befruchtung, Fremdschwangerschaft und genetische Beratung – Eine Handreichung der Evangelischen Kirche in Deutschland zur ethischen Urteilsbildung.

———. 1987b. Zur Achtung vor dem Leben: Maßstäbe für Gentechnik und Fortpflanzungsmedizin: Kundgebung der Synode der EKD.

———. 1989. Sekretariat der Deutschen Bischofskonferenz. Gott ist ein Freund des Lebens: Herausforderungen und Aufgaben beim Schutz des Lebens.

———. 2002. Im Geist der Liebe mit dem Leben umgehen: Argumentationshilfe für aktuelle medizin- und bioethische Fragen.

Körtner, U. H. J. 2011. PID: die evangelische Kirche findet keine klare Linie. *Evangelisch.de*, February 17, 2011. https://www.evangelisch.de/inhalte/103706/17-02-2011/pid-die-evangelische-kirche-findet-keine-klare-linie

Kreß, H. 2003. Ethischer Immobilismus oder rationale Abwägungen? Das Naturrecht angesichts der Probleme des Lebensbeginns. In *Streitfall Biomedizin: Urteilsfindung in christlicher Verantwortung*, ed. R.A. Anselm and U.H.J. Körtner, 111–134. Göttingen: Vandenhoeck & Ruprecht.

———. 2006. Gesundheitsschutz und Embryonenschutz in ethisch rechtlicher Abwägung. Bericht der Bioethik-Kommission Rheinland-Pfalz zum Revisionsbedarf von Embryonenschutz- und Stammzellgesetz. *Ethik in der Medizin* 18: 92–99.

———. 2009. Reproduktionsmedizin und embryonale Stammzellforschung – aktueller Problemstand und gesetzlicher Regelungsbedarf. *Ethik in der Medizin* 2 (21): 135–143.

Krüger, M., and J. Gollnick. 2010. Zur Präimplantationsdiagnostik nach dem Grundsatzurteil des Bundesgerichtshofes. *Der Gynäkologe* 43 (11): 955–958.

Kuhlmann, A. 1996. *Abtreibung und Selbstbestimmung: Die Intervention der Medizin*. Frankfurt: S. Fischer.

Leist, A. 1990. *Eine Frage des Lebens: Ethik der Abtreibung und künstlichen Befruchtung*. Frankfurt: Campus.

Liedhegener, A. 2003. Veränderte politische Option? Kirche und Katholizismus im politischen System der Bundesrepublik Deutschland seit 1989/90. In *Religion – Staat – Politik. Zur Rolle der Religion in der nationalen und der internationalen Politik*, ed. M. Brocker, H. Behr, and M. Hildebrandt, 237–262. Wiesbaden: Westdeutscher Verlag.

Merkel, R. 2002. *Forschungsobjekt Embryo: Verfassungsrechtliche und ethische Grundlagen der Forschung an menschlichen embryonalen Stammzellen*. Munich: Deutscher Taschenbuch Verlag.

Minkenberg, M. 2000. Zivilreligion, Bürgerreligion, Staatsreligion: Anmerkungen zur deutschen Debatte um das rechte Verhältnis von Religion und Politik. In *Religion und Politik. Zwischen Universalismus und Partikularismus*, ed. H.-D. Meyer, M. Minkenberg, and I. Ostner, 21–48. Opladen: Leske + Budrich.

Monsma, S.V. 2000. Kirche und Staat in den USA und der Bundesrepublik Deutschland: auf dem Weg zur Konvergenz? In *Religion und Politik: Zwischen Universalismus und Partikularismus*, ed. H.-D. Meyer, M. Minkenberg, and I. Ostner, 81–107. Opladen: Leske + Budrich.

Pressestelle der EKD. 2001. Der Schutz menschlicher Embryonen darf nicht eingeschränkt werden: Erklärung des Rates der EKD zur aktuellen bioethischen Debatte.

———. 2011. Stellungnahme des Rates der Evangelischen Kirche in Deutschland zur Präimplantationsdiagnostik (PID): "Deine Augen sahen mich, als ich noch nicht bereitet war..." (Psalm 139, 16).

Robbers, G. 2005. Staat und Kirche in der Bundesrepublik Deutschland. In *Staat und Kirche in der Europäischen Union*, ed. G. Robbers, 83–101. Baden-Baden: Nomos.

Robertson, J.A. 2004. Reproductive technology in Germany and the United States: an essay in comparative law and bioethics. *Columbia Journal of Transnational Law* 43 (189): 189–227.

Schöne-Seifert, B., et al. 1995. German-speaking countries and Switzerland. In *The encyclopedia of bioethics*, ed. W.T. Reich, vol. 3, 1627–1639. New York: MacMillan.

Sekretariat der Deutschen Bischofskonferenz. 2001a. Der Mensch: Sein eigener Schöpfer? Wort der Deutschen Bischofskonferenz zu Fragen von Gentechnik und Biomedizin.

———. 2001b. Stellungnahme der Deutschen Bischofskonferenz zur Präimplantationsdiagnostik (PID).

Spiewak, Martin. 2017. Embryonenschutzgesetz: Regelt das endlich! *Die Zeit* 18 (10): 2017.

Stammzellgesetz (StZG – Stem Cell Act). 2002. Act ensuring Protection of Embryos in connection with the importation and use of human embryonic stem cells. http://bundesrecht.juris.de/stzg/index.html. Accessed 18 October 2010.

Voigt, K.D. 2007. Religion und Politik. In *Ein Vergleich zwischen der Situation in Deutschland und den USA*, ed. D. Pruin and R. Schieder, 17–33. Münster: LIT Verlag.

Willems, U. 2002. Religionspolitik in der Bundesrepublik Deutschland 1945–1999: Die politische Regulierung der öffentlichen Stellung von Religion und Religionsgemeinschaften. In *Demokratie und Politik in der Bundesrepublik 1949–1999*, ed. U. Willems, 137–162. Opladen: Leske + Budrich.

———. 2004. Weltanschaulich neutraler Staat, christlich-abendländische Kultur und Laizismus: Zu Struktur und Konsequenzen aktueller religionspolitischer Konflikte in der Bundesrepublik. In *Religion und Politik: Zu Theorie und Praxis des theologisch-politischen Komplexes*, ed. M. Walther, 2303–2328. Baden-Baden: Nomos.

———. 2007. Kirchen. In *Interessenverbände in Deutschland*, ed. T. von Winter and U. Willems, 316–340. Wiesbaden: Verlag für Sozialwissenschaften.

Zachhuber, J. 2007. Introduction. In *Religion and politics in the United States and Germany. Old divisions and new frontiers*, ed. D. Pruin and R. Schieder, 7–16. Münster: LIT Verlag.

Chapter 14
Bioethics and Biopolitics in Switzerland: Stem Cell Research and Preimplantation Diagnostics in the Public Discourse

Monika Bobbert and Yvonne Zelter

Abstract This contribution analyzes Switzerland's public discourse about the Stem Cell Research Act (2004) and the current modification of the Law of Reproductive Medicine concerning preimplantation diagnostics. In Switzerland, citizens have direct influence on the shaping of new laws through plebiscites, referenda, and consultations. Official information material and summarizing protocols were examined in the present study.

The discourse about the moral status of the embryo in which both, proponents and opponents, sought the "human dignity argument," as well as the effort of Switzerland toward a middle compromise position in comparison with European countries, is striking in the two publically discussed changes of law in regard to bioethical questions. Focusing on Christian churches reveals a common rejection of preimplantation diagnosis, while stem cell research was discussed much more controversially.

1 Introduction

Looking back at the last 15 years of public discussion of bioethical issues in Switzerland, two legal changes considering "The beginnings of human life" attract attention: the Stem Cell Research Law[1] introduced in 2004 which also regulates therapeutic cloning and the change of the Reproductive Medicine Act in regard to preimplantation diagnostics.[2] Both laws are based on two central questions,

[1]Cf. Federal Law of Embryonic Stem Cell Research (Stammzellenforschungsgesetz StFG) from December 19th 2003 (as of January 1st 2014).

[2]Currently valid: Federal Law of Reproductive Medicine (Fortpflanzungsmedizingesetz FmedG), from December 18th 1998 (as of January 1st 2013), AS 2000 305; in the future: FMedG as of December 12th 2014, BBI 2015 6301).

M. Bobbert (✉) · Y. Zelter
Faculty of Catholic Theology, University of Münster, Münster, Germany
e-mail: m.bobbert@uni-muenster.de

M. Weiberg-Salzmann, U. Willems (eds.), *Religion and Biopolitics*,
https://doi.org/10.1007/978-3-030-14580-4_14

referencing each other: When does human life begin and what duties to protect exist toward embryonic stem cells and embryos? In addition, both laws are based on the biomedical possibility to cultivate and analyze stem cells and embryos in vitro, i.e., in the laboratory. The actors and arguments of the discussions on stem cell research and preimplantation genetic diagnosis are traced and ethically commented in the following. Other publicly debated bioethical issues during this period cannot be addressed in the context of the present contribution.[3]

In Switzerland, ethical and legal issues of biomedicine are characterized by a particular form of political participation as the citizens have—more directly than in other European countries—the possibility to participate in the design of new regulations and laws via plebiscites, referenda, and consultations. The objective is, in particular through the procedural element of "consultation," that rules and laws are based on a broad consensus.[4]

In the following, the discussion processes around these two laws are illustrated by the historical development, the political and social actors, and by presenting their central arguments. The vote dossiers of the referendums, the reports of consultations, and other information of the Swiss Federal Office for Health (BAG) were used as sources.[5] The results of the analysis of the present contribution and their commentary from an ethical perspective relate only to the sources mentioned and, thus, are limited. In addition, only those actors are represented in the consulted sources, who participated in the form of written reports. And finally, the analysis only refers to documents of the Federal Government. The consultation reports of the BAG not only summarize submitted statements but also select and interpret positions and arguments of consultation participants, i.e., social groups, associations, and institutions. Thus, this chapter will not reflect the wider public debate in Switzerland. The selection of the sources rather illustrates a picture shaped by the Confederation's point of view that reports on socially relevant issues. Beyond the scope of this contribution, it is to examine the documents' underlying presuppositions of "political philosophy," "ethics," "religion," and other premises that ran the interpretation and selection of positions and arguments.

[3]We thank Raina Schreitz and Anna Kohlwey for the translation of this contribution into English.

[4]Cf. for further reading Linder (2005, 241ff).

[5]The following sources were used for the debate about stem cells: Schweizerische Eidgenossenschaft, Abstimmungsdossier (2004); Brunner and Riedo, Externe Evaluation (2011); Bundesamt für Gesundheit, Vernehmlassungsbericht (2002); Bundesamt für Gesundheit, Vernehmlassungsbericht (2004).

The following sources were used for the debate about reproductive medicine: Schweizerische Eidgenossenschaft, Abstimmungsdossier (2015); Schweizerische Eidgenossenschaft, Abstimmungsdossier (2016); Bundesamt für Gesundheit: Vernehmlassungsbericht (2010); Bundesamt für Gesundheit, Vernehmlassungsbericht (2012); Bundesamt für Gesundheit, Vernehmlassungsbericht (2017), Bundesamt für Gesundheit, Faktenblatt Fortpflanzungsmedizingesetz (2015).

Since this anthology highlights the relationship between biopolitics and religion, special attention is given to the analysis of the Swiss documents on the opinions of religious communities. The authors of the present contribution reconstructed and structured the debate on the basis of the selected sources from an ethical perspective. Due to hermeneutic reasons, there is an ethical-evaluative presupposition underlying the analysis and the systematization of the public debate as it appears in the source texts, without whom a problem-oriented order of the statements of political and social actors would be impossible. However, this part of the analysis is mainly descriptive. Finally, the authors of this chapter present their tracing of the political and social actors and arguments to a short commentary from an ethical point of view that feeds from their own observations and the knowledge of relevant literature on those two issues of applied ethics.

2 Characteristics of the Political System of Switzerland

The political scientist Adrian Vatter points out three unique features of the political system of Switzerland[6]: Switzerland neither has a common language, ethnicity, culture, nor religion and, thus, illustrates the heterogeneity of Europe in miniature. The direct democracy could also be considered as a "modern laboratory"[7] which is granted a broad research interest. To understand the following explanations of the creation of laws, the legislative procedure of Switzerland will be explained shortly: In the initiation phase, different actors can provide impetus for a new law. As second step, the consultation process is followed, in which institutions, organizations, cantons, parties, and individuals can take a stand in writing. Then the comments and incorporated changes in the draft will be evaluated, and modifications will be included. This draft is then passed to the National as well as the Council of States for consulting. If there is no agreement, there is an attempt to settle differences. If the Council reaches an agreement, a final vote on the legislative text follows. The law is published and can be put to referendum during a period of 100 days by at least 50,000 signatures.[8] When such a referendum takes place, the citizens receive a vote dossier or a so-called Abstimmungsbüchlein,[9] which presents the Federal Council's concerns associated with the act, as well as its arguments. On the other hand, the vote dossier contains the objections of referendum actors.[10]

---

[6]Cf. Vatter (2016, esp. 34–36).

[7]Vatter (2016, 35).

[8]Cf. Bundeskanzlei (2017, 34–35).

[9]"Voting booklet".

[10]Cf. for further information about direct democracy, Linder, Wolf (2005, 241–300).

3 The Stem Cell Research Act

3.1 Current Legal Situation

The Law on Stem Cell Research "determines the conditions under which human embryonic stem cells may be obtained from supernumerary embryos and used for research purposes."[11] According to the law, it is forbidden to create embryos for research purposes; to intervene in their genetic material; to form clones, chimera, or hybrid; to develop a parthenote, or to import and export them. Furthermore, it is prohibited to use surplus embryos for any purpose other than the extraction of stem cells; to import or export them; to win stem cells after the seventh day of developing from an embryo; or to transfer surplus embryos to a woman.[12] The gratuitousness of access, the information of the couple before the so-called embryo adoption, the independence of the persons involved, the approval of the stem cell extraction by the BAG, as well as the permit requirements for research projects to improve extraction procedures from supernumerary embryos must be met.[13] For a permit, again the aspired research project must satisfy the scientific and ethical requirements listed below and may only be conducted, if knowledge cannot be achieved in any other way.[14] After the extraction of stem cells, the embryo must be destroyed, a report has to be sent to the BAG, and the results of the research project have to be made publicly accessible.[15] The scientific and ethical requirements of the research projects are defined as follows: "A research project with embryonic stem cells is allowed only, if (a) the project should provide significant findings; (b) equivalent results cannot be obtained in any other way; (c) the project complies with the scientific quality requirements; d) the project is ethically defensible."[16]

3.2 History of Origins

The discussion about a new Stem Cell Research Act (StFG) erupted in 2001, when the Swiss National Fund agreed to support a research project with human embryonic

[11]StFG (2003), Art. 1,1; translated by R. Schreitz (Original text in German: "... legt fest, unter welchen Voraussetzungen menschliche embryonale Stammzellen aus überzähligen Embryonen gewonnen und zu Forschungszwecken verwendet werden dürfen.")

[12]Cf. StFG (2003), Art. 3.

[13]Cf. StFG (2003), Art. 4–8.

[14]Cf. StFG (2003), Art. 8.

[15]Cf. StFG (2003), Art. 9.

[16]StFG (2003), Art. 12; translated by R. Schreitz (Original text in German: "Ein Forschungsprojekt mit embryonalen Stammzellen darf nur durchgeführt werden, wenn a) mit dem Projekt wesentliche Erkenntnisse erlangt werden sollen: 1. im Hinblick auf die Feststellung, Behandlung oder Verhinderung schwerer Krankheiten des Menschen, oder 2. über die Entwicklungsbiologie des Menschen; b) gleichwertige Ergebnisse nicht auf einem anderen Weg erlangt werden können; c) das Projekt den wissenschaftlichen Qualitätsanforderungen genügt; und d) das Projekt ethisch vertretbar ist.")

stem cells from abroad. The Department of the Interior was to regulate on behalf of the Federal Council an "Embryonenforschungsgesetz" on how embryonic stem cells and supernumerary human embryos in the future should be dealt with. In the course of the debate, it was decided to rename the act in "Stammzellenforschungsgesetz" (StFG), because surplus embryos were to be used only for stem cell preparation. The first consultation process was opened on May 22, 2002, and offered the first draft of the law. Of the 121 comments received (12 political parties, 5 associations from the economy, 11 religious organizations, 10 from the fields of medicine/pharmaceutics/ genetics, and 40 individuals), 71 players from politics and society welcomed the draft law, 36 refused it. A second version of the law text was adopted on December 12, 2003, by Parliament with 103 to 57 votes; the Council of States stated an overwhelming majority of 35 to 1 vote. In the second consultation process, which was opened in late January 2004, 67 comments were received. They were broadly classified into 4 groups: 33 were in favor of the draft, 18 refused it, 8 statements were neither clearly for it nor against it, and 8 renounced a statement.

Against the law, the "Komitee gegen das Stammzellenforschungsgesetz"[17] took the referendum with 46,000 signatures. Also the "Arbeitsgruppe Stammzellenforschung der Schweizerischen Vereinigung Ja zum Leben"[18] collected 29,500, and the "Basler Appell gegen Gentechnologie"[19] collected 11,600 signatures. Officially, a selection of counter-arguments by the referendum initiators was rendered at the end of the vote dossier, which has the right to inform the general public on the occasion of a referendum by publishing a summary of the briefing paper. The authors of the present contribution have allocated short headings following the aspects mentioned in the dossier on the floating text[20]:

- Ban of turning embryos into purposes
- Species argument in reference to human dignity
- Missed opportunity of embryos' adoption

[17]"Committee against the Law on Stem Cell Research": The organizations of the committee included the "Schweizerische Hilfe für Mutter und Kind, Human Life International Switzerland, Schweizerische Gesellschaft für Bioethik, Vereinigung Katholischer Ärzte der Schweiz, Arbeitsgemeinschaft Evangelischer Ärztinnen und Ärzte" and the organization "Qui a la vie romand".

[18]Working Group stem cell research of the Swiss Association "Yes to Life": The Organization (Stiftung Ja zum Leben) sees itself as "a non-profit public foundation of civil law. It commits to the protection of human life and the rights of unborn children" (translation by R. Schreitz).

[19]"Basel Appeal Against Gene Technology" ("Basler Appell gegen Gentochnologie") was founded in 1988, and has about 1000 members and claims: "[. . .] no patents on life, no release of genetically manipulated organisms, no genetic interaction with the animal, democratic control of the research in gene and reproduction technology, participation of the population for genetic engineering of large projects, no genetic selection and genetic manipulation in humans." (Translated by R. Schreitz.) Homepage of "Basler Appell gegen Gentechnologie: Dafür stehen wir ein", retrieved May 4, 2017, from: http://www.baslerappell.ch/verein/

[20]Cf. Schweizerische Eidgenossenschaft, Abstimmungsdossier (2004, 27).

- Greater success in research with adult stem cells
- Peer pressure on couples to request for "embryo adoption"

The selection of central supportive arguments of the Federal Council for the law at the end of the consultation dossier can be described on the basis of the authors' short headers[21]:

- Hope for sick people through development of new therapies
- Advances in medicine due to research
- Damage to the research location Switzerland in case of permanent ban
- Preferential worthiness of research over destruction of embryos
- Controlled conditions instead of prohibition
- A clear demarcation in the law through a ban on cloning and ban on the creation of embryos for research purposes
- Honesty: research in their own country instead of benefit or depending on the progress of other countries

This summary of the critics' important arguments and the reasons of the Federal Council's endorsement of the Stem Cell Research Act at the end of the consultation dossier represented only a part of the debate. Therefore, a somewhat wider insight into the political debate is given below, taking into account the other sources underlying this post in addition to the consultation dossier.[22]

At this point, it should be noted first that 66.4% of the Swiss were in favor in the referendum on November 28, 2004, of the proposed law, which then came into effect on March 1, 2005. An external evaluation, to be carried out 5 years later, was one of the requirements of the act. On behalf of the BAG, an evaluation was conducted in 2010 by the "Agency Evaluation and Research,"[23] which consisted of three questions: Implementation and enforcement, effectiveness and expediency of the STFG, and international comparison.[24] Although deficiencies had been identified, in particular with regard to the coordination of the STFG and the Reproductive Medicine Act, there was no need for revision.[25] In terms of an international comparison with Sweden (liberal legislation), Switzerland (middle position) hardly achieved results within 5 years[26]: "At the research policy level, the late start in hESC research compared to Sweden is noticeable in the fact, that the research still has few independent publications, hardly has any own stem cell lines, few junior researchers and (still) no therapeutic application projects compared to original research with

[21]Cf. Schweizerische Eidgenossenschaft, Abstimmungsdossier (2004, 28–29).

[22]Cf. Sect. 3.3. of this chapter.

[23]In German terms: Bundesamt für Gesundheit, Fachstelle Evaluation und Forschung.

[24]Cf. Brunner and Riedo, Externe Evaluation (2011, 7).

[25]Cf. Brunner and Riedo, Externe Evaluation (2011, 5).

[26]Cf. Brunner and Riedo, Externe Evaluation (2011, 18).

adult stem cells. In addition, little effort to actively promote hESC research within the framework of the Federal Government's innovation policy has been made."[27]

3.3 Ethical Questions and Arguments in the Debate

The authors of the present contribution identified and structured the following ethical issues and arguments from the public debate on the introduction of the Stem Cell Research Act[28]:

3.3.1 Questionable Benefit of Research with Embryonic Stem Cells

In some places it had been fundamentally questioned whether embryonic stem cell research is necessary at all.[29] For example, the initiators of the referendum drew attention to the fact that research with adult stem cells is likely to achieve a higher success rate than research with embryonic stem cells.[30] Major findings in research with embryonic stem cells were not obtained in the last 20 years. The proponents purported that it was impossible to compare research with embryonic stem cells with research with adult stem cells, and a research ban would harm Switzerland as a research location.[31] According to the consultation dossier, the Federal Council had also lobbied for this kind of research because the hope was to be able to replace damaged cells of people with diabetes, myocardial infarction, Parkinson's disease, or a spinal cord injury.[32] Apparently, in the public debate, little importance was attached to the possible success or lack of it, of embryonic stem cell research. Rather, the "therapy argument" was very strong.

3.3.2 Switzerland's Legal Situation Compared to Other Countries, in Particular Germany and Sweden

In the discussion over the approval of embryonic stem cell research, it can be observed that a comparison with other European countries is frequently drawn.

---

[27]Brunner and Riedo, Externe Evaluation (2011, 18); translated by R. Schreitz. (Original text in German: "Auf forschungspolitischer Ebene macht sich der späte Start in die hESC-Forschung im Vergleich zu Schweden dahin gehend bemerkbar, dass die Forschung noch wenig eigenständige Publikationen, kaum eigene Stammzelllinien, geringen Nachwuchs und (noch) keine therapeutischen Anwendungsvorhaben vorweisen kann im Vergleich zur angestammten Forschung mit adulten Stammzellen. Zudem sind Bemühungen zur aktiven Förderung der hESC-Forschung im Rahmen einer Innovationspolitik des Bundes wenig ausgeprägt.")

[28]"Public debate" has been reconstructed by a synopsis and structuring of the above mentioned sources (see footnote 5).

[29]Cf. Bundesamt für Gesundheit, Vernehmlassungsbericht (2002, 7).

[30]Cf. Schweizerische Eidgenossenschaft, Abstimmungsdossier (2004, 27).

[31]Cf. Bundesamt für Gesundheit, Vernehmlassungsbericht (2002, 4).

[32]Cf. Schweizerische Eidgenossenschaft, Abstimmungsdossier (2004, 28).

This becomes very clear by looking at the external Swiss evaluation by the BAG in 2011. The new Stem Cell Research Law of Switzerland is presented as a law of the "middle" between a relatively restrictive legislation in Germany and Sweden's very liberal legislation.[33] This argument is not ethical in the strict sense. Nevertheless the "middle position" seems to be implicitly valued positively in the public debate in Switzerland.

3.3.3 Requirements of the Import and Export of Embryos and Embryonic Stem Cells

The question of the import and export of embryonic stem cells is asked in particular by law advocates: They argue that an import is ethically inconsistent while there is an existing ban on research using Swiss embryonic stem cells. The same applies to the recourse to research with embryonic stem cells from abroad if the research in their own country remains banned.[34] Before admitting stem cell research, it should first be clarified whether a permit for the import and export should be required and what criteria have to be fulfilled.[35]

3.3.4 Changed Purpose

The licensing opponents responded that with the embryonic stem cell research, the original purpose of a human embryo, namely, "life," would merely become "research." This changed purpose settlement shows an objectification of the embryo.[36] Moreover, they pointed out the risk of the exploitation of the female body if ovum were to be grown for research purposes.

3.3.5 Embryos, Embryonic Stem Cells, and the Paramount Norm of Human Dignity

The licensing opponents repeatedly referred to the aspect of human dignity in arguing the case for the ban of exploiting the embryo[37]: The moral status of the embryo does not depend on research interests or the fact of supernumerary embryos from reproductive medicine's therapies. On the one hand, from the moment of the fertilization of the egg cell, human dignity and, thus, the protection of human life are

[33]Cf. Brunner and Riedo, Externe Evaluation (2011, 18).

[34]Cf. Schweizerische Eidgenossenschaft, Abstimmungsdossier (2004, 29).

[35]Cf. Bundesamt für Gesundheit, Vernehmlassungsbericht (2002, 30–31), Bundesamt für Gesundheit, Vernehmlassungsbericht (2004, 17).

[36]Vgl. Bundesamt für Gesundheit, Vernehmlassungsbericht (2002, 6).

[37]Cf. Schweizerische Eidgenossenschaft, Abstimmungsdossier (2004, 27).

to be ensured.[38] On the other hand, other participants of the consultation considered the use of the concept "human dignity" in the context of human embryos as inadequate. A "respect model" which argues for respect toward early human life would be more favorable.[39]

3.3.6 Possibility of the Adoption of Supernumerary Embryos

In the course of the debate on the instrumentalization of human embryos, the opponents referred to the possibility of "adoption." The supernumerary embryos resulting in reproductive medicine could correspond to its purpose of "life" through an "adoption."[40]

3.3.7 Slippery Slope Arguments

According to some critics, an approval of the embryonic stem cell research would open the floodgates for other problematic techniques, including "therapeutic" cloning in the future.[41] This had already been advocated by some participants of the consultation procedure as an alternative to the development of treatments based on embryonic stem cells. However, it was frequently pointed out that such authorization would be contrary to international standards.[42] Some also saw the future approval of genetic interventions on the embryo as a subsequent risk.[43]

3.3.8 Compatibility with the Reproductive Medicine Act

The preimplantation genetic diagnosis (PGD or PID in German), which was regulated in the Reproductive Medicine Act (FMedG), was still prohibited at the time of the debate on the Law on Stem Cell Research. Thus, some participants of the consultation process warned against the incompatibility with the Reproductive Medicine Act: An approval of embryonic stem cell research with simultaneous prohibition of PGD would be inconsistent.[44] The National Council had spoken several times about maintaining the ban on the PGD.

[38]Cf. Bundesamt für Gesundheit, Vernehmlassungsbericht (2002, 6), Bundesamt für Gesundheit, Vernehmlassungsbericht (2004, 5).

[39]Cf. Bundesamt für Gesundheit, Vernehmlassungsbericht (2002, 16).

[40]Cf. Schweizerische Eidgenossenschaft, Abstimmungsdossier (2004, 27).

[41]Cf. Bundesamt für Gesundheit, Vernehmlassungsbericht (2002, 6).

[42]Cf. Bundesamt für Gesundheit, Vernehmlassungsbericht (2002, 19).

[43]Cf. Schweizerische Eidgenossenschaft, Abstimmungsdossier (2004, 27).

[44]Cf. Bundesamt für Gesundheit, Vernehmlassungsbericht (2002, 6).

3.3.9 Assigning Embryos to Specific Research Projects

The question whether a parent's consent to the release of embryos should be connected with specific research projects was discussed heavily. Those who voted in favor added that this would diminish the risk of commercialization. Opponents claimed that assigning embryos to specific research projects would not make sense from a scientific point of view.[45]

3.3.10 Criteria to Consider when Advising Parents

In the second consultation process, it was focused on the concrete implementation, including the decision of the parents: Might the doctors be biased?[46] Are couples who have supernumerary embryos pressurized by the possibility of the release of embryos and/or embryonic stem cells for research? Should perhaps at least a certain decision deadline be guaranteed? The idea is discussed whether a handout with ethical issues—for example, by the National Ethics Commission (NEK)—should be drawn up and presented to the parents as a basis for decision-making.[47]

3.4 Statements of Religious Communities

At the first consultation phase, a community of the Jewish faith, four Catholic organizations/institutions, four evangelical organizations/institutions, a nondenominational church association, and a theological faculty of a university took part. Out of the 11 participants in the consultation process, 5 welcomed the law, and 6 opposed it. The distribution is not assigned to the two Christian denominations. The Swiss Federation of Jewish Communities welcomed the stem cell research and called for an extension to the "therapeutic" cloning.[48] The latter has been rejected by all other church organizations (Table 14.1).

Of the four Catholic organizations/institutions, the Swiss Association of Catholic Women Farmers and the Christian Catholic Church of Switzerland (Old Catholics) plead for the draft, while the Swiss Association of Catholic Women and the Swiss Bishops' Conference spoke against the template. The Swiss Bishops' Conference was the most prominent institution opposing the bill. The bishops justified their decision-making on different levels: The criticism in general is aimed against the instrumentalization of the embryo.[49] In addition, the consent of the parents is not

[45]Cf. Bundesamt für Gesundheit, Vernehmlassungsbericht (2002, 10).

[46]Cf. Bundesamt für Gesundheit, Vernehmlassungsbericht (2004, 9).

[47]Cf. Bundesamt für Gesundheit, Vernehmlassungsbericht (2004, 10).

[48]Cf. Bundesamt für Gesundheit, Vernehmlassungsbericht (2002, 9).

[49]Cf. Bundesamt für Gesundheit, Vernehmlassungsbericht (2002, 6).

Table 14.1 Draft of the Act to the first consultation in 2002[a]

PRO	CONTRA
Christian Catholic Church of Switzerland	Swiss Federation of Catholic Women
Swiss Federation of Jewish Communities	Association of Protestant Physicians
Swiss Theological Society	Opinion and Discussion Forum of Reformed Protestants
Swiss Association of Catholic Women Farmers	Protestant Woman Federation of Switzerland,
Association of Evangelical Non-Denominational Churches and Communities	Swiss Bishops' Conference
	Federation of Swiss Protestant Churches

[a]Cf. Bundesamt für Gesundheit, Vernehmlassungsbericht (2002, 3) (translated by R. Schreitz). Institution's nomenclature in German: Pro: Christkatholische Kirche Schweiz, Schweizerischer Israelitischer Gemeindebund, Schweizerische Theologische Gesellschaft, Schweizerischer Verband Katholischer Bäuerinnen, Verband Evangelischer Freikirchen und Gemeinden; Contra: Schweizerischer Katholischer Frauenverbund, Arbeitsgemeinschaft Evangelischer Ärzte, Meinungsbild und Diskussionsforum Evangelisch-Reformierter, Evangelischer Frauenverbund Schweiz, Schweizer Bischofskonferenz, Schweizerischer Evangelischer Kirchenbund

sufficient; rather it constitutes a heteronomy of the individual by a third party.[50] The therapy possibilities are also a distant prospect and do not justify research with embryos and embryonic stem cells at the present time. In addition, a debate on the approval of embryonic stem cell research only during the summer months was much too short.[51] Finally, the claim that scientific and ethical standards would be respected must be specified.[52]

The Swiss Association of Catholic Women Farmers welcomed the clear framework conditions and, thus, a high legal certainty.[53] The Swiss Association of Catholic Women expressed great skepticism considering the intended approval of embryonic stem cell research and, if introduced, demanded a narrowly defined scope, as well as a review of the proposals by the NEK.[54] Of the four Protestant organizations/institutions, one welcomed the law, and three refused. The Swiss Federation of Protestant Churches legitimately considered a research with embryos and embryonic stem cells only ethically consistent if the research objective could not be achieved by other means. Many alternatives were currently available.[55] The Association of Protestant Physicians emphasized clearly that this would be an instrumentalization of human life, which cannot be agreed to.[56] Far fewer religious communities took part in the second consultation. Only the Swiss Federation of Protestant Churches and, from the Catholic side, Justitia et Pax and the Swiss Bishops' Conference participated (Table 14.2).

[50]Cf. Bundesamt für Gesundheit, Vernehmlassungsbericht (2002, 26).
[51]Cf. Bundesamt für Gesundheit, Vernehmlassungsbericht (2002, 7).
[52]Cf. Bundesamt für Gesundheit, Vernehmlassungsbericht (2002, 22).
[53]Cf. Bundesamt für Gesundheit, Vernehmlassungsbericht (2002, 4).
[54]Cf. Bundesamt für Gesundheit, Vernehmlassungsbericht (2002, 13, 15, 28).
[55]Cf. Bundesamt für Gesundheit, Vernehmlassungsbericht (2002, 11).
[56]Cf. Bundesamt für Gesundheit, Vernehmlassungsbericht (2002, 6).

Table 14.2 Draft of the Act in the second consultation in 2004[a]

PRO	CONTRA
Federation of Swiss Protestant Churches	Justitia et Pax Swiss Bishops' Conference

[a]Cf. Bundesamt für Gesundheit, Vernehmlassungsbericht (2004, 3). Pro: Schweizerischer Evangelischer Kirchenbund; Contra: Justitia et Pax, Schweizer Bischofskonferenz

After much effort to modulate it, the Swiss Federation of Protestant Churches supported the second draft, as this one would be more considerate of fundamental ethical values and principles unlike before[57]: So the draft argues clearly for a written handout for parents, which would set out the ethical positions and arguments. However, it is still criticized that a reasonable amount of time for the parents must be guaranteed, and a request should be refrained at all if the couples were stuck in a crisis.[58] The Swiss Bishops' Conference stresses again that research with embryonic stem cells could be justified only if knowledge could not be achieved in other ways.[59] The Swiss Bishops' Conference and Justitia et Pax demanded effective control despite their general refusal in the case of the authorization and set the framework in a manner similar to the Swiss Bishops' Conference.[60]

3.5 *Other Participants of the Consultation Process*

There is was a clear division between supporters and opponents of the Stem Cell Research Act: Leading associations of the business[61] (100% consent), medical professional associations and academies, university faculties of medicine and of the sciences, as well as research institutions (approximately 80% consent) faced the opponents of the abovementioned law, which were composed by political parties, religious communities, and citizens' initiatives.[62] Of the parties, 70% rejected the first draft and 60% the second one. The majority of the cantons, in particular the institutions of health care, argued relative consistently for the law. The NEK highlighted positively that the dignity of the embryo and the interests of couples were protected in the second draft of the law.[63]

---

[57]Cf. Bundesamt für Gesundheit, Vernehmlassungsbericht (2004, 4).

[58]Cf. Bundesamt für Gesundheit, Vernehmlassungsbericht (2004, 10).

[59]Cf. Bundesamt für Gesundheit, Vernehmlassungsbericht (2004, 18).

[60]Cf. Bundesamt für Gesundheit, Vernehmlassungsbericht (2004, 15).

[61]Term as used in the Vernehmlassungsbericht (2002, 3) des BAG.

[62]Cf. Bundesamt für Gesundheit, Vernehmlassungsbericht (2002, 3).

[63]Cf. Bundesamt für Gesundheit, Vernehmlassungsbericht (2004, 4).

3.6 Commentary from an Ethical Point of View

The reported range of arguments and counter-arguments is to be commented from an ethical point of view. On the one hand, it deals with inconsistencies or unanswered questions of the Swiss debate, on the other hand with selected ethical issues of embryonic stem cell research, as reflected in the relevant medical and social ethical reflections of the past decades.[64] In the debate, the proponents repeatedly pointed out that research with embryonic stem cells gives rise to new hope for the sick. The hope was further defined by the Federal Council, with regard to certain diseases such as diabetes, myocardial infarction, paraplegia, or Parkinson's disease.[65]

Both, the Federal Council and top business associations[66] and universities, referred to the need for research with embryonic stem cells.[67] A ban would harm Switzerland as a research location. An external evaluation, which was carried out 5 years later, examined the law on the basis of the following criteria: (1) implementation and execution, (2) effectiveness and expediency, and (3) international comparison (especially with Germany and Sweden). Among other things, the results were stated as follows: "At the research policy level, the late start in hESC research compared to Sweden is noticeable in the fact, that the research still has few independent publications, hardly has any own stem cell lines, few junior researchers and (still) no therapeutic application projects compared to original research with adult stem cells. In addition, little effort to actively promote hESC research within the framework of the Federal Government's innovation policy has been made. The expediency of the law in ethical terms is very much linked to the relevant legal bases."[68]

Now, the BAG is commissioned to evaluate the suitability of the law and not the expediency of the research. In this sense it is an "instrumental" evaluation with the leading question: Does the law provide a good basis for embryonic stem cell research? Looking more closely at the abovementioned quotation from the evaluation report, however, a mixture of this instrumental perspective with a perspective on the concrete research success is shown: For example, it is noted that research has

[64]See for the ethical and legal debate among others Devolder, Ethics of Embryonic Stem Cell Research (2015), Düwell, Bioethik (2008), Körtner and Kopetzki (Ed.), Stammzellforschung ethisch und rechtlich (2008); Bender et al. (Eds.), Grenzüberschreitungen kulturell, politisch, religiös (2005); Hilpert (Ed.), Forschung contra Lebensschutz (2009).

[65]Cf. Sect. 3.3.

[66]Terminus cf. Bundesamt für Gesundheit, Vernehmlassungsbericht (2002, 3).

[67]Cf. Schweizerische Eidgenossenschaft, Abstimmungsdossier (2004, 28–29), Bundesamt für Gesundheit, Vernehmlassungsbericht (2002, 4).

[68]Brunner and Riedo, Externe Evaluation (2011, 18); translated by R. Schreitz (Original text in German: "Auf forschungspolitischer Ebene macht sich der späte Start in die hESC-Forschung im Vergleich zu Schweden dahin gehend bemerkbar, dass die Forschung noch wenig eigenständige Publikationen, kaum eigene Stammzelllinien, geringen Nachwuchs und (noch) keine therapeutischen Anwendungsvorhaben vorweisen kann im Vergleich zur angestammten Forschung mit adulten Stammzellen. Zudem sind Bemühungen zur aktiven Förderung der hESC-Forschung im Rahmen einer Innovationspolitik des Bundes wenig ausgeprägt.")

"(still) no therapeutic applications"[69] so far. This extended claim to the evaluation is not surprising in the sense that ultimately an instrumental evaluation of the Stem Cell Research Law cannot be carried out without the anticipation of concrete successes. Thus, the evaluation of the BAG cannot avoid questions about the appropriateness and effectiveness of this type of research, despite the existence of assessing the appropriateness of the Stem Cell Research Act. This observation is relevant to the opponents' serious doubts whether the research objective to develop new therapies for incurable diseases via embryonic stem cells is achievable at all. The opponents had been very skeptical about the results of therapeutic research in the near future, concerning already known research results with embryonic stem cells. Also the promise of healing is important for the consideration of risk and benefit, because not only the goals but also the consulted means, as the research with embryonic stem cells, affect the calculation. It may be possible to justify a problematic means by the prospect of a pioneering result, helpful for many people.

However, if the research success becomes increasingly unlikely, i.e., if there is no realistic prospect of therapies using embryonic stem cells to treat disorders such as Parkinson's disease or certain forms of cross-paralysis, which originally had been regarded as promising in the medium term, then the problem of means is the only remaining one. If there are no positive research results, the ethical debate changes with new tracings. One might also ask why a country comparison is important for Switzerland, why the comparison with Sweden and Germany was chosen, and why a "middle position" seems to be desirable for Switzerland. This argument has repeatedly appeared in the proceedings but lacks justification. From an ethical point of view, a compromise or a position between two extremes is not good per se. However, the question as to which initial and end points are concerned, and what justifies a middle position, is decisive. Certainly, there is an attempt to reach consensus in Swiss society—e.g., through the consultation processes and popular votes—to find regulations which are supported by a broad majority. However, neither the reference framework (e.g., the German-speaking area, political Europe, or all nations globally) is prescribed for the external policy, nor is it a broad consensus or an intermediate position within a reference frame which was selected with good ethical reasons.

Another frequently used argument in the discussion is that—from an ethical point of view—controlled conditions are considered better than exporting embryonic stem cells from abroad without control. But without specification of conditioning criteria and without arguing why this form of control would defuse the given ethical problems or turn them for the better, a "controlled option" cannot be considered as a good solution per se. From an ethical point of view, it must be shown why something which is problematic per se becomes more acceptable by introducing conditions.

[69]Brunner and Riedo, Externe Evaluation (2011, 18); translated by R. Schreitz (Original text in German: "(noch) keine therapeutischen Anwendungsvorhaben vorweisen.")

Many proponents added that research with supernumerary embryos is better than their destruction. The idea of not throwing something away but putting it to use otherwise seems logical at first glance. However, human embryos are not things and therefore a comparison with usefulness and sustainability considerations of everyday life, and ecology is not reasonable. The question of an "exploitation" or purpose which is justifiable from an ethical point of view is still not answered. Eventually an already existing ethical problem, the question of the creation of surplus embryos and the resulting additional ethical problems, is faded out. The question whether it is justifiable to produce embryos that are not implanted should be focused on again.

Finally, in the debate it can be observed that, from an ethical point of view, important arguments, which had only been put forward by individual actors, were not taken up—for example, the risk of a heteronomy of embryos through a third party and their interests—as soon as they are no longer used for the purpose of reproduction. Furthermore, it is quite disputed to what extent a couple may dispose of their embryos in whatever way, for example, when it comes to the question of the purpose.

Another aspect must be mentioned in regard to religious communities and their positions in regard to bioethics: In the cases analyzed, Christian religious communities had no uniform position; in fact, moral dissent already existed within religious communities or Christian confessions. Empirical reports on strong religious moral persuasions are frequently queried and may be used as information about a social group of political interest. It must, however, be considered that quite often different moral views are represented on the empirical level of moral beliefs within a religious community. Moreover, moral dissent can have different reasons, and dealing with it may vary within various religions. Instead of only reporting on moral persuasions, it would be more helpful to look for the ethical reasons which are given on behalf of religious statements.

From a metaethical point of view, it is significant, whether Christian, Jewish, or Muslim ethics are based on a cognitivist or noncognitivist understanding of ethics. In terms of a cognitivist understanding of ethics, morality and ethics are distinguished, and thus, moral beliefs and moral practice are descriptive, and ethical reflection with reasonable ethical norms and judgments is prescriptive. In light of the author's understanding of theological ethics as "autonomous morality in the Christian context,"[70] which refers to philosophical concepts and methods, including rational reasons, the need to exploit ethically relevant issues, internal scientific disagreements (e.g., relating to pluri- and totipotency), and the examination of reasons and objections on this basis relating to embryonic stem cell research is shown.[71] In this respect, not the position itself but rather the reasons and objections of religious communities or denominations are of interest.

[70]This is the authors' concept of Christian ethics and also a major consensus among ethicists within Christian theology; cf. in few words Bobbert, Christlich motiviert—vernünftig begründet (2016), see in detail Bobbert and Mieth, Das christliche Proprium der Ethik (2015).

[71]Cf. Bobbert, Urteile in der (Bio-)Medizin- und Pflegeethik sind "gemischte" Urteile (2015), 299–306.

The argument of the proponents of the law to provide legal certainty is indirectly relevant from an ethical point of view. Rules of procedure and legal certainty may be interesting from a perspective of social ethics but have again to be consulted and justified in relation to its material content. The same applies to the "negotiating argument," though ethically not directly relevant, that stricter requirements in the legislative procedure have been made and the acceptance of the second draft paid tribute to this concession.[72] This aspect refers to the quest for an inner Swiss compromise. Even if a consensus system is welcomed sociopolitically, it has yet to discuss how the two poles can be justified from an ethical point of view, as a compromise cannot be considered as correct and good in general.[73]

4 Preimplantation Genetic Diagnosis

4.1 Current Legal Situation

The Reproductive Medicine Act (FMedG) of 1998 was modified in 2017 in favor of approval of preimplantation genetic diagnosis (PGD or PID in German).[74] Article 5 of the FMedG sets the conditions of the admissibility of an examination of the embryos produced in vitro prior to its implantation and its subsequent selection by gender or other characteristics. The PGD is allowed if:

> (a) the risk that an embryo with a hereditary predisposition to a serious disease settles in the uterus, cannot otherwise be averted; (b) it is likely that this severe disease occurs before the age of 50; (c) there is no effective and appropriate therapy available to combat the serious illness; and (d) the couple declares in writing to the doctor that the risk according to (a) is unreasonable.[75]

Furthermore, the law states that a PGD may only be performed, if the couple had given written consent for such a diagnostics, after having been informed by the physician.[76] This also applies for the reactivation of conserved germ cells, as well as

[72]Cf. Bundesamt für Gesundheit, Vernehmlassungsbericht (2004, 4).

[73]Cf. Mieth, Christliche Überzeugung und gesellschaftlicher Kompromiß (1984).

[74]Applicable law: FMedG (1998); the statements refer to the fourth law edition, which was proposed and voted for in the referendum 2016 (Swiss Confederation, Voting Dossier 2016), and which will come into force in 2017: FMedG (2014).

[75]FMedG (2014), Art. 5a (2); translation by R. Schreitz (Original text in German: "a) die Gefahr, dass sich ein Embryo mit einer vererbbaren Veranlagung für eine schwere Krankheit in der Gebärmutter einnistet, anders nicht abgewendet werden kann; b) es wahrscheinlich ist, dass die schwere Krankheit vor dem 50. Lebensjahr ausbrechen wird; c) keine wirksame und zweckmäßige Therapie zur Bekämpfung der schweren Krankheit zur Verfügung steht; und d) das Paar gegenüber der Ärztin oder dem Arzt schriftlich geltend macht, dass ihm die Gefahr nach Buchstabe a nicht zumutbar ist.")

[76]Cf. FMedG (2014), Art. 5b/6a.

carrying out a multiple pregnancy, if multiple embryos are implanted.[77] The number of embryos produced is increased from a maximum of 3–12. A granting authority takes the place of a supervision of the procedure.[78] The improper and therefore legally invalid use of the embryo is stated as follows[79]:

- Development outside of the woman's body, although there is possibility of implantation
- Surrogacy
- Interventions in the human germline
- Selection of the germ cells to other properties
- Making a clone, hybrid, or a chimera

4.2 *History of Origins*

In 2001 the PGD was prohibited in accordance with the FMedG Article 5, paragraph 3. The regulation in 2001 and the related prohibition have not ended the discussion, but the PGD has remained in the public debate.[80] In total three consultation phases (2009, 2011, 2017) and two referendums (14.06.2015 and 05.06.2016) have taken place. In the last referendum, 62.4% of the Swiss voted for the Fourth Template for Amending the Reproductive Medicine Act.[81] Starting in autumn 2017, the PGD will be permitted in Switzerland under certain conditions.

The first draft of the amendment limited the PGD to "medical use" in the sense that it would have only been allowed for couples, who carry genetically inherited diseases, which are classified as severe.[82] These couples could get a child in the natural way but—due to a genetic disease in the family—would undergo in vitro fertilization (IVF) in order to first examine fertilized embryos, which are then implanted after the preimplantation diagnosis. Childless couples who performed artificial insemination (IVF) to have a baby with the help of reproductive medicine were not admitted. This changed in the course of public discourse and modification of the law. Eventually also involuntarily childless couples who underwent IVF treatment could be allowed to access to PGD—at least to the aneuploidy procedure, i.e., the study of embryos in the laboratory on chromosome disorders.

Of the 92 comments received in the first phase of the consultation process (19.02. to 18.05.2009), 22% rejected the law in principle, 28% agreed, and 50% were for the PGD but against the actual drafts of the bill. The latter claimed, in particular, that the

[77]Cf. FMedG (2014), Art. 5b.

[78]Cfl. FMedG (2014), Art. 12.

[79]Cfl. FMedG (2014), Art. 30–36.

[80]Cf. e.g. the report of Ackermann (2005).

[81]Cf. Schweizerische Bundeskanzlei, Volksabstimmung vom 05.06.2016.

[82]In Germany, the PID is limited to this group of couples.

conditions of the bill were still too strict: It is held at the rule of three, i.e., the transfer of a maximum of three embryos, as well as the cryopreservation and the aneuploidy screening, i.e., the study of embryos on chromosome disorders was still not allowed.[83] A second consultation (29.06 to 30.09.2011) received 87 submissions. 18% voted in favor of the text, 61% in general voted in favor but still demanded changes, and 21% still opposed the law. Again, options for amendment concerned the extension of the rule of three and the approval of aneuploidy screenings and approval of HLA-typing of embryos for the purpose of a subsequent cell- or tissue donation for a sick sibling (so-called savior baby).[84]

In December 2014, the Federal Council and Parliament adopted the draft of the FMedG authorizing the PGD. To introduce the law, a vote on the amendment to Article 119 of the Constitution had to be made first. Therefore, in the future no longer only so many oocytes should be produced as "can be implanted immediately" but "[...]only so many oocytes may be developed into embryos outside the body of the woman, as are necessary for the reproduction process."[85] The vote on the amendment of Article 119 for the change of the FMedG took place on June 14, 2015. The Federal Council submitted the following arguments for approval in the consultation dossier[86]:

- Enabling fertility despite difficult circumstances
- Protecting couples confronted with difficult choices
- Prevention of reproductive tourism
- Respect for human dignity and the protection of the embryo

The Swiss people approved the amendment of the Federal Council with 61.9%, so the vote was considered to be paving the way for the introduction of the PGD in the FMedG. Against the legislative proposal, the "Zusammenschluss für eine solidarische Gesellschaft ohne Normierungszwänge—Vielfalt statt Selektion"[87]

---

[83]Cf. Bundesamt für Gesundheit: Vernehmlassungsbericht (2010, 3).

[84]Cf. Bundesamt für Gesundheit, Vernehmlassungsbericht (2012, 4–5).

[85]Bundesverfassung der Schweizerischen Eidgenossenschaft (BV) on the 18. April 1999 (On the 12th of February 2017), AS 1999 2556, Art. 119 (Fortpflanzungsmedizin und Gentechnologie im Humanbereich), 2c; translated by R. Schreitz. (Original text in German: "sofort eingepflanzt werden können", sondern: "Es dürfen nur so viele menschliche Eizellen ausserhalb des Körpers der Frau zu Embryonen entwickelt werden, als für das Fortpflanzungsverfahren notwendig sind.")

[86]Cf. Schweizerische Eidgenossenschaft, Abstimmungsdossier (2015, 12–13).

[87]Association for a solidarity-based society without standardization constraints—diversity instead of selection; The Alliance included: AGILE.CH—Die Organisation für Menschen mit Behinderung, anthrosana—Verein für anthroposophisch erweitertes Heilwesen, appella Telefon- und Onlineberatung, Behindertenseelsorge Katholische Kirche im Kanton Zürich, biorespect—Wir hinterfragen Biotechnik, Vereinigung Cerebral Schweiz, Dialog Ethik, Evangelische Frauen Schweiz, Frauen Ambulatorium pract.med. Theres Blöchlinger, Gruppenpraxis Paradies—ein Ort von Frauen für Frauen, indieme Schweiz, Parentela CH, Procap für Menschen mit Handicap, Schweizerischer Hebammenverband Selektion Bern, selbstbestimmung.ch, Selbsthilfe Intersexualität, Schweizerischer Katholischer Frauenbund, Verband für Anthroposophische Heilpädagogik und Sozialtherapie Schweiz.

took the referendum with 60,000 signatures. The arguments of the referendum committee "PID stoppen" were[88]:

- Risk of losing solidarity with disabled people through a change in the law
- No medico-reproductive advantage for infertile couples by PGD (cf. relevant studies)
- Facilitating a broad social debate through a "no" to changes in the law

On June 05, 2016, the people voted with 62.4% of all votes for the PGD. In addition, an adjustment of the reproductive medicine regulation was required in the future law introduction of PGD.[89] In an optional last consultation process (September 26, 2016, to January 09, 2017),[90] the participants gave their comments and opinions on the concrete execution of the law.[91]

4.3 *Ethical Questions and Arguments in the Discussion*

The pro and contra arguments compiled at the conclusion of the vote dossier, to be communicated to the general public as part of the voting process, were quite short—considering the complex medico- and socio-ethical discussions in the past two decades.[92] But also, in public debates in Switzerland, as they refer to the appropriate sources, the following questions and ethical arguments were put forward in the dispute over the admission of PGD:

4.3.1 Preimplantation Genetic Diagnosis for Couples with Severe Hereditary Diseases and also for Involuntarily Childless Couples?

Couples with a serious hereditary disease in the family are a group, which has prior experience with a particular disease. The probability of inheriting a specific genetic

[88]"Stopping PID"; Cf. Schweizerische Eidgenossenschaft, Abstimmungsdossier (2016, 37).

[89]Fortpflanzungsmedizinverordnung (FMedV) of the 4 December 2000 (On the 1 January 2017), AS 2000 3068.

[90]Bundesamt für Gesundheit, Vernehmlassungsbericht (2017).

[91]Within the framework of this contribution, the details of this consultation procedure cannot be mentioned.

[92]Cf., e.g., Francis (Ed.), Handbook of Reproductive Ethics (2017); Bobbert, Auswahlkriterien Präimplantationsdiagnostik (2014); Bobbert, Preimplantation Genetic Diagnosis as Non-Equality (2005); Maio et al., Kinderwunsch und Reproduktionsmedizin (2013); Haker, Ethik der Pränatal- und Präimplantationsdiagnostik (2011); Hildt and Mieth (Eds.), IVF—Medical, Social, Ethical Evaluation; Graumann, Präimplantationsdiagnostik und embryonale Stammzellforschung (2004); Kollek, Präimplantationsdiagnostik und weibliche Autonomie (2002); Schöne-Seifert, Präimplantationsdiagnostik und Entscheidungsautonomie (1999); Schweizerische Gesellschaft für Biomedizinische Ethik, Präimplantationsdiagnostik (2008).

disorder that has often to be examined by molecular genetic methods is given. To perform a PGD, couples who actually could have children naturally have to undergo artificial fertilization. This is a small group, and the decision concerning artificial insemination and PGD is not clear in the beginning.

There is also a group of a very different magnitude, namely, the involuntarily childless couples who undergo IVF treatment. To allow the PGD for them means introducing genetic screening into the reproductive medical procedure to exclude selected diseases, which society considers as problematic and deficient. Proponents of PGD for this group of couples argued that the PGD significantly reduced the risk of the transmission of hereditary diseases and thus purports a high moral goal, namely, health.[93] On the other hand, some opponents questioned the implicit understanding of health and disease. An approval of aneuploidy screening, i.e., testing for chromosomal abnormalities, was supposed to be indispensable as additional support for infertility treatment.[94] A PGD was said to increase the success rate (i.e., the percentage of embryos brought to term), since miscarriages were less frequent after a PGD. However, the opponents argued that according to studies, infertile couples do not get children more often using PID. [95] In addition, there were still no medical research results about possible long-term damages.[96] Several participants of the consultation process demanded accompanying research to examine possible long-term consequences.

4.3.2 Lower Risks for Mother and Child?

Proponents argued that the PGD significantly increased the probability of a complication-free pregnancy.[97] In addition, a "single-embryo transfer" would be a relief for many women by extending the rule of threes per cycle to 12 (through the amendment of Article 119 Federal Constitution) and the possible selection of the "healthiest" embryos. Thus, the probability of a multiple pregnancy could be reduced significantly.[98] The rule of threes and the prohibition on cryopreservation were set aside in the course of the proceedings. In the first phase of the consultation process, the Federal Council held it;[99] in the following process, the majority of consultation participants supported a removal of the rule of threes—not least, the National Ethics Commission (NEK).[100]

---

[93]Cf. Schweizerische Eidgenossenschaft, Abstimmungsdossier (2015, 12).

[94]Cf. Bundesamt für Gesundheit: Vernehmlassungsbericht (2010, 5).

[95]Cf. Schweizerische Eidgenossenschaft, Abstimmungsdossier (2016, 37).

[96]Cf. Bundesamt für Gesundheit, Vernehmlassungsbericht (2017, 18).

[97]Cf. Schweizerische Eidgenossenschaft, Abstimmungsdossier (2016, 34).

[98]Cf. Schweizerische Eidgenossenschaft, Abstimmungsdossier (2015, 7).

[99]Cf. Bundesamt für Gesundheit, Vernehmlassungsbericht (2010, 5).

[100]Cf. Bundesamt für Gesundheit, Vernehmlassungsbericht (2010, 14).

4.3.3 Admission and Duration of Cryopreservation of Embryos

Already in the first consultation process, proponents of cryopreservation stated that their claim was only the "applicable medical standard."[101] An introduction of PGD without the opportunity of cryopreservation would not be reasonable due to current medical standards.[102] The duration of the cryopreservation was also discussed: 5 years were adopted as standard, but an extension of another 5 years was possible.[103]

4.3.4 Enabling a Difficult Decision Before the Beginning of a Pregnancy: PGD Instead of PND?

In the discussion the PGD has often been compared to the prenatal diagnosis (PND): It was stated that PND is allowed and that conspicuous genetic findings often resulted in the termination of pregnancy. It was argued that it would be inconsistent to prohibit the PGD and to allow the PND, if a couple did not want a child with a severe genetic disorder.[104] When asked about the PGD control, individual participants in the first and second consultation added that the control requirements for the PGD should not exceed those of the PND.[105]

4.3.5 Approval of the Preimplantation Diagnosis in Connection with "Severe" Genetic Diseases?

There disagreement whether a specific list of diseases should determine admission to the PGD or what kind of criteria could mark a "severe" hereditary disease.[106] Some participants of the first consultation argued that there was no definition for the term "severe disease." Mentioning specific diseases was said to be similar to eugenic methods.[107] In the course of the third consultation process, discussions were led by the Federal Council, which defined a "severe" disease: intensive pain, severe restrictions of motor skills, dependence which go beyond childhood, severe limitations in cognition or severe mental disease, serious limitations of affected regulation, and continuous dependency on large equipment (e.g., oxygen-supply devise).[108]

---

[101]Cf. Bundesamt für Gesundheit, Vernehmlassungsbericht (2010, 15).

[102]Cf. Bundesamt für Gesundheit, Vernehmlassungsbericht (2010, 15).

[103]Cf. Bundesamt für Gesundheit, Faktenblatt Fortpflanzungsmedizingesetz (2015).

[104]Cf. Bundesamt für Gesundheit, Vernehmlassungsbericht (2010, 7); Schweizerische Eidgenossenschaft, Abstimmungsdossier (2016, 35).

[105]Cf. Bundesamt für Gesundheit, Vernehmlassungsbericht (2010, 11); Bundesamt für Gesundheit, Vernehmlassungsbericht (2012, 6/15).

[106]Cf. Bundesamt für Gesundheit, Vernehmlassungsbericht (2010, 7).

[107]Cf. Bundesamt für Gesundheit, Vernehmlassungsbericht (2010, 4).

[108]Cf. Bundesamt für Gesundheit, Faktenblatt Fortpflanzungsmedizingesetz (2015).

4.3.6 Preimplantation Diagnosis as Discrimination Against People with Disabilities?

On the part of the PGD opponents, the accusation was repeatedly made that an introduction of PGD questions solidarity with people with a disability.[109] The Federal Council, as a supporter of the law, countered that most disabilities were not genetically derived and thus, the PGD would not discriminate generally against people with disabilities.[110]

4.3.7 Extending the Selection Boundaries Through Introduction of PGD

The legislative proposals in the consultation processes provided a possible list of "severe" genetic diseases regarding the selection of embryos, but for some participants the boundaries were defined too strictly. They demanded the approval of PGD under certain constraints—e.g., approval of the NEK—which means a procedural control; and they demanded allowance of HLA matching, to select a donor child from several embryos which would be suitable for a sibling.[111] It was proposed that the HLA typing be examined in each individual case, and thus, would not necessarily mean an instrumentalization of human life.[112] Critics feared the expansion of the selection criteria on gender, or other desired characteristics, and also included the aspect of HLA matching to the latter.

4.3.8 Discrimination Against Women Through Egg Donation?

In the discussion about the introduction of PGD, the question of egg donation was also discussed. In the second consultation procedure, 17 participants added that the ban on egg donation constituted discrimination against women. Reference was made to the Court of Justice of the European Court of Human Rights (2010). In addition, it was claimed that egg donation is an internationally accepted medical practice that in particular benefits women who lose their fertility, e.g., due to cancer.[113]

4.3.9 Avoiding Reproductive Tourism

In the popular vote in 2015 as well as in the year 2016, the Federal Council led the argument of reproduction tourism. Due to the permission of PGD in all neighboring

[109]Cf. Schweizerische Eidgenossenschaft, Abstimmungsdossier (2016, 37).

[110]Cf. Schweizerische Eidgenossenschaft, Abstimmungsdossier (2016, 39).

[111]Cf. Bundesamt für Gesundheit, Vernehmlassungsbericht (2010, 8).

[112]Cf. Bundesamt für Gesundheit; Vernehmlassungsbericht (2012, 13).

[113]Cf. Bundesamt für Gesundheit, Vernehmlassungsbericht (2012, 23).

European countries (under various constraints), the possibility should also be offered to affected couples in Switzerland.[114]

4.3.10 Endangering Human Dignity

Both, law proponents and opponents, discussed the concept of human dignity, but the argument was used in different ways. The opponents of the law purported that the PGD was not compatible with human dignity.[115] But the Federal Council, as supporter of the PGD, also referred to human dignity: "It is, however, forbidden to select embryos on the basis of gender, physical characteristics or their suitability as a stem cell donor for sick siblings. In addition, the number of embryos that can be developed per treatment remains limited. Thus, the Federal Council considers human dignity and the protection of the embryo to be safeguarded."[116]

4.3.11 Structural Framework

In the deliberations on the introduction of the PGD, the framework conditions were also discussed. The number of PGD centers Switzerland should have was one factor. In the sense of a quality assurance, some participants in the consultation process in 2010 and 2012 suggested that the number of centers should be reduced from between five and ten to between one and three.[117] It was also discussed whether the health insurance companies should pay for the costs of the proceedings: Some participants of the consultation process demanded that the costs should be covered by health insurance on the grounds that a PND and a subsequent pregnancy termination would also be borne. There was also the argument that social inequality would otherwise be strengthened, and women or couples from certain social strata would be disadvantaged.[118]

---

[114]Cf. Schweizerische Eidgenossenschaft, Abstimmungsdossier (2015, 13); Schweizerische Eidgenossenschaft, Abstimmungsdossier (2016), 36. Cf. this factually inaccurate aspect below; see Sect. 4.3.

[115]Cf. Bundesamt für Gesundheit, Vernehmlassungsbericht (2010, 4).

[116]Schweizerische Eidgenossenschaft, Abstimmungsdossier (2015, 13); translated by R. Schreitz (Original text in German: "Es bleibt aber verboten, Embryonen aufgrund des Geschlechts, anderer Körpermerkmale oder aufgrund der Eignung als Stammzellenspender für kranke Geschwister auszuwählen. Zudem bleibt die Zahl von Embryonen begrenzt, die pro Behandlung entwickelt werden dürfen. Damit sind aus Sicht des Bundesrates die Menschenwürde und der Schutz des Embryos gewahrt.")

[117]Cf. Bundesamt für Gesundheit, Vernehmlassungsbericht (2010, 16); Bundesamt für Gesundheit, Vernehmlassungsbericht (2012, 24).

[118]Cf. Bundesamt für Gesundheit, Vernehmlassungsbericht (2010, 17); Bundesamt für Gesundheit, Vernehmlassungsverfahren (2012, 24).

4.4 *Statements of Religious Communities*

In the first consultation process, only the Swiss Federation of Jewish Communities had no reservations about the bill. The Swiss Catholic Women's League agreed in general terms to the submission, but submitted amendments—among others the deletion of further fields of application.[119] Furthermore, the consultation should be holistic,[120] and an expert commission should decide on each PGD.[121] The rule of threes had to be annulled.[122] The number of PID centers should be kept to a minimum (Table 14.3).[123]

The Swiss Bishops' Conference and the Swiss Federation of Protestant Churches were of the opinion that the PGD is contrary to human dignity and the right to life. The former also feared further expansions with eugenic purpose.[124] The rule of threes should be maintained and the number of PGD centers limited (Table 14.4).[125]

The second consultation process showed a clear position of the Christian churches against the PGD bill. Only the Swiss Federation of Jewish Communities welcomed the law on all points.[126]

Article 119 of the Federal Constitution of the Swiss Confederation (BV) was rejected by all Christian organizations.[127] In this context, the Swiss Evangelical Alliance called for a "constitutional legal clarification of the status of embryos."[128] They also called for mandatory accompanying research.[129] The Swiss Evangelical

Table 14.3 The draft law in the first consultation process of 2010[a]

PRO	CONTRA
Swiss Federation of Jewish Communities Swiss Catholic Women's League (with reservations)	Swiss Bishops' Conference Federation of Swiss Protestant Churches Swiss White Cross Association of Evangelical Non-Denominational Churches and Parishes in Switzerland

[a]Translated by R. Schreitz. (Nomenclature in German: Pro: Schweizerischer Israelitischer Gemeindebund, Schweizerischer Katholischer Frauenbund (mit Vorbehalten); Contra: Bioethikkommission der Schweizerischen Bischofskonferenz, Schweizerischer Evangelischer Kirchenbund, Schweizerisches Weißes Kreuz, Verband Evangelischer Freikirchen und Gemeinden in der Schweiz.)

[119]Cf. Bundesamt für Gesundheit, Vernehmlassungsverfahren (2010, 8).

[120]Cf. Bundesamt für Gesundheit, Vernehmlassungsverfahren (2010, 10).

[121]Cf. Bundesamt für Gesundheit, Vernehmlassungsbericht (2010, 12).

[122]Cf. Bundesamt für Gesundheit, Vernehmlassungsbericht (2010, 14).

[123]Cf. Bundesamt für Gesundheit, Vernehmlassungsbericht (2010, 16).

[124]Cf. Bundesamt für Gesundheit, Vernehmlassungsbericht (2010, 4).

[125]Cf. Bundesamt für Gesundheit, Vernehmlassungsbericht (2010, 15/16).

[126]Cf. Bundesamt für Gesundheit, Vernehmlassungsbericht (2012, 8, 11, 15).

[127]Cf. Bundesamt für Gesundheit, Vernehmlassungsbericht (2012, 8).

[128]Bundesamt für Gesundheit, Vernehmlassungsbericht (2012, 8); translated by R. Schreitz (Original text in German: "(verfassungs-) rechtliche Klärung des Status' von Embryonen.")

[129]Cf. Bundesamt für Gesundheit, Vernehmlassungsbericht (2012, 24).

Table 14.4 The draft law in the second consultation of 2012[a]

PRO	CONTRA
Swiss Federation of Jewish Communities	Federation of Swiss Protestant Churches Association of Evangelical Non-Denominational Churches and Parishes in Switzerland The United Methodist Church Swiss Evangelical Alliance Swiss Bishops' Conference

[a]Pro: Schweizerischer Israelitischer Gemeindebund; Contra: Schweizerischer Evangelischer Kirchenbund, Verband evangelischer Freikirchen und Gemeinden in der Schweiz, Evangelisch-methodistische Kirche, Schweizerische evangelische Allianz, Schweizerische Bischofskonferenz

Alliance advocated consultation by an independent person[130]; the Swiss Bishops' Conference generally regarded the indicative regulations as too liberal.[131]

4.5 *Other Participants in the Consultation Process*

The results of the first two consultation processes prove a clear distribution of business associations on the one hand and the academies, professional associations, universities, and hospitals on the other.[132] The first bill, while maintaining the tripartite scheme and the prohibition on cryopreservation, was defeated by over 90%, although PGD had been advocated.[133] However, the majority of business associations, academies, professional associations, universities, and hospitals agreed to the second template, although they had reservations. In both processes the strongest criticism was about the term "further organizations and companies." 40% of the political parties also opposed introducing the PGD in the second consultation phase.[134]

4.6 *Commentary from an Ethical Point of View*

In the legislative procedure and the accompanying debates, law proponents as well as opponents referred repeatedly to the norm of "human dignity."[135] The Federal Council considered the law consistent with human dignity, as can be seen in the vote dossier:

[130]Cf. Bundesamt für Gesundheit, Vernehmlassungsbericht (2012, 15).

[131]Cf. Bundesamt für Gesundheit, Vernehmlassungsbericht (2012, 13).

[132]Cf. the vote tables, in: Bundesamt für Gesundheit, Vernehmlassungsbericht (2010, 6); Bundesamt für Gesundheit, Vernehmlassungsbericht (2012, 7). The tables show clear distribution pro or contra PGD in the respective groups.

[133]Cf. Bundesamt für Gesundheit, Vernehmlassungsbericht (2010, 6).

[134]Cf. Bundesamt für Gesundheit, Vernehmlassungsbericht (2012, 7).

[135]In the debate about the dignity and protection needs of the embryo it was often referred to Article 119 of the Federal Constitution from 2015. Because several embryos should be produced in the procedure, Article 119 in has been modified.

> It remains forbidden to select embryos on the basis of gender, other physical characteristics or on their suitability as a stem cell donor for a sick sibling. In addition, the number of embryos that can be developed per treatment remains limited. Thus, the Federal Council considers human dignity and the protection of the embryo to be safeguarded.[136]

The conclusion is to question, as it is unclear, whose human dignity is safeguarded. It seems as if it were aimed at human embryos in general and less at the individual embryo. Human dignity relates to individuals and is significant regardless of external objectives. In ethics and law, the "nasciturus" already has human dignity.[137] Respecting human dignity includes not being discriminated against. In performing the PGD diagnosis, embryos are discarded through external selection criteria. Not only are the abovementioned prohibited selection criteria problematic from an ethical point of view but also the distinction between disease and health[138] as well as between severe and minor disease or between a high versus low probability of the manifestation of a disorder, such as a hereditary increased probability of cancer or dementia.[139] An appropriate use of the moral and legal term of human dignity would be necessary to avoid a not justified combination of this central ethical and legal norm with discriminating selection.

It was surprising that—this is a feature of the Swiss PGD debate—in connection with the discussion about the PGD, a dispute about the legal admissibility of egg donation came up, because there is no connection to the PGD from an objective point of view. Unlike as was represented in the second consultation phase in 2011, egg donation is not an internationally accepted practice. Although women's eggs or ovarian tissue are preserved in the case of cancer treatment to enable these women to have children at a later date, these are their own eggs and not the so-called "donated" eggs.[140]

It can also be noticed that liberalization has taken place in the course of the law procedure: Cryopreservation and aneuploidy screening were considered as prohibited in the first draft of the law; in addition, it adhered to the rule of threes. In contrast to previous debates in other European countries, the Swiss debate did not continuously set out that there are two different groups of couples, where the PGD can be applied. Ultimately, it adopted a regulation which included both groups and,

---

[136] Cf. Schweizerische Eidgenossenschaft, Abstimmungsdossier (2015, 13); translated by R. Schreitz. (Original text in German: "Es bleibt aber verboten, Embryonen aufgrund des Geschlechts, anderer Körpermerkmale oder aufgrund der Eignung als Stammzellenspender für kranke Geschwister auszuwählen. Zudem bleibt die Zahl von Embryonen begrenzt, die pro Behandlung entwickelt werden dürfen. Damit sind aus Sicht des Bundesrates die Menschenwürde und der Schutz des Embryos gewährleistet.")

[137] Cf. for ethical and judicial contributions Joerden et al. (Eds.), Menschenwürde und Medizin (2013).

[138] Cf. Bobbert, Krankheitsbegriff und prädiktive Gentests (2012).

[139] Cf. Bobbert, Entscheidung über Auswahlkriterien bei Pränataldiagnostik (2014); Bobbert Pre-Implantation Genetic Diagnostics (2005).

[140] Cf. Goeckenjan et al., Medical and ethical issues of fertility preservation in female cancer patients (2017).

thus, allowed PGD diagnosis for the large number of involuntarily childless couples who undergo the IVF procedure. Although the public controversy over the aneuploidy screening was referring to these two different groups, it is at least questionable as to how many discussion participants had basic knowledge about human genetics. This included knowing, for example, about the difference between molecular and chromosomal genetic dysfunctions and the difference between heritable mutations and spontaneously emerging dysfunctions of the genetic constitution, which is necessary to decipher the group of hereditary predisposed couples, which were often represented in radio and television broadcasts by individual stories, from all couples undergoing artificial insemination due to infertility.[141]

With this broad approval to the PGD, today Switzerland has no middle position in Europe, as it has been aspired to in the context of embryonic stem cell research, but one of the most liberal PGD regulations in Europe.

From an ethical point of view, important objections were left unacknowledged[142]: Each embryo is unmistakable already at an early stage determined in its species, continuity, and individuality and has the potentiality of development. Nevertheless, in using PGD, the embryos do not have the same chance of being implanted. Rather, it is decided by a profession, by a designated panel, or by society—by means of externally targeted characteristics, to implant a certain embryo instead of another: So the embryos are discriminated in an ethically problematic sense.[143] The objection of the unequal treatment on the basis of social evaluations or the problem of illegal discrimination by society was hardly mentioned in the debate. In addition, it was left open why limits can be drawn for selection criteria such as gender and race, but not for other socially undesired characteristics.

Within reproductive medicine, couples, physicians, and ultimately the PGD criteria approved by a society have a gate-keeper function.[144] The social practice of parenting—until now being determined by unconditional acceptance—changes through the preconditioning of a child into a parenting that includes elements of control and power. This social problem of selection is to be distinguished from a striking result after a genetic prenatal diagnosis (PND) in an existing pregnancy, when a woman gets into an individual conflict situation.

Another change can be observed in the procedure: Only in the second referendum in 2016, in which the Federal Council added a passage in defense against the accusation the PGD would lead to a discrimination of disabled people:

> The goal of the PGD is to prevent the transmission of a serious genetic disease from the parents to the child or to make a pregnancy without complications possible. Most disabilities

---

[141]There are also counterexamples, e.g., an ARENA show of the April 29, 2016 to the new reproductive medicine law; Cf. Arena (2016).

[142]See e.g for the ethical debate among others Düwell (2008), Ethik in der Medizin (1999), Francis (2017), Graumann (2004), Haker (2011), Hildt and Mieth (1998), Kollek (2002), Mathwig (2009), Schöne-Seifert (1999), Zimmermann-Acklin (2012).

[143]See the dispute about acceptable criteria of desirability or not desirability which can also not be solved by the sole distinction between health and disease or severe versus mild disease: Bobbert, Krankheitsbegriff und prädiktive Gentests (2012).

[144]Cf. Bobbert, Präimplantationsdiagnostik (2014).

> are not genetically determined but rather have other causes: Therefore, there will not be a society without disabilities in the future. People with disabilities belong to our society. The Federal Council has its disability policy to actively make sure that these people can participate in society on an equal footing.[145]

The late consideration on the possible discrimination against people with disabilities is surprising. Maybe the interests of people with disabilities lacked a strong political lobby? In addition, some central objections to the PGD, which were represented in the European ethical debate, were hard to hear in the Swiss debate: What selection criteria could be justified in detail with good reasons at all? Who is allowed to decide about such criteria in social practice? What are the consequences if these problematic decisions for selection criteria are delegated to a board and, in other words, are only delegated to a process?

Also the difficulty of distinguishing between disease and health (e.g., in relation to hereditary microsomia or sex chromosome-related dysfunctions such as Turner or Klinefelter's syndrome) or of differentiating between "severe" and "light" illnesses was hardly discussed. This would have been necessary, as the PGD—in contrast to a PND—in connection with an existing pregnancy represents a social selection which is allowed or limited by professional representatives or members of boards, whereas in the context of pregnancy and PND, a woman makes decisions according to her individual responsibility or conscience.

Often, the PGD is compared with the PND, and the elimination of "inconsistency"—in one case a ban and in the other an admissibility—is required. In addition, it is said that the PGD protects the couple from fatal decisions, i.e., a late termination of pregnancy after abnormal findings through the PND. Argumentative objection to this comparison was rarely represented in the Swiss debate.[146] There were several other academic discussions about ethics, which were represented in public debates to a lesser extent[147]: The missing identification of parents with the embryo as their future child leads to a delay of responsibility for parenthood. Due to abstract operations and due to a variety of embryos, there is a tendency to regard early embryos as objects. The legal regulation of the PGD turns responsibility over to the medical profession or society and, thus, the power of selection. The physical and mental capacity of the individual pregnant women is no longer crucial in the case of PID. Rather, an external "authority" sets the social standards by which embryos can

[145]Cf. Schweizerische Eidgenossenschaft, Abstimmungsdossier (2016, 39); translated by R. Schreitz (Original text in German: "Das Ziel der PID ist es, die Übertragung einer schweren genetischen Krankheit von den Eltern auf das Kind zu verhindern oder eine Schwangerschaft ohne Komplikationen zu ermöglichen. Die meisten Behinderungen sind nicht genetisch bedingt, sondern haben andere Ursachen: es wird deshalb auch in Zukunft keine Gesellschaft ohne Behinderung haben. Menschen mit einer Behinderung gehören zu unserer Gesellschaft. Der Bundesrat sorgt mit seiner Behindertenpolitik aktiv dafür, dass diese Menschen gleichberechtigt am gesellschaftlichen Leben teilhaben können.")

[146]Cf. Zimmermann-Acklin, Präimplantationsdiagnostik (2012); Mathwig, Präimplantationsdiagnostik (2009).

[147]See alongside comprehensive medico-ethical publications. Cf., e.g., Zimmermann-Acklin, Präimplantationsdiagnostik (2012) or Mathwig, Präimplantationsdiagnostik (2009), Bobbert (2014) particularly for contributions published in Switzerland which were critical of PGD.

be implanted and therefore develop. Although the Swiss Federal Council decided against a specific list of diseases, the selection according to set characteristics (see Sect. 3.3), which have been called exemplary, are problematic.

The comparison with other countries was repeatedly used as a prominent argument in the process. It remains ultimately unclear why an international comparison is used as a category for assessing an ethical question. In fact, medical tourism of pregnant women in European foreign countries was mentioned more often. Similar empirical phenomena can be observed in the context of other legally prohibited or ethically problematic practices in Switzerland or European Union states, such as organ trafficking, without it being less problematic from an ethical point of view or without abolishing governing law on grounds of travel practice.

Overall a mixture of patterns of reasoning can be observed. So the opponents of the "rule of threes" led the argument of the "single-embryo transfer" as a means of preventing fetocide. However, this argument is used incorrectly, because a "single-embryo transfer" does not necessarily require the production of several embryos. A "single-embryo transfer" can also be performed with only one embryo produced in the Petri dish. The medical dispute about the possibilities of increasing the likelihood of a pregnancy should be carried out openly. But factors such as the woman's age, the sophistication of the reproductive-medicine centers, and the expertise of its fertility experts are more relevant than those rare cases, in which a PID can diagnose genetic disorders that are incompatible with life.

5 Conclusion

The Stem Cell Research Act is now over a decade old and is no longer discussed in Swiss society in principle but still in the concrete application on research project proposals. In contrast, the discussion on the future Reproductive Medication Law, which allows PGD to a limited extent, lasted from 2001 to 2017. During this long period, an opening in the PGD occurred, with regard to preimplantation diagnosis, the causes of which might be revealed in a comprehensive media and discourse analysis in the future.

In both legislative amendments, the discussions about the moral status of the embryo outside the uterus were striking. Both, the proponents and the opponents of the two laws, often sought the "human dignity" argument, without any legal or ethically relevant clarification of human dignity and the associated protection obligations. The ethical question remained open as to whether couples could freely dispose of their embryos in terms of reproduction and research. If couples are given external limits and permission, the criteria and conditions set or accepted by society should be justified from an ethical point of view. There were no good reasons for these central issues. Nor has it been publicly justified why Switzerland is striving for an average compromise position in European comparisons on bioethical issues.[148]

[148]It has to be mentioned, that this strategy can be observed in debates of European countries.

In both laws, similar outlined, fundamental questions and debates can be observed. Looking at the relationship between religion and politics in the two legislative processes, different positions were made clear: The introduction of PGD was rejected by the majority of the Christian religious communities, whereas the stem cell research law was discussed much more controversially. A thorough confessional assignment to Catholic or Protestant admission or rejection cannot be ascertained on the basis of the evaluated sources, in which numerous church associations expressed themselves. It can only be inferred from the sources that the Swiss Bishops' Conference clearly rejected both laws and the Swiss Federation of Jewish Communities in both cases expressed itself in favor of liberal legislation.

The discussion on the law development in Switzerland concerning new biomedical procedures and techniques, such as stem cell research and preimplantation diagnostics, was traced by a defined stage of the legislative procedure which was decisive for configuration of judicial norms. More differentiated questions and empirical studies on the opinion-building processes with regard to new developments in biomedicine could thus follow this investigation in order to perform a comprehensive history of origins of law approach. Although the medial and political mediation and enforcement processes are quite relevant for application-related ethics, investigating the "is" and "was" is assigned to the interdisciplinary dialogue with empirically working social and historical sciences. Further analyses could connect to the results developed in the context of this contribution.

With regard to biomedical questions, good reasons and objections come to the fore from a normative ethical point of view, since ethics asks for a correct and equitable regulation of social coexistence, i.e., the "moral ought." Questions of normative ethics or, in other words, of what is morally right or wrong and how vulnerable human beings have to be protected were discussed. More or less good reasons were introduced into the public discourse. Open questions and desiderata of discourse will mostly be stated in regard to new biomedical findings. Thus, even if a period of public discourse has come to an end by voting for a specific legal regulation, the ethical discourse can continue—also regarding the ethical pros and cons of the applicable law.

References

Ackermann, Sibylle. 2005. Ethische Diskussion der Präimplantationsdiagnostik. Eine Darstellung ihrer Entwicklung, der ethischen Haupt- und Nebenargumente und der ausstehenden Fragen (im Auftrag der NEK), Fribourg. http://www.nek-cne.ch/fileadmin/nek-cne-dateien/Externe_Gutachten/gutachten_ackermann.pdf. Accessed 17 July 2017.

Arena: Abstimmungs-Arena (2016). Änderung Fortpflanzungsmedizin, 29.4.2016. https://www.srf.ch/sendungen/arena/abstimmungs-arena-aenderung-fortpflanzungsmedizin-2. Accessed 17 July 2017.

Bender, Wolfgang, Christine Hauskeller, and Alexandra Manzei, eds. 2005. *Grenzüberschreitungen. Kulturelle, religiöse und politische Differenzen im Kontext der Stammzellforschung weltweit*. Münster: Agenda.

Bobbert, Monika. 2005. Ethical problems of preimplantation genetic diagnosis. *ELSA Newsletter* 2, 21.
———. 2012. Krankheitsbegriff und prädiktive Gentests. In *Das Gesunde, das Kranke und die Medizinethik: Moralische Implikationen des Krankheitsbegriffs*, ed. Markus Rothhaar and Andreas Frewer, 167–194. Stuttgart: Franz Steiner.
———. 2014. Präimplantationsdiagnostik. Wer darf über die Auswahlkriterien entscheiden? In *Bioethica*, vol. 3, 110–111.
———. 2015. Urteile in der (Bio-) Medizin- und Pflegeethik sind gemischte Urteile. In *Ethik in den Wissenschaften*, ed. Regina Ammicht Quinn and Tom Potthast, 299–306. Tübingen: IZEW.
———. 2016. Christlich motiviert – vernünftig begründet. Bildungs- und forschungsethische Perspektiven. *Schweizerische Kirchenzeitung* SKZ 3, 23–27.
Bobbert, Monika, and Dietmar Mieth. 2015. *Das christliche Proprium der Ethik. Zur moralischen Perspektive der Religion*. Luzern: Exodus.
Brunner, Beat, and Verena Riedo. 2011. Stammzellenforschungsgesetz (StFG) – Externe evaluation. *Zusammenfassender Bericht, Bern* 28 (1): 2011.
Bundesamt für Gesundheit. 2002. Bericht über die Ergebnisse des Vernehmlassungsverfahrens zum Vorentwurf eines Embryonenforschungsgesetzes. Bern.
———. 2004. Vernehmlassungsbericht zum Entwurf einer Stammzellenforschungsverordnung. Bern.
———. 2010. Änderung des Fortpflanzungsmedizingesetzes (Zulassung der Präimplantationsdiagnostik). Ergebnis der Vernehmlassung (19. Februar bis 18. Mai 2009). Bern.
———. 2012. Entwurf zur Änderung von Artikel 119 BV sowie des Fortpflanzungsmedizingesetzes (Präimplantationsdiagnostik). Ergebnisse des Vernehmlassungsverfahrens (29. Juni bis 30. September 2011). Bern.
———. 2015. Faktenblatt Fortpflanzungsmedizingesetz. https://www.bag.admin.ch/bag/de/home/themen/mensch-gesundheit/biomedizin-forschung/fortpflanzungsmedizin/rechtsetzungsarbeiten-fortpflanzungsmedizin/volksabstimmung-2016.html. Accessed 26 June 2017.
———. 2017. Bericht über die Ergebnisse des Vernehmlassungsverfahrens zur Revision der Fortpflanzungsmedizinverordnung (FMedV). Bern.
Bundeskanzlei. 2017. Der Bund kurz erklärt. www.bundespublikationen.admin.ch. Accessed 29 May 2017.
Devolder, Katrien. 2015. *The ethics of embryonic stem cell research*. Oxford: Oxford University Press.
Düwell, Marcus. 2008. *Bioethik. Methoden, Theorien und Bereiche*. Stuttgart: Verlag J.B. Metzler.
Ethik in der Medizin. 1999. Themenheft Präimplantationsdiagnostik 1.
Francis, Leslie, ed. 2017. *The Oxford handbook of reproductive ethics*. Oxford: Oxford University Press.
Göckenjan, Maren, Bettina Böttcher, Ludwig Wildt, and Thomas Strowitzki. 2017. Medical and ethical issues of fertility preservation in female cancer patients. In *Ethics and oncology. New issues of therapy, care, and research*, ed. Monika Bobbert, Beate Herrmann, and Wolfgang U. Eckart, 49–66. Freiburg: Karl Alber.
Graumann, Sigrid. 2004. Präimplantationsdiagnostik, embryonale Stammzellforschung und das Regulativ der Menschenwürde. In *Biomedizin und Menschenwürde*, ed. Matthias Kettner, 122–144. Frankfurt am Main: Suhrkamp Verlag.
Haker, Hille. 2011. *Hauptsache gesund? Ethische Fragen der Pränatal- und Präimplantationsdiagnostik*. München: Kösel.
Hildt, Elisabeth, and Dietmar Mieth, eds. 1998. *In Vitro Fertilisation in the 1990s: Towards a Medical, Social and Ethical Evaluation*. Farnham: Ashgate.
Hilpert, Konrad, ed. 2009. *Forschung contra Lebensschutz? der Streit um die Stammzellforschung*. Freiburg: Herder.
Joerden, Jan C., Eric Hilgendorf, and Felix Thiele, eds. 2013. *Menschenwürde und Medizin. Ein interdisziplinäres Handbuch*. Berlin: Duncker & Humblot.

Kollek, Regine. 2002. *Präimplantationsdiagnostik: Embryonenselektion, weibliche Autonomie und Recht*. Tübingen: Francke Verlag.
Körtner, Ulrich H.J., and Christian Kopetzki, eds. 2008. *Stammzellforschung: ethische und rechtliche Aspekte*. Wien: Springer.
Linder, Wolf. 2005. *Schweizerische Demokratie: Institutionen, Prozesse, Perspektiven*. 2nd ed. Bern: Haupt.
Maio, Giovanni, Tobias Eichinger, and Claudia Bozzaro, eds. 2013. *Kinderwunsch und Reproduktionsmedizin. Ethische Herausforderungen der technisierten Fortpflanzung*. Freiburg: Karl Alber.
Mathwig, Frank. 2009. *Präimplantationsdiagnostik: Vernehmlassungsantwort des Rates des Schweizerischen Evangelischen Kirchenbundes zur Änderung des Fortpflanzungsmedizingesetzes*. Bern: Schweizerischer Evangelischer Kirchenbund.
Mieth, Dietmar. 1984. Christliche Überzeugung und gesellschaftlicher Kompromiß. In *Der ethische Kompromiß (Studien zur theologische Ethik 11)*, ed. Helmut Weber, 113–146. Freiburg: Herder.
Schöne-Seifert, Bettina. 1999. Präimplantationsdiagnostik und Entscheidungsautonomie. Neuer Kontext, altes Problem. *Ethik in der Medizin* 11: S87–S98.
Schweizerische Bundeskanzlei. Volksabstimmung vom 05.06.2016. https://www.admin.ch/ch/d/pore/va/20160605/index.html. Accessed 27 June 2017
Schweizerische Eidgenossenschaft, Abstimmungsdossier. 2004. Volksabstimmungen vom 28.11.2004. Erläuterungen des Bundesrates, Dritte Vorlage, Stammzellenforschungsgesetz. Bern, 24–39.
———, Abstimmungsdossier. 2015. Volksabstimmung vom 14.06.15. Erläuterungen des Bundesrates, Bundesbeschluss über die Änderung der Verfassungsbestimmung zur Fortpflanzungsmedizin und Gentechnologie im Humanbereich (Präimplantationsdiagnostik). Bern, 4–23.
———, Abstimmungsdossier. 2016. Volksabstimmung vom 05.06.16. Erläuterungen des Bundesrates, Vierte Vorlage, Änderung des Bundesgesetzes über die medizinisch unterstützte Fortpflanzung (Fortpflanzungsmedizingesetz FMedG). Bern, 32–47.
Schweizerische Gesellschaft für Biomedizinische Ethik. 2008. Präimplantationsdiagnostik, in: *Bioethica Forum*, 1(2).
Stiftung JA zum Leben: Wir über uns. http://www.ja-zum-leben.de/de/seiten/wir_ueber_uns. Accessed 4 May 2017.
Vatter, Adrian. 2016. *Das politische System der Schweiz*. 2nd ed. Baden-Baden: Nomos.
Verein "Basler Appell gegen Gentechnologie". *Dafür stehen wir ein*. http://www.baslerappell.ch/verein/. Accessed 4 May 2017.
Zimmermann-Acklin, Markus. 2012. Sollte die Präimplantationsdiagnostik (PID) unter den gleichen Voraussetzungen zulässig sein wie die Pränataldiagnostik (PND)? Gutachten zur rechtlichen Regelung der Präimplantationsdiagnostik aus ethischer Sicht. Fribourg, 13.10.2012. https://www.bag.admin.ch/dam/bag/de/dokumente/biomed/fortpflanzungsmedizin/zimmermann-acklin-vergleich-pid-pnd.download/zimmermann-acklin-vergleich-pid-pnd.zimmermann-acklin-vergleich-pid-pnd. Accessed 17 July 2017.

Chapter 15
Morality Policies: How Religion and Politics Interplay in Democratic Decision-Making

Nathalie Schiffino

Abstract Throughout history, there have been connections between religion and politics in Belgium. Abortion, euthanasia, same-sex marriage, and research on embryo are examples of life and death issues that still trigger vivid debates. They may even challenge the stability of a political system, especially when it undergoes a federalization process. Political parties are key players to find a compromise. The churches and the secular actors accept the public regulation as long as their values are not at stake. Therefore, policy change regarding morality issues in Belgium mainly focuses on policy instruments rather than on core beliefs. Such a strategy by decision-makers leaves the traditional cleavages untouched. It ensures social quietness during the framing and the implementation of the public policies. In Belgium, demonstrations related to morality issues are not in breach of the law as they are in France or Spain, for instance. Even the Belgian Catholic Church partially outpaces the Roman Catholic Church.

1 Introduction

Morality policies are policies that entail a tight interlace between polity, politics, and values. Morality policies result from political decisions after debates about what a society values (or, on the contrary, rejects) in order to promote the community life. Religion and secularism are one vector of such debates, decisions, and policies. The way members of a society experience religion and secularism influences the framing, the content, and the instruments of the policies. The policy-making is thus influenced by the polity (what a community values), the politics (the settings of the political system), and the issues at stake (morality as a specific policy field).

Linking policies and politics, this chapter pays attention to morality issues regarding life and death: abortion, euthanasia, same-sex marriages and adoption, but mainly assisted reproductive technologies and human embryonic stem cells. Biopolitics explicitly refers to the policies adopted in such fields. Religious,

N. Schiffino (✉)
Université Catholique de Louvain, Mons, Belgium
e-mail: nathalie.schiffino@uclouvain.be

M. Weiberg-Salzmann, U. Willems (eds.), *Religion and Biopolitics*,
https://doi.org/10.1007/978-3-030-14580-4_15

morality, and biopolitical issues are thus at the core of the chapter. It investigates why and how morality debates are triggered and tackled throughout the policy-making. In a democratic political system, the promotion of freedom and rights allows for the expression of opposed conceptions about community life, including about biopolitical issues. In divided societies, the expression of opposed conceptions may challenge the stability of the political system. If so, how can we explain that the political system meets the challenge, overcomes the puzzling policy-making process, and survives?

Two levels of explanations are presented here. The first one relates to the structures: the settings of the political system explain the general frame for the definition and the implementation of morality policies. The cleavage structure is a key explanation for the origin of the debates. The type of party system and the politicization of the morality issues are other key explanations for the outbreak of the debates. The second level of analysis targets the actors: organizations (such as Catholic institutions or political parties) and individuals (such as a clergyman or a deputy) play a major role in defining or implementing morality policies. Policy entrepreneurs act for policy change in the morality field, promoting it. Veto players try to make it more sluggish. In the parliamentary arena, the same decision-makers may even participate in successive policies (for the regulation of euthanasia, embryo research, assisted reproductive technologies, and same-sex marriage), even during several legislatures.

The chapter focusses on Belgium as an empirical field and a laboratory for theoretical generalizations. The Belgian case is often cited in scientific literature. It is compared to countries such as France, Italy, or the Netherlands, depending on the object of comparison. The characteristics of its political system and the positions of the Catholic Church (which is the dominant faith in Belgium) give Belgium an added value for comparison among (European) countries, especially where the regulation of morality issues is concerned. The Belgian policy-making of morality issues is puzzling because it combines regulatory reluctance and permissiveness, with the Catholic Church and the Social Christian parties being the most obvious yet moderate opponents.

In 1830, the Belgian state was built upon an institutional balance between religion and secularism. Actually, the whole architecture of the political system was determined by the stability to be found at the intersection of three cleavages. The division between religion and secularism is historically the first cleavage. It was experienced as soon as the constituent assembly in 1830 brought together the two existing factions, the Catholics and the liberals. The deep conflicts between the French-speaking community and the Flemish community developed from the decade 1840. The last line of cleavage in Belgium tore apart the socialists and the liberals with regard to socioeconomic issues, crucially in the 1880s, the Social Christian being traditionally centrist in this contest. As shown by Easton (1965), a political system needs to adapt to disturbance in order to survive. With regard to the Belgian political system, the crises caused along the cleavages were systematically solved by a quest for compromises.

Lijphart (1999) demonstrated that deeply divided societies avoid the risk of unresolvable opposition through consensus democracy. In terms of institutions and practices, this entails a segmented pluralism (Lorwin, 1971). Each pillar attracts a certain political party and the associations related to it. Pillars are experienced at organizations' level as well as at individuals' level. For instance, a member of the Catholic pillar would vote for the Catholic party, be registered with a Catholic health insurance company, attended a Catholic school, was a member of a Catholic scout troop, and so forth. The segmented pluralism entails a representative dynamic, not a participatory one, with elites embracing accommodating strategies (Lijphart, 1975). Due to these political settings, mixing societal divisions and institutional consensus, Belgium is the archetype of a cleaved and consociational society, where crises catalyze political decisions (Schiffino, 2003).

In their seminal work entitled *Cleavage Structures, Party Systems, and Voter Alignments*, Lipset and Rokkan decode the political party system that derives from the cleavages' intersection. The organization of the party system, and the understanding of how political parties oppose each other, is crucial to understanding the management of religious, morality, and biopolitical issues. As we will discuss below, it is of prime importance to explain the way political parties politicize the policy-making of such issues. Indeed, it is the cornerstone of the agenda setting, the framing, and the implementation of the related policies. Such politicization does not operate ex nihilo. It depends on the societal environment. In the biomedical policy sector, for instance, doctors, universities, and the Catholic Church are actors whose arguments matter to political parties.

The remainder of the chapter is organized into three sections. Each section provides a big picture about morality policies and afterwards some details regarding assisted reproductive technology (ART) and embryo research. The first section about the national relationship of religion and politics explains the origin of the interplay between the Catholic Church and the political actors, within the pillars that are created by the three cleavages. It is theoretically oriented, with a strong reference to the seminal work by Lipset and Rokkan. It takes into account the evolution of the political system, mainly regarding the current federalization process. The federalization process appears to be a low order explanation, the weight of the political parties being higher. The second section specifies the theoretically oriented explanations by focusing on the biopolitical and morality debates as they develop in the field. It provides a summary of the main political decisions that Belgian parliamentarians have adopted regarding morality issues. It is factually oriented, and it explains that morality decisions are both permissive and procedural. The third section is devoted to the religious actors and their arguments. Examples and scientific literature are drawn on to explain the role of key actors. The Catholic Church and the Social Christian political parties are major players in policy-making, but weighting differently.

2 The National Relationship of Religion and Politics

2.1 Polity Matters: Different Pillars Value Different Policies

Lipset and Rokkan (1967) rely on Parsons' systemic approach (1951) to theoretically explain how western societies divide and how western political parties structure. Lipset and Rokkan highlight four main lines of deep divisions. They label them "subject vs dominant culture," "church vs government," "primary vs secondary economy," and "workers vs employers and owners" (Lipset and Rokkan, 1967: 14). The second cleavage is of major interest in the purpose of this chapter. It is characterized by "the conflict between the centralizing, standardizing, and mobilizing *Nation-State* and the historically established corporate privileges of the *Church*" (Lipset and Rokkan, 1967: 14—italics by Lipset and Rokkan). The church (mainly Catholic in western democracies at that moment) and the state dispute over education and morality issues. Each one develops organizations (i.e., political parties and associations among the civil society) to support their views and to mobilize the population.

Therefore, and more generally along the four lines of divisions, the cleavage structure transforms into a party system. The deep conflicts are translated into oppositions between political parties depending on how opponents surpass four thresholds: legitimation, incorporation, representation, and majority power. Along the "church vs government" cleavage, the opponents to the prevalent religious-based decisions create secular parties.

In Belgium, most of the seculars are to be found in the liberal and the socialist parties, being in opposition to the Social Christian party. These days they are the three traditional political families in the Belgian party system. In the 1960s and 1970s, each family divided internally along the linguistic cleavage, doubling the number of parties. Only very few other political parties succeed in joining the party system, notably the greens since the 1980s. They now form part of the four main political families in Belgium, each one duplicated into French and Flemish speaking. The greens claim to be pluralist along the "church vs government" cleavage, supported by both religious and lay members.

Among the European party systems as highlighted by Lipset and Rokkan, and within each political party, individuals tend to vote for a particular party and to align with positions along the cleavages. Education and charities were the major issue that opposed the state and the church during the nineteenth century and still during the twentieth century. The Catholic electorate supported the Social Christian party, promoting the influence and even the initiatives of the church in school and charities. The socialist and the liberal electorates voted for the respective parties in order to favor the state position. Since the mid-twentieth century, scientific progress and the changes in society have been pushing new issues onto the political agenda along the "church vs government" cleavage. Abortion, ART, stem cells or embryo research, euthanasia, same-sex marriage, and adoption become prominent moral issues to be tackled by the Social Christian, the liberal, the socialist, and later the green parties.

Illustrating Lipset and Rokkan's theory, historical facts provide a background of current decisions. Let us recall significant facts since the creation of the state until nowadays. As early as December 13, 1830, when the founding fathers drafted the constitution, Archbishop *de Méan* wrote a letter to the congressional representatives. He asked for the Catholic Church's to be recognized in the constitution. As head of the Catholic Church, the clergyman declared to those drafting the constitution that depriving Catholics of the freedom of religion would trigger crises within the new state (Schiffino and Jacob, 2015). It proved to be the case during the so-called school wars (the first from 1879 to 1884 and the other from 1952 to 1955). The church (i.e., the leading organizer of education) and the state (willing to develop its educational network) were in disagreement on the number of schools and the obligation to teach religion. Ultimately, the "school wars" lead to a historical compromise: the "school pact" in 1958. It was the political parties, who negotiated and concluded the agreement. In a second round, the agreement was translated into the law of 29 May 1959, which stabilizes principles and resources: it foresaw equal rights for schools, be they Catholic or secular; an increase in the state budget for the creation or support of (new) schools; and greater subsidization of the Catholic schools. The agreement is still in force today.

Since the 1960s, secularization has gone beyond educational issues. It is noticeable in Belgian society as a whole, through the decrease in attendance at mass at Catholic churches (from 29.4% in 1977 to 5% in 2009), religious marriages (from 77.7% in 1977 to 25.6% in 2009), religious burials (from 83.7% in 1977 to 58.4% in 2009), and christenings (from 85.2% in 1977 to 54.6% in 2009).[1] In 2010, the Eurobarometer investigated the issue of biotechnologies, and it states that 57% of the respondents claim to be Catholic, 5% other Christians, 5% Muslims, and 1% orthodox.[2] In 2011, the vast majority of Belgian citizens proclaimed to be religious believers, with 50% of them claiming to be Catholic and 2.5% other Christians, 5% Muslims, 0.4% Jewish, and 0.3% Buddhist.[3] In 2017, Gallup released results that would weaken such a stable trend. Its survey would indicate "China is the least believing country, where religious beliefs reach values inferior to 20%. China is followed by Czech Republic, Japan, Belgium, Sweden and Denmark as countries with lower belief levels."[4] Bearing in mind the figures, while we investigate the regulation of morality policies, we should not confuse the evolution of day-to-day

---

[1]Arcq, Étienne, Sägesser, Caroline (2011) Le fonctionnement de l'église catholique dans un contexte de crise, *Courrier hebdomadaire du CRISP*, 2112–2113: 66.

[2]European Commission, *Special Eurobarometer 341* – Wave 73.1, Biotechnology, Report, Fieldwork: January 2010 – February 2010, Publication: October 2010.

[3]Dobbelaere, Karel, Billiet, Jaak, Voyé, Liliane (2011), Religie en kerkbetrokkenheid: naar een sociaal gemargineerde kerk? in Abts, Koen, Dobbelaere, Karel, Voyé, Liliane, *Nieuwe tijden, nieuwe mensen. Belgen over arbeid, gezin, ethiek, religie en politiek*, Leuven, Lannoo Campus: 145.

[4]Religion prevails in the world. Gallup International. http://gallup-international.bg/en/Publications/2017/373-Religion-prevails-in-the-world (accessed: 17 August 2017).

practices (which undergo a secularization process) and the maintenance of core values (which highlights the survival of a cleaved political system).

Crises along the "church vs government" cleavage arise again on some occasions. In 1990, the Chamber of Representatives passed a law that partially decriminalized abortion, but King Baudouin refused to sign the law, appertaining to a clause of conscience. This launched a major political crisis. To find a way out of the deadlock, the constitutional solution was to declare the King temporarily (over 36 hours) unable to reign. At this time (in fact also during the 1980s and 1990s, until 1999), the prime minister of the federal coalitions was a member of the Flemish Social Christian party. The Council of Ministers, encompassing all ministers, sanctioned and promulgated the law in the King's place. Then, the assembled chambers voted in favor of the end of the King's inability to reign.

Subsequently, the Social Christian parties strategically avoided decisions about morality issues in the years thereafter. However, political upheaval is likely to happen regarding issues along the "church vs government" cleavage. Repeatedly, lastly also in 2016, the issue of teaching philosophy at primary schools triggers fierce debates. Socialist and liberal politicians regularly propose to change education programs in order to promote secular lessons instead of courses of religions where they apply.

Yet, since nearly half of a century, the major disturbance of the political system has been the implementation of a federalization process replacing the unitary state. From 1970, Belgium has faced dramatic changes in its institutional settings. Six revisions of the constitution pave the federalization process, on average once in 10 years, the last adopted in December 2011. The unitary state, with a limited delegation of power to provincial and local layers, gives way to a federal state where provinces and municipalities remain, but along with three regions and three communities. The regions and communities exert a normative power equal to the federal state; they own one government and one parliament each, but no judiciary structure as the federal state. Asymmetry characterizes the political system because regions and communities do not geographically overlap, nor do they exert the same competencies. In such a context, where does the new point of equilibrium regarding religion and morality issues settle? Does federalization affect the regulation of morality policies?

The federal state remains competent for dealing with religious matters. It is in charge of the recognition of religious communities (Catholicism, for instance) and secular organizations, the designation of their representatives, the payment of the ministers of religion (wages and retirement benefits), and the establishment of a regulatory framework regarding the religious communities and secular organizations. At the level of federated entities, the regions are in charge of the "organic legislation" about the places of worship (e.g., the mosques) and their recognition. The regions supervise such establishments, and they are responsible for organic legislation about public investment in such establishments too. At the federated level too, the so-called communities are responsible for education (the school programs) in the Catholic and secular schools, as well as for financing their buildings. In front of this puzzling distribution of competencies, it is important to note that morality laws and decrees are adopted at the federal level.

2.2 *The Location of ART and Embryo Research Policies Within the Political System*

Regarding biopolitics and ART policies more specifically, the federal state remains the leading actor. Three public actors are prominent in the governing structure. The federal parliament is responsible for the adoption of decisions and the federal minister of public health for their execution. The administration under the domain of the minister, called the "federal public service—public health," is one major implementer of the policies. These actors intervene in consultation with a series of bodies dominated by experts (jurists, practitioners, and ethicists), all located at federal level.

The Belgian advisory committee on bioethics (BACB) provides advice, about any morality issue including ART, on its own initiative or on the request of several institutions (governments, parliaments, hospitals, research institutes, higher-education establishments, and recognized local ethical committees attached to a university or a hospital). One college of medical doctors, specializing in reproductive health, monitors the activities of the ART centers. Besides, it records the relevant data under the so-called Belgian Register for Assisted Procreation (BELRAP). The federal commission for medical and scientific research on in vitro embryos (federal commission embryos, for short) supervises the research institutes and the hospitals in that specific area of activities. The federal commission embryos gather information about the research projects; it reports to the federal parliament once a year and may even visit the laboratories.

The regions manage the implementation rather than the definition of the morality policies. With regard to ART specifically, the regions are responsible for the accreditation and the control of ART laboratories.[5] The federal ministers of public health are in charge of the competency on their respective territory. They work with the administrative agencies under their supervision, such as the directorate for hospital care within the *Agence pour une Vie de Qualité* (AVIQ) in the Walloon region. Regions are entitled to fund some research too.

Belgium owns 34 accredited ART centers. The royal decree of 15 February 1999 divides them into two categories: A and B. Doctors of a type A center practice ovarian stimulation and ovarian puncture. The B centers receive the oocytes for fertilization and embryo transfer because they own an ART laboratory. The royal decree plans 1 A center per 700,000 inhabitants, 1 B center per university hospital,

---

[5]The federal government has adopted a royal decree for "setting the standards that the care programs 'reproductive medicine' must meet to be accredited" (title of the decree). See below about the adoption of this norm in 15 February 1999.

and 1 nonuniversity center per province[6] with at least 1 center in a public hospital. The geographical location of the 34 centers accredited under these standards is as follows: 2 A centers and 4 B centers are located in Brussels, 5 A centers and 6 B centers in Wallonia, and 9 A centers and 8 B centers in Flanders. They are dispatched following the religious and secular cleavage, along with their respective hospital. This is a consequence of the standards by the law, as well as a rationale to them.

Does federalization make a difference regarding the intertwining of policies and religion in the policy-making of morality issues? As pointed out by Simeon (2009), one main argument for federalization in a culturally divided society is to accommodate differences. That is the point in Belgium, but rather for communitarian reasons, than because of religious divisions. Little research addresses the question of morality policies (Knill, 2013) and even less the influence of federalization on biopolicy-making. Interestingly, Snow (2015) investigates a potential link between the way biopolicies are adopted and the type of federalism. Snow considers "the way in which policy-makers frame issues as properly belonging to one particular level of government – that is how political actors frame a policy field in terms of jurisdictional competence in addition to policy substance" (Snow, 2015: 124).

In Belgium, the argument is of little relevance because there is no choice: according to the constitution and the laws, issues are regulated at federal level. Only one conflict of competence occurred, and it dates back to 1982. That year, two parliamentarians presented a bill to the parliament of the *Communauté française*, in order to regulate several aspects of artificial insemination (AI). But the bill was never adopted. The reason is that, until nowadays, federated communities own no power to modify the civil code or the constitution, which is de jure necessary for regulating the vast majority of morality issues. Actually, the roles actors play within the political system prove to be crucial. It is the (cleaved) party system and the politicization of the issues by the political parties, more than the federal structure, which explain the policy-making of morality issues.

3 The Biopolitical and Morality Debates

3.1 Politics Matter: The Political Settings Induce Instrumental Policies

Since the 1990s, morality issues trigger debates among the Belgian decision-makers, and they lead to political decisions. Intriguingly, they do not generate fierce protests or mass mobilization such as in France in the 2000s regarding same-sex marriage, for instance. The Belgian bioethical and morality debates are characterized by both pioneering decisions and little objection. The "church vs government" cleavage remains, yet deep conflicts along it are largely anesthetized. This is due to the

[6]There are ten provinces in Belgium.

institutionalized pluralism and the consociational policy-making. A motto for such a policy-making is "live and let live." If one social segment (e.g., the Catholic pillar) is not threatened in its core values and practices, it lets the other ones (e.g., the lay segments) promote their own values and practices. In this perspective, public policies in the moral and bioethical fields aim at settling general agreements and frameworks. Laws and decrees define implementation procedures, barely broad principles that serve as guidelines.

Let us sketch the morality policies that have been adopted up to now, in the aftermath of debates. Belgium decriminalized abortion a long time ago, but only partially. The law of 3 April 1990 related to the termination of pregnancy modifies the penal code. Abortion remains a crime unless the woman is in situation of distress and she asks the doctor to terminate her pregnancy. The law prescribes conditions to avoid penal pursuit. The abortion must occur before the twelfth week of pregnancy, and the woman must be counselled, including receiving information about adoption, for example. The decriminalization of abortion has never been challenged since then, contrary to what happened in Spain in the 2000s for example. Yet, from January to June 2017, the federal coalition was put under pressure because several law proposals aimed at fully decriminalizing abortion (i.e., removing it from the penal code). In June 2017, the federal majority, including the Flemish Social Christians and the liberals, finally adjourned the debates to avoid internal conflicts.

Shortly after the Netherlands in 2001, Belgium was among the first countries to have euthanasia decriminalized, by passing the law of 28 May 2002. The law of 28 February 2014 revised it, extending the practice to minor patients. The Social Christian parties voted against such an extension. Scholtes, the spokesperson of the Belgian bishops, argues that "it endangers social bonds."[7] On the contrary, Biarent, the doctor directing the intensive care unit at the Children's University Hospital Queen Fabiola, considered that the law was necessary and that "it is not a secular law" but "a law by and for people who live their beliefs with a wider reflection, and who want to move forward."[8] The law revision also aimed at adjusting the implementation procedures to take into account reports by the federal commission for control and evaluation of euthanasia. Still, the principle of decriminalization of euthanasia is not disputed. "The law is there, there will not be a majority to come back totally on it. But it can be revised in a certain number of respects," declared De Graeve from the Flemish Social Christian party.[9] The law defines euthanasia as the practice by a third party who intentionally ends the life of a person at that person's own request. It states that it is not a crime anymore if the third party, actually the doctor, complies with the requirements of the law. The annual reports from the

[7]Quoted by Brener Perla (2014), L'euthanasie des mineurs: fallait-il légaliser?, *Regards*, 794 (no page numbers).

[8]Quoted by Brener Perla (2014), L'euthanasie des mineurs: fallait-il légaliser?, *Regards*, 794 (no page numbers).

[9]*Le Figaro*, 4 February 2016. That year, two sisters filed a complaint against the euthanasia of their third sister. The latter had requested euthanasia because of deep psychological suffering. A potential revision of the law is still being debated regarding such pathology.

control commission at parliament show a constant increase in declared euthanasia, from 24 in 2002 to 1807 in 2013.[10]

The country is also a pioneer in allowing same-sex marriages. In an astute move, associations and politicians decoupled the issue of same-sex marriage from the issue of parenthood. The law of 13 February 2003 paved the way for the marriage of homosexuals by modifying the civil code: "husband and wife" became "the spouses," and all gendered formulations relating to marriage were adapted. The French-speaking liberals asked the issue to be classified as ethics, in order to neutralize the party discipline.[11] Actually, their votes showed internal divisions, unlike the Flemish liberals (except one representative), the socialists, and the greens who all supported the law. The opponents argued that the aim of marriage is procreation. The proponents advocated that marriage was primarily about the creation of stable relationships. The French-speaking Social Christians voted against the law, or they abstained from voting. Surprisingly, Flemish Social Christians were supportive. They argued that the law did not cover parenthood. Adoption by same-sex couples came later, in the law of 18 May 2006. Its modus operandi is the same: adapting gendered sentences in the civil code. Both Social Christian parties voted against, and so did the French-speaking liberals, while the Flemish liberals showed internal divisions. Figures regarding homosexual marriages show a stable trend.[12] Between 2004 and 2014, 1123 men and 1016 women married same-sex partners on average per year.[13]

Regarding assisted reproductive technologies (ART), only a few European countries, including Belgium,[14] grant lesbians or single women access to ART. Sperm donation is allowed for lesbian couples and single women. In fact, access by gays depends on the guidelines of each ART center, since the legislation does not forbid it. More generally, and comparably to other European countries,[15] Belgium offers generous reimbursement by the national health plan for any couple and single person entitled to ART treatments under the conditions prescribed by the law.

---

[10]Commission fédérale de Contrôle et d'Évaluation de l'Euthanasie, *Seventh report to the legislative Chambers*, years 2014–2015, p. 14.

[11]The party discipline implies that the deputies stick to the programmatic axis of the political party when voting. (Bio)ethics or morality issues are among the only ones for which the discipline is not imposed on the deputies.

[12]Statistics Belgium (i.e., federal public administration), http://statbel.fgov.be (date of consultation: 6 December 2017).

[13]As a basis for comparison, 40,049 marriages took place in 2015 in Belgium and 39,878 in 2014 (Statistics Belgium, http://statbel.fgov.be (accessed: 6 December 2017).

[14]Präg, Patrick, Mills, Melinda (2015) Assisted reproductive technology in Europe. Usage and regulation in the context of cross-border reproductive care, *Families and Societies*, Working Paper Series, 43: 14.

[15]Präg, Patrick, Mills, Melinda (2015) Assisted reproductive technology in Europe. Usage and regulation in the context of cross-border reproductive care, *Families and Societies*, Working Paper Series, 43: 6.

Concerning surrogate motherhood, the legislative process is currently in transition. Surrogacy is not yet explicitly allowed by law, but it is not forbidden either (except the commercial purpose of it), and it actually depends on the guidelines of the centers concerned. Currently, adoption is required to transfer legal parenthood.

Regarding human embryonic stem cell research (HESCR), Belgium is considered a leading country too (Salter, 2007: 277). HESCR is allowed and guided by the law of 11 May 2003. The law authorizes therapeutic cloning, the creation of HESC, and their procurement from residual cells. The law forbids reproductive cloning, eugenics, and chimeras. It subjects the research projects to the authorization by local ethics committees. It creates the federal commission for medical and scientific research on embryos in vitro, also referred to simply as federal commission embryo (FCE). The FCE controls the research by the centers, the hospitals, and the universities at every stage, from the protocols at the beginning of the research until the final reports it receives. It directly reports to the federal parliament responsible for the FCE.

The law regarding HESCR was expanded to ART 4 years later, with the law of 6 July 2007. The third title of the law is specifically devoted to reproductive medicine. It binds the ART centers tightly to university hospitals, and it focusses on the conventions that these centers must systematically establish with the patients. It stipulates the terms and conditions for AI, in vitro fertilization (IVF), cryopreservation, insemination, and donation of supernumerary embryos. Transparency from the centers and informed consent of the patients are core notions in the law.

How were such policies adopted? What are the characteristics of the policy-making in terms of the interplay between politics and religion? In a similar way that Snow links frame and federalism (see above), Mucciaroni (2011) connects frame and morality policies as a peculiar type of policy. Mucciaroni suggests that morality policies are framed as such by the actors in order to put the priority on principles rather than on instruments. This is congruent with the scientific literature about morality policies that defines them as a "debate over first principles" (Mooney, 1999: 676). Yet what is striking in the Belgian case is that instruments or procedural dimensions of policy-making are used to go beyond the principle oppositions that tore the actors apart. Thanks to the major characteristic of the policy-making, we shed new light on the current knowledge about morality and ART policies.

In 1991 and 2002, the laws about abortion and euthanasia briefly defined each notion and then described the procedure to be respected in practice. Regarding ART, the royal decree of 15 February 1999 was adopted long before the law of 6 July 2007 that defines core notions[16] and then establishes the conditions for ART treatments. Even if the first chapter of the law is entitled "General Principles," its content is mainly instrumental (practices in the accredited centers only, age of the woman, e.g.). When research on embryos was regulated in 2003, the law also fixed the procedures to be followed by the practitioners, following a ban on reproductive

[16]The law of 6 July 2007 defines 22 notions, from assisted procreation to pairing. The law of 11 May 2003 defines only five notions: embryo, research, relating persons, researcher, and human reproductive cloning.

cloning (as formally in all countries[17]) and commercial purposes. Laws for same-sex couples explain the judicial procedures for marriage (2003) and adoption (2006). Significantly, none of the morality laws are preceded by preambles that would argue the necessity of the measures and set them in context, as with other types of policies (in the environmental field, for instance).

Meanwhile, these decisions are an integral part of morality policies as such. They result from debates on values including religious positions. They entail moral principles about life and death (e.g., allowing euthanasia, allowing postmortem insemination), family (e.g., integrating single parent in ART treatments), and sexuality (e.g., recognizing same-sex couples). These morality policies are even "manifest" in the sense by Knill (2013: 312) who defines "manifest morality policies" as "issues in which value conflicts constitute the standard mode of political decision-making." The point is that outputs of the policy-making are procedural norms. They are mainly connected to instruments and change in the settings of instruments (Hall, 1993).

In fact, the politicization of morality issues in Belgium mainly comes from secular actors. They act as policy entrepreneurs (Mintrom, 2013: 442, 444) and are the catalysts of policy change in the morality field, from legal void to procedural regulation. The subsequent policies focus on procedural content because the "church vs government" cleavage can potentially trigger conflicts so deep that compromises are sought to avoid confrontations. Instrumental agreements do not challenge the values of the different segments of societies. They establish frameworks that each segment applies following its core beliefs. Notably, article 5 of the ART law in 2007 states that the ART centers may invoke a conscience clause toward demands addressed by patients. In divided societies, decision-makers tend to circumscribe the framing and the expansion of a (morality) issue from principles to instruments, because the characteristics of the political system prevent finding one most commonly recognized solution around values, especially when sources of deep conflict remain.

Even the Belgian Advisory Committee on Bioethics (BACB) provides an illustration of such a complex functioning mode. It took up its work in 1993 with a twofold mission: to provide concrete advice to decision-makers on biomedical issues and to inform the public about these issues. In terms of composition, the BACB strikes a balance between transdisciplinary and political pluralism. Its 43 members represent the secular and confessional pillars and equally the French- and Dutch-speaking communities, as well as respect a gender balance. An equilibrium is also ensured between scientists and doctors on the one hand, philosophers, jurists, and social scientists on the other hand. Since its first opinion in May 1997, BACB has delivered 68 opinions; 22 opinions are directly related to ART issues, such as the use of preimplantation diagnosis or the donation of sperm and eggs. The opinions are rarely consensual. Even when they are, differences are added to the broad consensus.

[17]UNESCO, *National Legislation concerning Human Reproductive and Therapeutic Cloning*, Paris, July 2004, SHS-2004/WS/17, 28 p.

In opinion n°10, for instance, all members of the BACB agreed on excluding reproductive cloning; however, beyond this broad consensus, they suggest three different scenarios for the future political decisions: ban, moratorium, and democratic debates. Some BACB members do not want reproductive cloning at all, not now or later (ban); some BACB members think it is a matter of time before the issue is set again on the agenda (moratorium); and some BACB members suggest further democratic debates about this complex issue.

3.2 The Political Environment Surrounding ART and HESCR

From the 1960s until the 1980s, the policy field was almost not publicly regulated. Physicians self-regulated the development of AI (the main practice at that time). In 1975, the national council of the medical order introduced an article about AI into the medical deontology code.[18] In the 1980s, university hospitals took the initiative to settling local ethics committees. These committees aimed at advising practitioners about ethical aspects of care, including those related to individual cases. Care was conceived in a larger perspective than ART and HESCR activities, yet the creation of such committees was closely linked to the development of biomedicine and bioethics. In 1994, a royal decree legally recognized the hospital ethics committees as part of a general law regulating hospitals. The committees still play a day-to-day role in accompanying the ART centers.

In the 1990s, the political decision-makers began to adopt first norms. They first dealt with the institutionalization of the debates and with the instruments to settle. The cooperation agreement[19] of 15 January 1993 institutionalized the BACB. At that moment, the federal majority was composed of the four French-speaking and Flemish socialist and Social Christian parties, with the prime minister belonging to the latter. The liberals, who historically promote secular positions, stayed in the opposition. The BACB institutionalizes pluralism notably through its composition: by law, its members come equally from the three pillars. In 1996, the committee released its sixth opinion about the optimization of the supply and of the functioning of the in vitro fecundation (IVF) centers. The opinion constituted a basis for the royal decrees in 1999. These were not comprehensive laws about ART, rather they established the accreditation of the centers (with the obligation of annual reporting,

[18]Article 88 stipulated that AI with donor (AID) was limited to married couples, who must give their written and informed consent, while the sperm donor remained anonymous. Furthermore, the physician should investigate the motives of the married couples as well as the good health of the donor.

[19]A cooperation agreement is a binding policy instrument that reflects the consociative dynamic of the Belgian political system. The federal government, the regions, and the communities (or some of them, depending on the issue) formally negotiate the agreement in order to avoid conflicts between the public authorities and to increase their collaboration.

among others) and a physicians' college. However, their adoption is remarkable since they are the first formal and public decisions in the field of ART.

Two coalitions at the federal government from 1999 to 2003, and from 2003 to 2007, were able to define limited yet substantive content (the law on embryo research of 11 May 2003 and the law on ART of 6 July 2007) rather than procedures (as in the royal decrees in 1999). The adoption of the laws was made possible because only secular parties formed part of the federal coalition, for the first time since World War II. This opened a window of opportunity for a public and permissive regulation of ART. Secular parliamentarians endorsed their role of policy entrepreneurs. The Social Christian representatives globally acted as a counterbalance within the parliamentary opposition.

Morality debates usually take place in parliament (by government decision after the abortion crisis in 1991), and decisions take the form of parliamentarian votes (generally majority against opposition). Binding referendums are not allowed in Belgium. The judiciary is scarcely solicited about ART or HESC files (nor it is for the implementation of the other morality issues). The Belgian political regime is a partitocracy: the political parties are the main players in the legislative arena, inducing the politicization of morality debates. Nevertheless, a striking difference between the morality policy-making and other fields is that voters' discipline is relaxed. This makes religious actors, and even more Social Christian parties within the federal parliament, key players in policy-making.

4 The Role of Religious Actors and Their Arguments

4.1 Values Matter: Discretely Fighting for One's Pillar

The social quietness that accompanies the framing and the implementation of the legislative norms, preceded by agreements among the political parties, is explained by a second factor. The political culture (characterized by pluralism and consociationalism) is a structural first one. At the actors' level, the second factor is the role played by civil society organizations. In terms of policy analysis, they are generally the final beneficiaries of the public programs. In terms of political science, they may act as interest groups. They actively participate in the debates, without mobilizing several social segments in a strong protest, only in their own pillar.

This happens, for example, regarding euthanasia, with the opposition between the *Association for the Right to Die with Dignity and Euthanasia Stop*, the latter gathering religious ministers and professors from the different universities in Belgium. Regarding same-sex marriage and adoption, *Rainbow Wallonia* (the French-speaking federation of the lesbian, gay, and transsexual associations) faces no major opponent, but the doctrine of the Roman Catholic Church. In the same perspective, patients support the ART centers and HESCR, much more than they oppose them. They do not even appear within the associations listed by some insurance companies or university hospitals, unlike other patients' associations. Organizations among the civil society participate in the debates, they deliver press releases, some organize demonstrations, they serve as sound boxes, they prompt the media to comment on polls, and they are listened to by decision-makers, but the social segmentation hampers a global mobilization in one sole direction, for or against a morality issue.

Physicians and researchers form a peculiar group within civil society. In terms of policy analysis, they are most of the time the target groups of the public programs. Laws target them either individually or collectively through the hospitals and the medical centers. Following Schneider and Ingram's typology (1993), physicians and researchers are privileged target groups: they are powerful and positively constructed as deserving beneficial policies that other groups will hence also approve. Moreover, doctors are profiled by politicians as having direct linkages with public purposes. In addition, they benefit from policy tools that enable them to take action on a voluntary basis (Schneider and Ingram, 1993: 339). Regarding morality issues, they are solicited as experts, notably by the parliament's commissions. Practitioners preferably counsel the political representatives of their own pillar. Some are even elected representatives and members of the political parties. Physicians and researchers are well represented within political institutions, and they do not need to lobby to reach specific policy outcomes. They belong to all advisory bodies related to the issues. The law of 30 May 2005 even revises the law of 11 May 2003 to make the membership to the BACB and to the FCE compatible. In general, practitioners prefer autonomy over legislation that might prevent them from keeping the pace with scientific and medical progress. Their self-regulation follows the lines of cleavages and aligns with the structuration of the political system.

Hence, political parties, and especially some of their individual members, are the most vocal about the agenda setting and the framing of morality issues, albeit avoiding (intense) media coverage. They are active within the representative, cleaved, and consensual political system. Each party is the center of gravity of its pillar. The pillars altogether segment the whole society, with elites serving as pediment of the pillars, both superimposing each one and bridging them together. Each political party heading its pillar, it is the leading representative of its segment during any policy-making process, with a more or less active support of the related associations in the field. Political parties even disrupt the policy-making by representatives (De Winter and Dumont 2006). In such a context, discretion regarding controversial debates can be a key for successful decision-making.

The president of one of the two Social Christian parties explained that "that every week he confers with all governments, putting every decision to the test – the sign of a classic party chairman."[20] The same president considers that the political party is "a place of centralization, it is also a place of orientation, and it is a place where inevitably one has a much broader vision of all the stakes of society, than by being a minister who is stalled in his competences." The president of the other Social Christian party declared that his core business as the party leader was to "support

[20]This is an excerpt from interviews that were conducted for a larger research project, from October 2015 to May 2016.

organization but also externally to try to change something in society and thereby play a supporting role."

With regard to biopolitics, political parties stand for the values and the moral principles of their pillar while trying to find agreements about the (non)decisions to be made in order to avoid deep conflicts. Based on the experience of the abortion crisis in 1990, the Belgian political parties chose the parliament and the government as the main policy venues, avoiding a conflict expansion (Schattschneider, 1960). Engeli et al. (2013: 348) have suggested that "government participation of confessional parties makes a great difference for policy development." From the comparison of ART regulation by the Belgian and the Italian decision-makers, scholars also have demonstrated that "a governmental coalition adopts a public policy that fits with the policy image of key actors, including the church if necessary" (Schiffino et al., 2009: 581). The withdrawal of the church or, on the contrary, its involvement clearly influences the policy-making and its outputs, leading to more permissive or restrictive decisions in the biopolitics field (Schiffino et al., 2009). The impact of the Catholic Church on morality policies has thus been assessed, and it has even been proven to be an actor of permissive policies at least because of its sluggishness (Knill et al., 2014: 853). In this case, if the Catholic Church does not hamper permissive laws in the morality field, it can decelerate the rhythm of their adoption.

In Belgium, the focus is set on the Catholic Church, more than on other confessions, for several reasons. It is the first and the main religious family in Belgium. It exists since the creation of the state. More than a half of the population identifies with Catholicism or at least to Christianity. It is the only religion that has provided support for the creation of a (dedicated) political party, the socialist and the liberal families being secular parties. The Social Christian segment, be it French or Flemish speaking, traditionally has deputies and ministers in the legislative and executive arenas who determine the policy-making.

However, there is permanent tension between the doctrine of the Roman Catholic Church and its effective influence on Belgian morality policies. First, several facts demonstrate that the doctrine is not systematically nor strictly fueled through diverse channels of communication because actors take distance themselves from it. Second, the Belgian Catholic Church has historically adopted a rather progressive stand on many issues along the lines of cleavages.[21] It has not acted as a veto player, i.e., an actor "whose agreement is necessary for a change of the *status quo*" (Tsebelis, 1999: 591).

The evolution of the name of the French-speaking Social Christian party provides a first example of the distance between the doctrine of the Roman Catholic Church and the Belgian actors. The party first evolved from the name "Catholic party" in the nineteenth century to be the so-called Social Christian party (with the French acronym *PSC*) after World War II. In 1947, the *PSC* formed a coalition with the socialist party, and the government adopted a strong social program. In the twenty-

---

[21]Christian Laporte, L'Eglise belge à la pointe du combat social et sociétal, cinquante ans après "Populorum Progressio", *La Libre Belgique*, 20 February 2017.

first century, the *PSC* went further with in-depth reforms. It opted for the name "Humanist Democratic Centre" (*cdH* in French) in 2002. For Milquet, the president of the party when the congress adopted the new status and the new name (18 May 2002), the rationale is explicit: "People think that we are bound to religion, that we are narrow-minded. It is time to clarify this."[22] A tiny minority of the *PSC* created the "French-speaking Christian Democrats" (*CDF*) because they firmly opposed the abandonment of the Catholic axis. This secession was considered as an epiphenomenon, and the *CDF* never succeeded in reaching the legal 5% threshold necessary to have representatives in the parliaments.

Let us take a second example. In 2015, a course about citizenship was added to the program at primary schools, along with the courses of religion and moral. The decision appeared as a compromise in the stormy debate about a course of philosophy at primary schools. The Social Christian minister at the *Communauté française* was behind the decision. The minister, Milquet, distanced herself from the Belgian Catholic Church. She agreed that religious education in public schools should be decreased from 2 hours teaching to 1. In June 2016, all Belgian bishops took note of the decision, without vocally opposing it. They addressed an open letter to all parents: "We are pleased that Catholic religious education is always offered in schools (...) Now that decisions have been made, let's move forward and have full confidence in the competence of teachers and professors of Catholic religious education. Their courses contribute to the greater good of your children." From 2014 to 2016, the Belgian Catholic Church did not circulate press releases about morality issues, but one explicitly about the importance of either religious or moral education at primary schools (on October 2016). The press release was cosigned by the ministers of the protestant, orthodox, Muslim, and Jewish faiths. It promoted the support of parents (90% of them, according to the press release) to maintain religious or moral education at school: "Today, parents and teachers are asking for these courses to be taught in the best possible way, for the quality of the teaching of both religious education and of citizenship." The rationale behind this joint initiative was the implementation problems of courses on citizenship (too busy schedules, too late teachers' course allocation, etc.).

In the spring of 2017, when secular parties and the pluralist greens supported legislative bills to completely decriminalize abortion, we find a third illustration of the progressive stand of the Belgian Catholic Church, searching for an equilibrium between, on the one hand, the Roman Catholic doctrine and, on the other hand, the societal evolution and the need for compromise. Indeed, one of the Belgian bishops, Delville, declares that "It is important that abortion remains a crime, but at the same

[22]Quoted by the public Belgian media, RTBF, https://www.rtbf.be/info/belgique/detail_partis-politiques-changer-de-nom-pour-changer-quoi?id=9134743 (accessed: 9 August 2017).

time, we must understand that there are concrete and difficult situations, and the legislator must also solve them."[23]

While morality issues returned to the forefront of societal debates, the Belgian Catholic Church expressed some consistent positions. On May 2017, the bishops declared that they did not agree with euthanasia for nonterminal psychiatric patients. On December 2017, they rebuked ART issues by agreeing with the speech by Testart[24] who warns against transhumanism in ART and HESC practices. They shared his view: "In 30 years, we have gone from the prohibition of embryo sorting to the acclimatization of these technologies."[25]

Meanwhile, the Catholic Church positions itself in second place, behind the Social Christian parties that are the leading decision-makers, even more in a partitocratic political system. The Social Christian parties have been part of the federal coalitions since 1945.[26] They have largely contributed to nondecisions in the morality field. On March 1992, after the abortion crisis, the governmental agreement[27] stated that consensus would prevail at both governmental and parliamentary levels for matters such as bioethics, which did not explicitly form part of the governmental agreement. During the electoral campaign for the elections in June 1999, the president of the Flemish Social Christian party repeated that no bioethical issue would be added in a governmental agreement while its party formed part of a coalition. On July 1999, before the assembled chambers, the government declared: "In seeking answers to ethical questions and challenges, the government makes a break with the past. Instead of restricting them to the need for governmental solidarity and consensus amongst the parties of the majority, parliament will assume its responsibility, on the basis of individual awareness."[28] The Social Christians were then relegated to the opposition. From 1999 to 2007, they sluggishly limited decisions to procedures. In 2011, a task force delivered a report about faiths in Belgium (a religious albeit not a bioethical issue). However, we find no official tracking of a follow-up of its recommendations by the federal minister of justice. In 2014, the electoral programs of the parties show some reluctance in regulating religious issues. The Flemish Social Christian party underlines the importance of religious heritage on the one hand and of pluralism on the other. It "acknowledges the value that religions provide to the polity." It recommends "no general ban on

---

[23]Quoted by the public Belgian French-speaking media, RTBF, 9 April 2017.

[24]Professor Jacques Testart is the French doctor who led the birth of the first test-tube baby in 1982.

[25]Catholic Church Belgium. http://www.cathobel.be (accessed: 6 December 2017).

[26]Only two historical examples of secular governments (without the social Christians) exist: firstly, during the "Royal Question" (1945–1947) and, secondly, during the second "School War" (1954–1958).

[27]The governmental agreement is a fundamental text that the parties agree on after the elections: when they decide which of them should form part of the coalition and they establish their schedule for the whole legislature.

[28]Chamber of the Representatives, Declaration of the federal government, extraordinary session, legislative document, 20 / 1 – 1999 (S.E.), 14 July 1999, p. 8.

religious or philosophical symbols in education and in public services."[29] Actually, its program focuses on the coexistence of diverse religions. The French-speaking Social Christian party raises the issue of a reform of the faiths: criteria for the recognition of religions, organization of representative bodies, harmonization of the salaries for ministers of religion, and data on the number of ministers of religion. It promotes a reform about religious education. The party generally argues that it "does not want to lock up the debate on values in the private sphere" and it promotes "the appropriation by everyone of his/her own religious or philosophical identity" as well as "independent judgement and exercising free thinking."[30] Meanwhile, in 2017, the federal government (with the CD&V on board) has frozen the renewed debates about abortion.

Adding the internal divisions inside political parties helps deepen the analysis. It allows avoiding simple dichotomies between religious actors and secular actors. In Belgium, being Catholic does not mean per se being conservative. Catholics as well as the seculars have both left and right wings. A major reason for the Social Christian parties not to regulate morality issues or ART while they are coalition leaders, or even coalition members, is to protect their internal cohesion. If the parties adopt a radical position on bioethical matters, they reduce themselves de facto to a conservative nucleus. If they soften their position, they lose a traditionalist part of their electorate. "The *PSC* I knew was not a left-wing party. It was a center party with a left centrist current and a centrist current. It was a centrist party in the full sense of the word," said Deprez, who chaired the party from 1981 until 1996.[31]

4.2 The Progressive Social Christians Outpace the Roman Catholic Church Regarding ART and HESCR

In the 1980s (when IVF was being developed), the Roman Catholic Church categorically condemned AI (the type of ART then most widely practiced). Its argument refers to an unnatural form of procreation. In the light of technological developments in the field of ART, the Roman Catholic Church has adopted a slightly more open, though still very restrictive position.[32] The starting point for the church's doctrine on ART is the belief that human life begins at conception (i.e., the moment of fertilization) and must be protected as such. The embryo must thus enjoy the rights and level of protection of a human being. Consequently, the Roman Catholic Church opposes recourse to ART outside of matrimony (as it opposes procreation in general

[29] CD&V, *Innestocongres*, 15–16–17 November 2013, p. 107.

[30] cdH, *Electoral program for the European, Legislative, and Regional elections 2014*, p. 48.

[31] *La Libre Belgique*, 14 June 2014.

[32] A comprehensive account of the Roman Catholic Church's position on ART can be found in a document published in 1987 by the church's congregation for the doctrine of the faith and entitled "Instruction on respect for human life in its origin and on the dignity of procreation."

outside of marriage), since it would conflict with the right of the child to be born into a family. ART in which fertilization occurs outside the body, i.e., IVF followed by embryo transfer, are not acceptable to the Roman Catholic Church because the (often-produced) excess embryos are frequently destroyed. ART using donated gametes are also unacceptable from the Roman Catholic Church's point of view since they deny the newborn's right to a clear legal status and identity. The only form of ART that the church does not categorically exclude is AI with the husband's sperm.

Some distance from the doctrine is noticeable in the work of the Catholic universities (*Université catholique de Louvain*, French-speaking, and *Katholieke Universiteit te Leuven*, Flemish) that allow the creation of embryos and therapeutic cloning by their practitioners, despite the explicit opposition by a bishop. In 2002, a vivid debate ensued involving the bishop *Léonard*[33] and the *Université catholique de Louvain* (*UCL*). It was reported in the media. The clergyman considered that part of the university work did not align with the Roman Catholic Church's doctrine. He referred to the medical practices of the university hospital related to ART and to birth control, abortion, or euthanasia. In 2010, when the bishop became the primate of Belgium, a debate emerged inside the university about the episcopal conference no longer being its organizing power. A professor launched a petition arguing: "We, doctors practicing in UCL clinics, consider the ideas and remarks made by Monsignor Leonard as incompatible with the values essential to the deontological practice of medicine, namely non-discrimination and respectful support of patients in their disease, whatever it may be. We therefore ask for the resignation of Bishop Leonard from all the functions he performs within our university." The governance of the university currently preserves its historical roots: the archbishop and the French-speaking bishops officially exert the organizing power. Yet the responsibility of the university management rests with the board of directors, not the clergy who now only provide advice. The university officially claims this hybridity: "An institution as rich and complex as UCL calls for governance that takes into account its history and traditions, the diversity of its activities and indispensable subsidiarity."[34]

During the legislative process about stem cell research, therapeutic and reproductive cloning, ART, and preimplantation genetic diagnosis, doctors and researchers from all the Belgian universities spoke at the federal parliament as experts. Some were invited because of their leading practices, in their capacity as director of a center, or because they served either as president or as member of advisory bodies. Professor Devroey from the *Centrum voor Reproductieve Geneeskunde* at the *Vrije Universiteit Brussel* (VUB, secular, and Flemish) was a

[33]Léonard is a traditionalist minister of faith. Pope Benoit XVI appointed him as the Belgian archbishop. Before the extension of euthanasia to minors, Léonard called for a day of fasting and prayers. He denounces the "arrogant democracy, which transforms a voted law into absolute truth" (quoted by *7 sur 7*, 24 February 2012). Other clergymen disagree with his statements, notably the cardinal Danneels.

[34]University website: https://uclouvain.be/fr/decouvrir/gouvernance-0.html (accessed: 6 December 2017).

pioneer of the intracytoplasmic sperm injection (ICSI) technique. He considers the legislative framework as "a model for Europe" that brings "juridical security and prospects."[35] His view is shared by Professor Comhaire, endocrinologist. Before it was forbidden by law in 2007, Dr. Comhaire was practicing sex selection on embryos in his private clinic. In parliament, he contended that "the choice of sex can increase the happiness of a family." His argument is that "sexuality and reproduction remain purely private issues." Professor Van Steirteghem is the well-known co-pioneer of the ICSI technique. Belonging to the VUB as well, he insists on "the need to create embryos for research if there are no alternatives and with the consent of donors." He is convinced that "embryonic stem cells are crucial for medical progress." Professor D'Hooghe, from the KUL Fertility Center, confirms a stricter stand when declaring "research on embryos is only a potential option when all other means have been excluded." Another member of this Catholic university, Professor Cassiman, argues that HESCR is "a therapeutic driver but that evidence is lacking when it comes to its application." Skepticism also dominates the discourse by a Catholic French-speaking scholar, Professor Verellen-Dumoulin from the Human Genetic Center at the UCL. As a geneticist, she welcomes the legislative framework, but thinks "the creation of embryos for using them as therapeutic cells might be adverse to health." Trends emerge from the auditions at the parliament. They confirm not only the division of practitioners along substantive lines but also their joint support toward laws that would provide guidelines while preserving the doctors' autonomy. This is congruent with Schneider and Ingram's theory about the social construction of target groups.

The policy-making of ART and HESCR provides enlightening examples of both the misalignment of the Belgian parliamentarians with the Roman Catholic doctrine and their alignment with the traditional line of cleavages, the Social Christian deputies adopting stricter positions than the secular. In their legislative bills, the French-speaking Social Christian deputies want to promote several principles.[36] They insist on guaranteeing "the respect and protection of the uniqueness of each human person." They also advocate the "prohibition of deciding the genetic characteristics of a child." They consider "the conception of the human person as an end and not as a means." They also insist on "the gendered nature of reproduction human." They believe that preimplantation diagnosis still "raises ethical questions" which allow for restrictions. They ban any cloning including for therapeutic purposes. Among the specialized senate commission for bioethical issues, the Flemish Social Christian representatives generally adopt a more relaxed position. For instance, they agree on the creation of embryos for research purposes. However, at the Chamber of Representatives, they also declare that their party as a whole

[35]All quotes in this paragraph come from the Chamber of Representatives, plenary session, legislative document CRIV 50 PLEN 353, 2 April 2003.

[36]See J. Milquet et al., Senate, legislative document, 877/1, 12 February 1998. J. Milquet, Chamber of Representatives, legislative document, 157/1, 13 October 1999.

considers that "human life, from its premises to its end, deserves respect."[37] The deputy member of the party, Goutry, proclaims "scientific progress must not be blocked by ethical considerations. Belgium has useful instruments and can count on the expertise of academics and scientists."[38] He suggests a ban on commercial purposes, which is not a big deal since all parties agree on it. The French-speaking Social Christian party tends to adopt firmer positions. When the law of 2007 was on the parliament agenda, the party declared itself to be in favor of a regulation, but it also refused to vote for it if amendments were not taken into account. For example, the *cdH* condemns the postmortem insemination. And it does not grant single parents ART treatment. He argues: "the birth of a child must not be a personal project but a couple's project."[39] The law of 6 July 2007 bypasses such proposed restrictions.

5 Conclusion

The Belgian policy-making is influenced by both the Belgian Catholic Church's policy image on morality issues and its sluggish position on morality policies. The weight of other actors, and above all the political parties and the practitioners who all favor a public regulation, is higher. But to avoid deep conflict about core values, the public regulation aims at setting a general framework rather than privileging specific principles to the detriment of others. Pluralism and consociationalism prevail. That is why there is a distance between the overall position of the Roman Catholic Church on morality issues and the way it is experienced in Belgium. ART regulation provides a convincing illustration of the religious actors' nuanced positions and their arguments.

All practitioners are in favor of a regulation of both ART and research embryo. They demand it, in order to avoid a legal vacuum. Such a void could be detrimental to them in case of criminal prosecution. But they argue for a legislation in tune with the scientific progress and the practitioners' autonomy. Their argument is audible since they are a privileged target group which has proved its efficiency in self-regulation. Patients associations are not vocal in the ART field. The conventions that they conclude within the ART centers seem to meet their expectations. The French-speaking Social Christian party shows a paradox: it has distanced itself from the Catholic Church, and yet it remains the most visible opponent of the content of the ART law and of HESCR law. It promotes a more traditional model of family and a more natural mode of procreation,[40] close to the historical discourse of the Catholic

[37]Chamber of Representatives, legislative document, CRIV 50 PLEN 353, plenary session, 2 April 2003, p. 10.

[38]Chamber of Representatives, legislative document, CRIV 50 PLEN 353, plenary session, 2 April 2003, p. 10.

[39]Chamber of Representatives, legislative document, DOC 512567/004, report, p. 22.

[40]Hébette Martine, 9 June 2006, Senate, legislative document, 3-1440.

Church. The Flemish Social Christian party adopts a less restrictive position, especially during the parliament debates. They argue to be confident in the excellence of the medical community.

Morality policies in Belgium are puzzling. Religion and politics have been intertwined since the creation of the state; but so is secularism. The state is divided by fundamental cleavages that mobilize core values and beliefs in opposite directions. To reduce the vital pressure that result from the bitter conflict, the political system searches for agreements. In the morality field, agreements on opposed religious and moral principles are difficult to find. Agreements can be found at another level, an instrumental one. As far as principles are concerned, each social segment sustains the political party of its pillar in defending the quiet respect of the core values. Permissive regulation is acceptable, provided there is the possibility to invoke conscientious objections. In the ART and HESC policy sector, such regulation enables each center and hospital to practice in respect with its guidelines. Generally, no major clash is visible on morality issues. Thanks to the role played by civil society organizations and the political parties, policy-making takes into account the interplay between religion and politics, but it balances it rather than overweighting it. It is the precondition for a peaceful community life. This is the case for Belgium. And it may be inspiring for every democratic political system, even with different settings, if democracy is the measurement standard.

References

De Winter, Lieven, and Patrick Dumont. 2006. Do Belgian parties undermine the democratic chain of delegation? *West European Politics* 29 (5): 957–976.

Easton, David. 1965. *A systems analysis of political life*. New York: Wiley.

Engeli, Isabelle, Christoffer Green-Pedersen, and Lars Thorup Larsen. 2013. The puzzle of permissiveness: Understanding policy processes concerning morality issues. *Journal of European Public Policy* 20 (3): 335–352.

Hall, Peter. 1993. Policy paradigms, social learning, and the state: The case of economic policymaking in Britain. *Comparative Politics* 25: 275–296.

Knill, Christoph. 2013. The study of morality policy: Analytical implications from a public policy perspective. *Journal of European Public Policy* 20 (3): 309–317.

Knill, Christoph, Caroline Preidel, and Kerstin Nebel. 2014. Brake rather than barrier: The impact of the Catholic Church on morality policies in Western Europe. *West European Politics* 37 (5): 845–866.

Lijphart, Arend. 1975. *The politics of accommodation: Pluralism and democracy in the Netherlands*. Berkeley, CA: University of California Press.

———. 1999. *Patterns of democracy, government forms and performance in thirty-six countries*. New Haven, CT: Yale University Press.

Lipset, Seymour Martin, and Stein Rokkan. 1967. Cleavage structure, party systems, voter alignments: An introduction. In *Party systems and voter alignments: Cross national perspectives*, ed. Seymour Martin Lipset and Rokkan Stein, 1–64. London: Free Press.

Lorwin, Val. 1971. Segmented pluralism: Ideological cleavage and political cleavage in the smaller European democracies. *Comparative Politics* 3 (2): 141–144.

Mintrom, Michael. 2013. Policy entrepreneurs and controversial science: Governing human embryonic stem cell research. *Journal of European Public Policy* 20 (3): 442–457.
Mooney, Christopher. 1999. The politics of morality policy. *Policy Studies Journal* 27 (4): 675–680.
Mucciaroni, Gary. 2011. Are debates about "morality policy" really about morality? Framing opposition to gay and lesbian rights. *Policy Studies Journal* 39 (2): 187–216.
Parsons, Talcott. 1951. *The social system*. London: Routledge and Kegan Paul.
Salter, Brian. 2007. The global politics of human embryonic stem cell science. *Global Governance* 13 (2): 277–298.
Schattschneider, Elmer Eric. 1960. *The semi-sovereign people: A realist's view of democracy in America*. New York: Holt, Rinehart, and Winston.
Schiffino, Nathalie. 2003. *Crises politiques et démocratie en Belgique*. Paris: L'Harmattan.
Schiffino, Nathalie, and Steve Jacob. 2015. Uncertainty and cleavages at stakes: Do the Belgian constitutions of 1831 and 1993 stabilize political power? In *Behind a veil of ignorance? Power and uncertainty in constitutional design*, ed. Louis Imbeau and Steve Jacob, 121–133. New York: Springer. Studies in Public Choice.
Schiffino, Nathalie, Célina Ramjoue, and Frédéric Varone. 2009. Biomedical policies in Belgium and Italy: From regulatory reluctance to policy changes. *West European Politics* 32 (3): 559–585.
Schneider, Anne, and Helen Ingram. 1993. Social construction of target populations: Implications for politics and policy. *The American Political Science Review* 87 (2): 334–347.
Simeon, Richard. 2009. Constitutional design and change in federal systems: Issues and questions. *Publius: The Journal of Federalism* 39 (2): 241–261.
Snow, Dave. 2015. Explaining a policy failure: Jurisdictional framing, federalism, and assisted reproductive technologies in Canada. *Canadian Public Policy* 41 (2): 124–136.
Tsebelis, George. 1999. Veto players and law production in parliamentary democracies: An empirical analysis. *American Political Science Review* 93 (3): 591–608.

Printed by Printforce, the Netherlands